D1295863

Nuclear Power and Its Environmental Effects

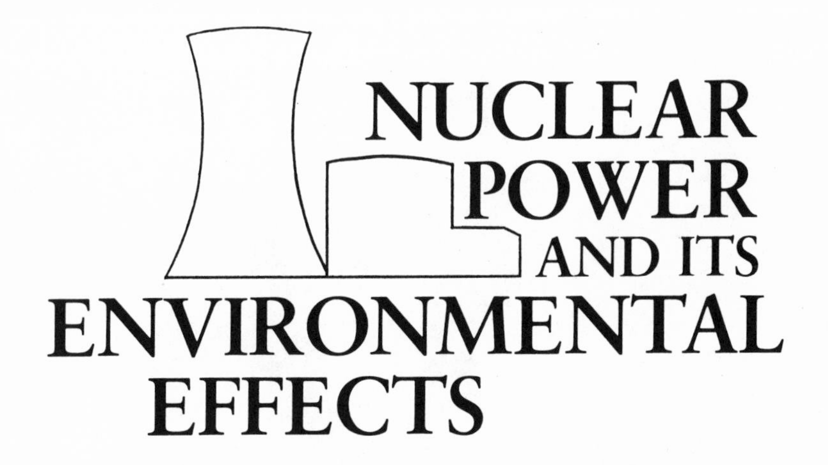

NUCLEAR POWER AND ITS ENVIRONMENTAL EFFECTS

Samuel Glasstone
Walter H. Jordan

AMERICAN NUCLEAR SOCIETY
La Grange Park, Illinois

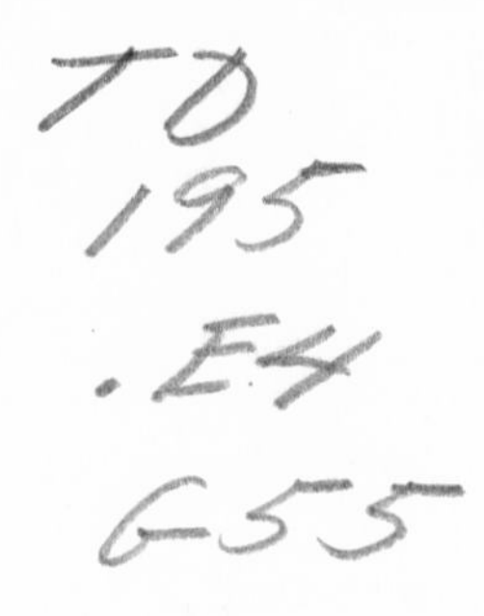

American Nuclear Society Order Number: 690006
International Standard Book Number: 0-89448-022-7 (hardbound)
0-89448-024-3 (softbound)
Library of Congress Catalog Card Number: 80-67303

American Nuclear Society
555 North Kensington Avenue
La Grange Park, Illinois 60525

Second Printing, 1981.

Printed in the United States of America
Designed by Christopher J. FitzGerald

Contents

List of Tables

Publisher's Foreword

In publishing this book, the American Nuclear Society, a multidisciplinary scientific society with a membership of some 12 000 persons worldwide, departs somewhat from its customary role as a publisher of technical texts directed to the interests of engineers and scientists in the nuclear field. With the publication of *Nuclear Power and Its Environmental Effects,* the Society reaches out to the broader audience constituted by those among the general public who would be serious students of what has become a topic of great importance, of some controversy, and of needless confusion. Authors Glasstone and Jordan, writing from a wealth of practical experience over many years in the nuclear field, must be commended for their clear, dispassionate exposition. Written with the general reader in mind, this book also serves as a convenient summary of safety measures and environmental concerns for the professional.

Like all published works of the American Nuclear Society, *Nuclear Power and Its Environmental Effects* has received a rigorous review. In their Preface, the authors have acknowledged the organizations that have reviewed the text and contributed information. The publisher recognizes, in addition, the following Argonne National Laboratory personnel for their special assistance: James E. Carson, Robert M. Goldstein, William Hallett, Barbara L. Reider, and Jan van Erp. The Society is also grateful to R. C. Greenwood (EG&G Idaho Inc.), R. M. Jefferson (Sandia National Laboratories), W. Meyer (University of Missouri-Columbia), N. M. Schaeffer (Radiation Research Associates), and Robert F. Pigeon (U.S. Department of Energy) for their guidance in bringing this book to print.

NORMAN H. JACOBSON
Manager, ANS Publications

Preface

At the beginning of 1980, about 70 nuclear power plants were licensed to operate in the United States and they produced about one-eighth of the total electric power generated. At the same time, some 90 additional plants were under construction. It is possible, therefore, that by the end of the decade nuclear power stations will be generating roughly one-fifth of the electricity consumed in this country.

In its report entitled "Energy in Transition, 1985-2010," published in 1980, the Committee on Nuclear and Alternative Energy Sources of the National Academy of Sciences concluded that coal and nuclear power are the only economic alternatives for the large-scale generation of electricity during the next 20 to 30 years. Furthermore, in spite of his known lack of enthusiasm for nuclear power, President Carter has stated that: "We cannot shut the door on nuclear energy." However, the future of the nuclear option in the United States (and elsewhere) is greatly dependent on the public's perception of its environmental and safety aspects.

In this book we have tried to describe the environmental effects associated with nuclear power operations from the mining of uranium to the final disposal of waste products. The generation of nuclear energy is unique in the respect that it is accompanied by the formation of radioactive materials. These materials emit radiations that can have adverse health effects. Consequently, special emphasis has been placed on the possible release of radioactive substances to the environment and the potential effects of the associated radiations. In doing so, we have explained some of the measures taken to ensure safe operation of nuclear power plants and prevent the escape of dangerous amounts of radioactive materials in case of an accident.

The general public has been made all too well aware of the potential hazards of radiation, especially after the accident at the Three Mile Island nuclear power station in March 1979. But little effort has been made to place the effects of radiation in perspective. The report of the President's (Kemeny) Commission on this accident pointed up the difficulty experienced by the news media in presenting such information as was available in a form that would be understood by the public; this difficulty was particularly acute in the reporting of information on radiation releases. Clearly, there is need for a better understanding of radiation and its consequences; we think that this book can make a useful contribution to that understanding.

We wish to express our thanks to the many people who have helped us in one way or another. Several members of the Argonne National Laboratory, the Oak

Ridge National Laboratory, and the Battelle Pacific Northwest Laboratories reviewed various stages of the manuscript and made many suggestions for which we are grateful. We also wish to thank the Department of Energy (and its predecessors, the Atomic Energy Commission and the Energy Research and Development Administration) for partial financial support. In particular, we are greatly indebted to E. J. Brunenkant, L. J. Deal, R. F. Pigeon, and W. W. Schroebel, former or present staff members of these agencies, whose encouragement and help made this book possible.

SAMUEL GLASSTONE
WALTER H. JORDAN

Oak Ridge, Tennessee
April 1980

1

The Generation of Electric Power

NUCLEAR STEAM SUPPLY SYSTEM

The first step in the generation of electricity, in nuclear power plants and in conventional plants using fossil fuels (i.e., coal, oil, or natural gas), is the production of steam by heating water.[a] There is a fundamental difference, however, in the ways in which the heat is produced. In a conventional power plant, heat is liberated by burning a fossil fuel, but in a nuclear installation the heat is released by a process called *nuclear fission*, as explained in Chapter 2.

The device in which fission heat is produced in a controlled manner from the nuclear fuel is called a *nuclear reactor* or, in brief, a *reactor*. The heat is then utilized either directly by boiling water within the reactor or indirectly by transferring the heat to water in a *steam generator* (heat exchanger) outside the reactor. In any case, the basic purpose of the reactor and its associated equipment is to produce steam at as high a temperature and pressure as is possible under the circumstances. Consequently, the term *nuclear steam supply system* is commonly used to describe the reactor and other components that serve to produce steam by utilizing nuclear fission heat. Once the steam has been produced, the subsequent operations are basically the same in both nuclear and fossil-fuel plants. The steam drives a *turbine*, which is connected to an *electrical generator*. In the turbine, part of the heat content of the steam is converted into mechanical energy of rotational motion. Then, in the generator, this energy is converted into electrical energy.

The exhaust steam leaving the turbine passes to a *condenser*, where cold water flowing through tubes causes the steam to condense to liquid water (Fig. 1-1). The condenser serves several functions. By condensing (and thus removing) the steam, the back pressure of the turbine exhaust is reduced; at the same time, the heat remaining in the steam is removed at the relatively low temperature of the condenser cooling water. As described shortly, this is an important aspect of turbine operation. Finally, the high-quality water used in producing steam is recovered by condensing the steam; the liquid condensate is then returned as *feedwater* to the steam generator.

[a]Gas-turbine and diesel generating plants are exceptional in this respect.

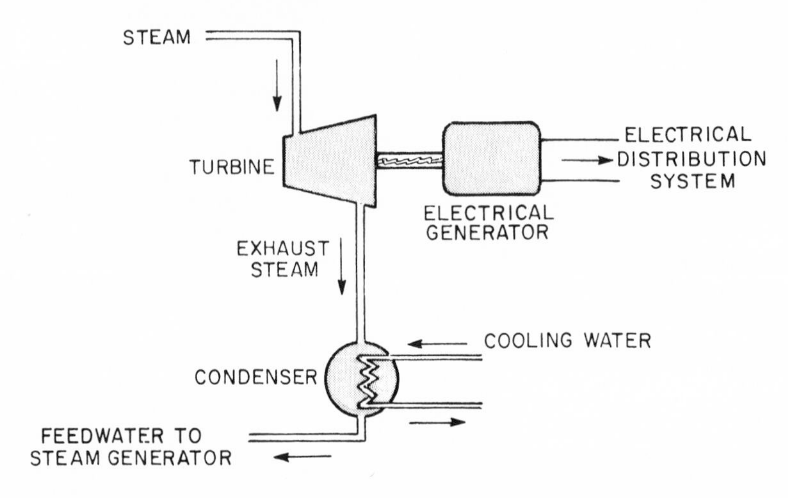

FIG. 1-1. General principles of electric power generation from steam.

THERMAL EFFICIENCY

It is a fundamental law of nature, to which no exceptions are known, that any machine, such as a turbine, that continuously converts heat (e.g., in steam) into mechanical work must operate in a series of stages. First, heat is taken up by a *working fluid* (e.g., steam) at an upper temperature; only part of this heat is converted into mechanical work in the machine, while the remainder is given up to a *sink* at a lower temperature (Fig. 1-2). In a turbine, the sink is the cooling water in the condenser. The working fluid, at the lower temperature, is now returned to the steam generator, where it is reheated. The steam at the upper temperature then repeats the cycle. In this way, the conversion of heat into work can proceed in a continuous manner.

The fraction of the heat supplied at the upper temperature that can be converted into mechanical work is increased by having the upper temperature of the working fluid as high as possible, while the lower temperature of the cycle is as low as possible. In practice, the lower temperature, which is that of the condenser cooling water, is always much the same. Hence, the efficiency with which a turbine, for example, can convert heat into work depends on the upper temperature—that is, on the temperature of the steam supply to the turbine. The efficiency of a properly designed turbine can thus be increased by increasing the steam temperature.

As a general rule, nearly all the mechanical energy of rotation supplied by a turbine to a generator is converted into electrical energy. Consequently, the conclusion reached above concerning the effect of steam temperature on the efficiency of converting heat to mechanical energy also applies to the net conversion of heat into electrical energy. The thermal efficiency of any electric power generating plant can be defined as the percentage of the heat energy supplied by the fuel in the boiler (or reactor) that is converted into electrical energy; i.e.,

$$\text{Thermal efficiency} = \frac{\text{Electrical energy generated}}{\text{Heat energy supplied}} \times 100.$$

Although the thermal efficiency is defined here in terms of heat supplied by the fuel rather than heat supplied to the turbine, it is still true that the thermal efficiency of a steam-electric power plant can be increased by increasing the temperature of the turbine steam.

In the great majority of nuclear power plants, steam enters the turbine at a lower temperature than in modern fossil-fuel plants; consequently, the former convert a smaller fraction of the heat energy supplied into electric power and discharge a somewhat larger fraction to be removed in the condenser. The discharged heat eventually is dispersed to the environment, and the resulting *thermal effects*, as they are called, can be ecologically important.

In most modern central-station plants using fossil fuels, steam temperatures are as high as 570 °C (1060 °F). The thermal efficiency of these plants—that is, the proportion of the heat supplied that is converted into electrical energy—is about 40 percent. In addition, roughly 10 percent is carried off with the combustion (or stack) gases, and 5 percent is dissipated within the plant. This leaves some 45 percent of the total heat to be removed in the condenser.

Nearly all nuclear reactor plants now operating or under construction employ ordinary (or light) water to remove the heat generated by fission; consequently, they are referred to as light water reactors (p. 26). In these reactors, the maximum steam temperature is around 285 °C (540 °F), and the thermal efficiences, on the average, are close to 33 percent. There are no stack losses; about 5 percent of the heat supplied is dissipated in the plant, however, leaving 62 percent to be removed in the condenser.

Although the gross thermal efficiency of a modern fossil-fuel plant can be 40 percent, as compared with 33 percent for the nuclear plant, the net efficiencies are closer when a coal-fired plant is considered. In the latter case, about 10 to 12 percent of the electricity generated is utilized in operating plant equipment, such as coal and ash conveyors, precipitators, and flue-gas scrubbers. The net efficiency, as measured by the electrical energy available for sale, thus is reduc-

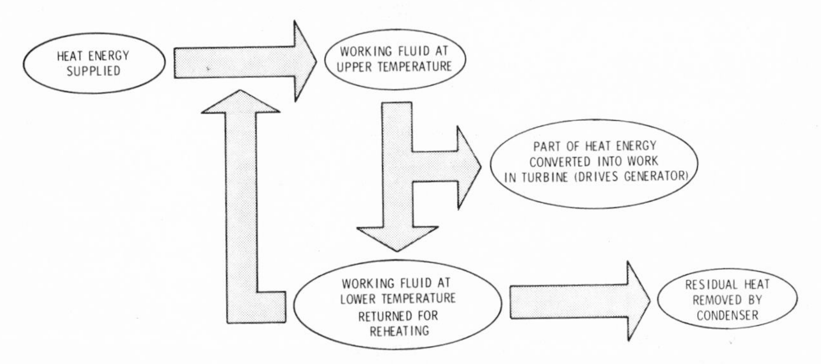

FIG. 1-2. Cycle of changes for continuous conversion of heat into mechanical work.

ed to 35 or 36 percent, whereas for a nuclear power plant, it is about 31.5 percent.

It should be understood that the lower steam temperatures (and lower efficiencies) of existing nuclear power plants, as compared with modern fossil-fuel plants, are not characteristic of nuclear energy, but rather of water-cooled reactors. In more modern reactor designs that use either liquid sodium or helium gas instead of water as the coolant to remove the fission heat, much higher steam temperatures (and higher efficiencies) should be possible.

ELECTRICAL CAPACITY OF A GENERATING PLANT

The rated or design capacity of a power plant is a measure of the maximum rate at which the plant can generate electricity. It is expressed in kilowatt (kW) units or, often more conveniently, in megawatts (MW), where 1 MW is equal to 1000 kW. Most existing fossil-fuel central-station power plants have capacities of less than 300 MW of electricity, but for modern installations, both fossil-fuel and nuclear, they are in the vicinity of 1000 MW (1 million kW) or more.

The demand for electricity is not uniform; it varies with the time of day and with the season of the year, as well as from one year to another. The average operating level of a power plant is commonly expressed in terms of the *plant factor*. This is defined as the ratio of the actual net output over a specified period of time to the net output if the plant had operated at its rated (or maximum) capacity during the same period, expressed as a percentage. Thus, in brief,

$$\text{Plant factor (\%)} = \frac{\text{Actual output}}{\text{Rated output}} \times 100$$

over a given time period. For most large power plants, the plant factor averaged over the whole year is from 60 to 70 percent.

The *quantity* (or amount) of electricity consumed or generated, as distinct from the *rate* at which it is generated, is commonly stated in terms of the kilowatt-hour (kWh) unit.[b] In fact, electric meters in the home and in industry measure electrical consumption in kilowatt-hours. A power plant with a capacity of 1000 MW (1 000 000 kW), for example, operating at a plant factor of 70 percent for 24 hours would generate (1 000 000) (70/100) (24) = 16.8 million kWh of electricity per day. On the average, this would satisfy the requirements (residential, industrial, and commercial) of a population of about 500 000 in the United States.

[b] The rate of electrical generation (in kilowatts or megawatts) can be compared to miles per hour on an automobile speedometer, whereas the quantity generated or consumed (in kilowatt-hours or megawatt-hours [MWh]) is equivalent to the mileage traveled as indicated on the odometer.

Bibliography—Chapter 1

The following list of general references covers the environmental aspects of electric power generation, with special emphasis on the use of nuclear power.

American Nuclear Society. *Nuclear Power and the Environment: Questions and Answers.* La Grange Park, Ill.: American Nuclear Society, 1980.

Bodansky, D. "Electricity Generation Choices for the Near Term." *Science* 207 (1980): 721.

Eichholz, G. G. *Environmental Aspects of Nuclear Power.* Ann Arbor, Mich.: Ann Arbor Science Publishers, 1976.

Foreman, H., ed. *Nuclear Power and the Public.* Minneapolis: University of Minnesota Press, 1970. See also paperback edition. New York: Anchor Books, 1972.

Hohenmeser, C.; Kasperson, R.; and Kates, R. "The Distrust of Nuclear Power." *Science* 195 (1977): 25.

International Atomic Energy Agency. *Nuclear Power and the Environment.* IAEA information booklet. Vienna: International Atomic Energy Agency, 1973.

______. *Proceedings of the Symposium on the Environmental Aspects of Nuclear Power Stations.* Vienna: International Atomic Energy Agency, 1971.

Jordan, W. H. "The Issues Concerning Nuclear Power." *Nuclear News* 14, no. 10 (1971): 43.

______. "Nuclear Energy: Benefits and Risks." *Physics Today* 23, no. 5 (1970): 32.

Karam, R. A., and Morgan, K. Z., eds. *Environmental Impact of Nuclear Power Plants.* New York: Pergamon Press, 1976.

National Academy of Engineering, Committee on Power Plant Siting. *Engineering for Resolution of the Energy-Environment Dilemma.* Washington, D.C.: National Academy of Engineering, 1972.

Nuclear Power and the Environment. Madison: University of Wisconsin Press, 1970.

"Public Health Risks of Thermal Power Plants." *Nuclear Safety* 14 (1973): 267.

Sagan, L. A., ed. *Human and Ecological Effects of Nuclear Power Plants.* Springfield, Ill.: Charles C. Thomas, Publisher, 1974.

U.S., Atomic Energy Commission. *Comparative Risk-Cost-Benefit Study of Alternative Sources of Electrical Energy* (WASH-1224). Washington, D.C., 1974. See also *Nuclear Safety* 17 (1976): 171.

______. *The Environmental and Ecological Forum 1970–1971* (TID-25857). A. B. Kline, Jr., Coordinator. Washington, D.C., 1972.

U.S., Congress, House of Representatives, Subcommittee on Energy and the Environment, Committee on Interior and Insular Affairs. *Oversight Hearings on Nuclear Energy—Overview of the Major Issues.* Washington, D.C., 1975.

U.S., Congress, Joint Committee on Atomic Energy. *Hearings on the Environmental Effects of Producing Electric Power,* Parts I and II. Washington, D.C., 1970.

______. *Selected Materials on the Environmental Effects of Producing Electric Power.* Washington, D.C., 1969.

U.S., Department of Energy. *Nuclear Reactors Built, Being Built, or Planned in the United States as of December 31, 1977* (TID-8200-R37). Washington, D.C., 1977.

U.S., Energy Policy Staff, Office of Science and Technology, Executive Office of the President. *Electric Power and the Environment.* Washington, D.C.: Government Printing Office, 1970.

2

Fundamental Principles of Nuclear Reactors

INTRODUCTION

The purposes of this chapter are first, to explain the basic scientific principles of the fission process and how this process is utilized for the release of nuclear energy; and, second, to outline some of the main characteristics of the chief nuclear reactor systems that make use of fission for generating electricity in the United States. The information presented here is intended to serve as a background for understanding the environmental problems associated with nuclear electric power plants.

CHARACTERISTICS OF ATOMS

Elements and Atoms

All naturally occurring substances on earth are made up from one or more of about 90 different kinds of simple materials known as *elements.* Among the common elements are gases, such as oxygen and nitrogen, which are the main components of air; solid nonmetals, such as carbon and sulfur; and various metals, including iron, zinc, and copper. There are also other elements—for example, mercury—that are normally liquids. A less familiar element, which has attained prominence in recent years because of its use as a source of nuclear energy, is the metal uranium.

The smallest part of any element that can exist, and that determines the characteristics of that element, is called an *atom* of that element. Thus, there are atoms of hydrogen, of iron, of uranium, and so on, for all the elements. The hydrogen atom is the lightest of all atoms, whereas atoms of uranium are the heaviest atoms found in nature in significant amounts. Still heavier atoms, such as those of plutonium, also important for the release of nuclear energy, have been artificially made, starting with uranium.

The Nucleus of the Atom

Even the heaviest (and largest) atoms are extremely small. About 100 million of the larger atoms placed side by side would extend for a length of one inch.

Every atom consists of a still smaller central region, or *nucleus*, surrounded by light particles known as *electrons*. The nucleus carries a positive charge of electricity, whereas the electrons have a negative charge. In the normal atom, the positive charges on the nucleus are exactly balanced by the negative charges of the electrons; therefore, the atom as a whole has no net electrical charge (Fig. 2-1). In other words, the atom is said to be electrically neutral.

The atomic nucleus itself is made up of even simpler fundamental particles called *protons* and *neutrons*. The nucleus of the light hydrogen atom is merely a proton, but the nuclei of all the other atoms contain both protons and neutrons. These two particles have almost the same mass, but differ in the following respect: the proton carries a single charge of positive electricity, whereas the neutron, as its name implies, carries no charge and is electrically neutral. It is, in fact, the protons that are responsible for the positive charge on the nucleus. However, both the protons and the neutrons contribute to the mass of the nucleus. Because the electrons have such a very small mass, the nucleus carries more than 99.9 percent of an atom's mass.

Chemical Energy and Nuclear Energy

The negatively charged electrons are retained in the atom as a result of the force of electrical (electrostatic) attraction between them and the positively charged nucleus. However, the neutrons and protons within the small but relatively heavy nucleus are held together by a very much stronger force of a different type. It is this great difference between electrostatic forces and nuclear forces that is responsible for the potency of nuclear energy sources, as compared with the more conventional sources of energy.

In the combustion of fossil fuels, heat is produced as a result of chemical reactions between the elements carbon and hydrogen in the fuel and oxygen in the air. These reactions involve only the electrons of the various atoms, the nuclei remaining unchanged. Thus the energy release is determined by the electrostatic forces.

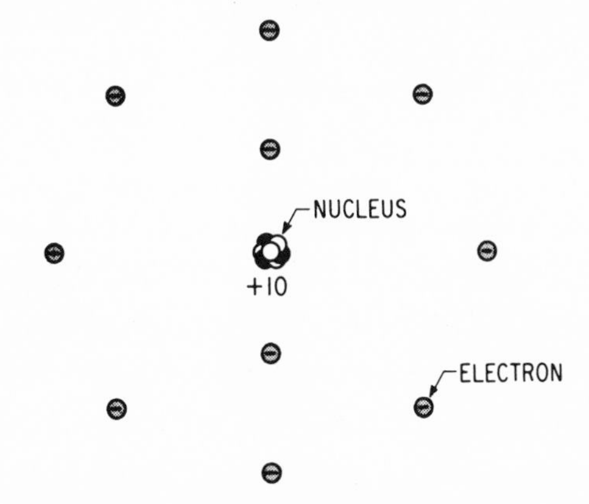

FIG. 2-1. Representation of the structure of an atom (not to scale). This particular (oxygen) atom has ten positive charges in the nucleus, which is surrounded by ten electrons.

In nuclear fission, on the other hand, the nucleus of a uranium (or other fissionable) atom splits up into smaller nuclei. Because there is a rearrangement of neutrons and protons, the strong nuclear forces are operative. The amount of energy released is then very much greater than from chemical (combustion) reactions with the same quantity of a fossil fuel. For example, the complete fission of one pound of the element uranium would release as much energy as the combustion of about 1300 (short) tons of coal or of 240 000 gallons (5700 barrels) of residual fuel oil. In metric units, 1 kilogram (kg) of uranium would produce the same amount of energy as about 2.6 million kg of coal or 910 000 litres (910 cubic meters [m^3]) of oil.

The Existence of Isotopes

The essential difference between atoms of different elements lies in the number of protons (or positive charges) in the nucleus; this is called the *atomic number* of the element. The nuclei of hydrogen atoms, as stated above, each contain one proton; helium atoms have two protons; uranium atoms, 92 protons; and plutonium atoms, 94 protons. Although all the nuclei of a given element have the same number of protons, they may have different numbers of neutrons. The resulting atomic species, which have identical atomic numbers but differ in their masses (i.e., total number of protons and neutrons), are called *isotopes* of the particular element.

An isotope of a given element is identified by its *mass number;* this is the sum of the numbers of protons and neutrons in the nucleus. For example, the element uranium occurs in nature mainly in two isotopic forms; the nuclei of one isotope contain 92 protons and 143 neutrons (mass number 235), whereas those of the other isotope contain 92 protons and 146 neutrons (mass number 238). Consequently, these two isotopes are referred to as uranium-235 and uranium-238, respectively. Both play important but different roles in the release of nuclear energy.

The term *nuclide* is frequently used to describe a species of atom (or element) characterized by the composition of its nucleus. Thus, one nuclide differs from another in that there is a difference in the number of protons and/or neutrons in their respective nuclei. The isotopes of a given element are nuclides that have the same number of protons, but differ in the number of neutrons in their nuclei.

RADIOACTIVITY

Radioactive Isotopes and Their Radiations

All but about 20 of the known elements exist in nature in two or more isotopic forms. Most of these natural isotopes are stable; that is to say, the nuclei do not change in any way over long periods of time. Some isotopes, however, have nuclei that are not stable; they continuously undergo changes by emitting radiations. These unstable isotopes are said to be *radioactive* and exhibit the phenomenon of *radioactivity.* The process of radioactive change is commonly referred to as *radioactive decay.* Some 40 (or so) radioactive isotopes (or *radioisotopes*) of 12 heavy elements, such as uranium, thorium, and radium, occur naturally on earth. In addition, more than a thousand radioactive species (i.e., *radionuclides*)[a] have been produced artificially by various nuclear reactions, including fission.

The subject of radioactivity is of special importance in connection with the release of nuclear energy by fission. The main reason is that the residues of the fission process, called *fission products*, constitute a complex mixture of more than 300 radionuclides.

Three kinds of radiation are associated with the more common types of radioactive decay. First, there are *alpha particles*, which consist of two protons and two neutrons and are identical with the nuclei of helium atoms. When emitted from a radionuclide, alpha particles have a high velocity and travel two or three inches in air before they are brought virtually to rest. They then pick up two electrons and thus become ordinary atoms of helium. In materials denser than air, alpha particles are stopped within much shorter distances. For example, they are unable to get through the outer layers of the human skin.

Beta particles represent the second type of radiation. These particles are actually electrons moving at very high speeds. But they are not those electrons that normally surround the nucleus. In the radionuclides that emit beta particles, a neuton in the nucleus changes spontaneously into a proton and an electron. The proton remains, but the electron is expelled immediately as a beta particle. Beta particles can travel several feet through air before they are absorbed; they can penetrate the human skin to a depth of a very small fraction of an inch.

The nuclei of nearly all radionuclides emit either an alpha particle or a beta particle. This emission is often accompanied by the third kind of nuclear radiation, *gamma rays*. Gamma rays are basically the same as x rays, but they generally have more energy. Gamma rays can travel great distances through air and can pass through appreciable thicknesses of denser material (Fig. 2-2). Thus, gamma rays are sometimes able to penetrate quite deeply into the body.

When a radioactive nucleus expels an alpha or beta particle, the numbers of both protons and neutrons in the nucleus are changed. Hence, the product nucleus, called the *decay product*, is that of a different element from the one undergoing radioactive decay. The decay (or daughter) product may or may not be radioactive. If it is radioactive, then it will change in turn into another decay product. After a number of stages of radioactive decay, a stable (i.e., nonradioactive) species is formed as the end product of the particular decay chain.

The expulsion of an alpha or beta particle often leaves the resultant nucleus with energy in excess of the normal value. The extra energy is then removed by the emission of gamma rays, but the numbers of protons and neutrons in the nucleus are unchanged. The nature of the element (and isotope) is thus unaffected by the emission of gamma rays; the nucleus that remains is simply in a lower energy state than it was initially.

[a]The terms *radionuclide* and *radioisotope* are often used interchangeably; however, radionuclide is more general, referring to any radioactive species (or nuclide), whereas radioisotope should be used only when the element is specified (e.g., radioisotopes of uranium).

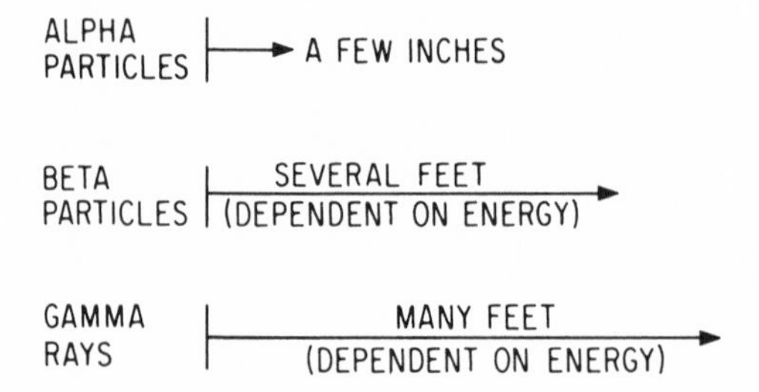

FIG. 2-2. Qualitative comparison of penetration of alpha and beta particles and gamma rays in the air.

Activity and Half-Life

The emission of radiations from a radioactive nuclide is a spontaneous process. The intervals between successive emissions of particles (or rays) are random in character, but a statistical average over a significant time period indicates a definite emission (or disintegration) rate that depends only on the nature and amount of the emitting species. If this statistical average is determined over an extended period, the rate of disintegration is found to decrease because the number of radioactive nuclei present is being depleted by their decay. The rate at which disintegration (i.e., particle emission) of a given quantity of radioactive material is occurring is frequently called the *activity* of the sample.

The time required for the activity of a given sample of a particular radioactive species to decrease to half of its original value is called the *half-life*. Every such species has its own characteristic half-life, which can be determined by observing how the rate of emission of radiation changes over a period of time. The half-lives of different radioactive species range from less than a millionth of a second to billions of years.

The half-lives of some radionuclides of special interest for the purpose of this book are given in Table 2-I. The first part of the table lists the substances, referred to as *fuel materials*, that are used either directly or indirectly for the release of nuclear energy by fission. The second part contains a few of the radioactive species that result from the fission process.

An important aspect of the half-life of a radionuclide is that no matter how much of the nuclide is considered, it always takes the same time for the activity (or rate of particle emission) to decrease to half of the initial value. The rate of decrease of radioactivity in this manner is illustrated in Fig. 2-3.

Suppose the activity of a certain amount of radioactive material at any time is represented by 100; this corresponds to zero time on the curve. Then, after one half-life, the activity will be down to 50; another half-life later, it will have decreased to 25; after three half-lives, the activity will be down to 12.5, and so on. Thus, it takes a little more than three half-lives for the activity of any quantity of radioactive material to decrease to 10 percent of its initial value. In

TABLE 2-I. Half-Lives of Some Radionuclides

Radionuclide	*Half-Life*
Fuel Materials	
Uranium-233	158,000 yr
Uranium-235	704 million yr
Uranium-238	4.47 billion yr
Thorium-232	14 billion yr
Plutonium-239	24,400 yr
Products of Fission	
Iodine-131	8 days
Krypton-85	10.8 yr
Tritium	12.3 yr
Strontium-90	28 yr
Cesium-137	30 yr

somewhat less than seven half-lives the activity will be down to 1 percent, and in ten half-lives, it will be only 0.1 percent.

Hazards of Radiations from Radioactive Substances

In their passage through matter, alpha and beta particles and gamma rays are able to remove one or more (negatively charged) electrons from atoms they encounter, thereby leaving positively charged residues called *ions*. Alpha, beta, gamma, and certain other radiations thus are said to be capable of causing *ionization*, and hence they are referred to as *ionizing radiations*. Exposure of living organisms to sufficiently large amounts of ionizing radiations can cause harmful effects.

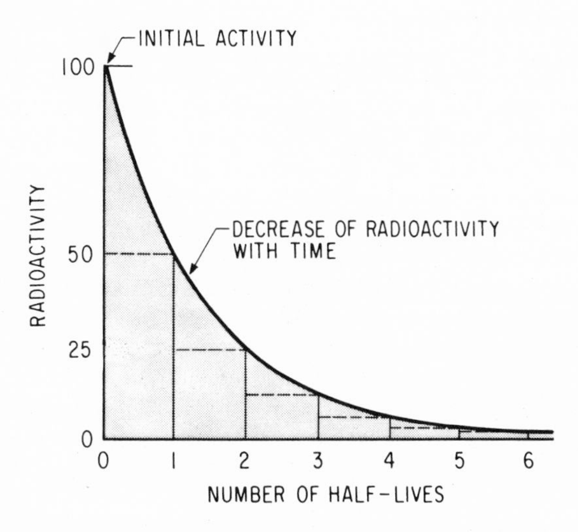

FIG. 2-3. Representation of the decay of a radioactive species with time.

Because they have such short penetration distances, alpha and beta particles from sources outside the body do not generally constitute a hazard. The skin can be injured if the source remains in contact with it for some time, but this would be an unusual situation. On the other hand, gamma rays, like x rays, can penetrate some distance into the body. Hence, they can represent a hazard from even external sources. It is not the gamma ray itself that can cause damage to tissue, but rather the high-energy electrons produced when gamma-ray energy is absorbed in the body. The potential injury depends primarily on the amount of energy absorbed in this manner.

If the radioactive source should enter the body (e.g., from food, water, or air), injury could arise from the alpha or beta particles, as well as from the gamma rays.[b] The extent of the potential hazard depends on many factors, including the following: the known tendency of the isotopes of a few elements to concentrate in specific tissues (e.g., iodine in the thyroid gland); the amount of radioactive material taken into the body and the rate at which it is removed by natural elimination processes and radioactive decay; and the nature and energy of the radiations emitted.

As a result of the operation of a nuclear power plant, radioactive fission products are formed. Essentially all of these products emit beta particles, often accompanied by gamma rays. As noted in later chapters, every effort is made to minimize the release of this radioactive material to the surroundings. Nevertheless, there is always some discharge of low-level radioactivity, carried out under carefully controlled conditions, in both gaseous and liquid effluents. The quantities of radioactivity discharged and their probable effects on man and the environment are important concerns of this book.

THE FISSION REACTION

Fissile and Fertile Nuclides

A nuclear species (or nuclide) that can be used to achieve the sustained release of nuclear energy by fission is said to be *fissile*. The only fissile nuclide existing naturally on earth is the lighter isotope of uranium—namely, uranium-235. This isotope is present to the extent of only 0.71 percent by weight in natural uranium; nearly all of the remainder consists of uranium-238. Thus, the fissile uranium-235 is by far the less abundant of the two main isotopes of uranium found in nature.

Uranium-233 and plutonium-239 are also fissile, but these substances do not occur naturally. They are obtained by nuclear reactions starting with thorium-232 and uranium-238, respectively, which exist in nature in substantial amounts (p. 16). Natural uranium contains about 99.3 percent of uranium-238, and natural thorium is almost exclusively thorium-232. These two nuclides cannot be employed directly for the sustained release of nuclear energy, but they

[b]The human body always contains significant amounts of beta-emitting radioisotopes of potassium and carbon, as well as small quantities of uranium, radium, and other alpha-particle emitters derived from natural sources (see Chapter 6).

can be converted into substances that can be so used. Consequently, thorium-232 and uranium-238 are referred to as *fertile* nuclides.

The Fission Chain

The nuclei of all elements, except those of the lightest isotope of hydrogen, contain neutrons. It is a relatively simple matter to release neutrons from certain nuclei and thus obtain *free neutrons.* If such a free neutron enters the nucleus of a fissile species, fission may occur. Actually, fission is not the only process that can take place when a fissile nucleus absorbs a neutron, but attention will be restricted here to the fission reaction.

The entry of a neutron into a fissile nucleus results first in the formation of a *compound nucleus.* The latter then splits almost instantaneously into two smaller (i.e., lighter) nuclei called *fission fragments.* At the same time, two to three neutrons, on the average, are set free, and there is a release of a large amount of energy. The fission process may thus be represented as follows:

Neutron + Fissile Nuclide ⟶ Fission Fragments + Neutrons + Energy

The important point is that the free neutrons produced in the fission reaction can serve to induce fission in more fissile nuclei, thereby releasing more neutrons. These, in turn, can cause further fissions, and so on. Thus, a single neutron could, in principle, initiate a chain of fission reactions accompanied by the continuous release of energy. In other words, because fission is caused by neutrons and results in the liberation of other neutrons, a self-sustaining energy release process is possible. High-energy neutrons can cause fission of the fertile nuclides uranium-238 and thorium-232, but a chain reaction cannot be maintained in them.[c]

Fission Energy

Except for a small amount that escapes, the energy released by fission in a nuclear reactor (p. 18) appears as heat. After a short time of operation, to permit a steady state to be attained, about 93 percent of the heat is generated at (or close to) the instant of fission. The remaining 7 percent (or so) arises from the absorption of the energy carried by the beta particles and gamma rays emitted in the radioactive decay of the fission products. An important consequence, as noted in Chapter 4, is that when a reactor is shut down after a period of operation, heat continues to be liberated by the decay of the accummulated fission products at a rate that decreases with time.

Calculations, which have been confirmed by actual measurements, indicate that complete fission of one pound (0.45 kg) of fissile material in a nuclear reac-

[c]Such nuclides are said to be fissionable, rather than fissile. The latter term is reserved for nuclides in which neutrons of all energies can cause fission and in which a fission chain can be maintained.

tor would result in the production of about 35 000 million British thermal units (Btu) of heat.[d] If this heat could be completely converted into electricity, one pound of fissile material would generate more than 10 million kWh. The conversion efficiency of most current nuclear power systems is roughly 33 percent (Chapter 1); hence, the actual quantity of electricity generated would be about 3.3 million kWh (i.e., 3300 megawatt-hours [MWh]). Consequently, when a 1000-MW electrical nuclear power plant operates at full capacity, about 5.3 oz (0.15 kg) of fissile material actually undergo fission per hour (i.e., roughly 8 pounds [3.6 kg] per day). For reasons that will be apparent shortly, the actual amount of fuel required to operate a nuclear reactor is considerably larger than the fissile material that is consumed. Nevertheless, the total weight of the nuclear fuel requirements is many thousand times smaller than the fossil fuel burned in a conventional generating plant with the same electrical capacity.

The Critical Mass

An average of between two and three neutrons are liberated in each act of fission, whereas only one is needed at each stage to maintain the fission chain. It would seem, therefore, that once the fission reaction is initiated in a mass of fissile material, it would readily sustain itself. However, this is not the case if the mass is too small, because not all neutrons produced in fission are available to carry on the fission chain. Some neutrons are lost in nonfission reactions, both in the fissile species itself and in any other materials that may be present. In addition, neutrons are lost by escape from the system in which fission is taking place.

The fraction of neutrons lost by escape can be reduced by increasing the size (or mass) of the system. The minimum quantity of fissile material that is capable of sustaining a fission chain, once it has been initiated by a source of neutrons, is called a *critical mass*. The magnitude of the critical mass depends on a wide variety of conditions. These include the nature of the fissile nuclide, its proportion in the fuel material, the geometry of the system, and the presence of substances that either absorb neutrons readily or affect the speed with which the neutrons are moving. For a specific reactor system, however, the critical mass of fissile material has a definite value.

If the quantity of fissile material in a reactor falls below that required for criticality under the existing conditions, the system is said to be *subcritical*. Although some fissions can occur in the subcritical state, fission chains gradually die out because not enough neutrons are available to maintain them. On the other hand, in a *supercritical* system, there is more fissile material than in a quantity that is just critical. The number of fission chains then increases at a rate depending on the degree of supercriticality.

The Fission Products

As noted earlier, the lighter nuclei formed in the fission reaction are known as fission fragments; they are also sometimes referred to as the *initial fission products*. There are about 40 different ways in which a fissile nucleus can split into

[d]Combustion of 907 kg (1 [short] ton) of coal would produce, on average, about 27 million Btu, and 0.16 m^3 (1 barrel) of fuel oil would produce about 5.8 million Btu.

two parts; consequently, there are some 80 or so different fission fragments produced. The different modes of fission are not equally probable, and the proportions of the individual fission fragments cover a considerable range.

Nearly all of the approximately 80 fission fragments are radioisotopes of familiar elements. Since they are radioactive, the fragment nuclei begin to decay immediately after they are formed. Some of the fission fragments have such short half-lives that they virtually disappear within a few seconds, whereas others persist for much longer times. In most cases, the decay products of the fission fragments are also radioactive, and they decay in turn. On the average, there are about four stages of beta-particle emission before a stable (nonradioactive) species results.

The general term *fission products* is applied to the complex mixture that arises from the radioactive decay of the fission fragments and their decay products. The exact composition of the fission product mixture—that is, the nature and proportions of the nuclides present—depends on the species undergoing fission and on the speed of the neutrons causing the fission. Regardless of their origin, however, the fission products formed in the roughly four stages of decay of the 80 fission fragments are a mixture of more than 300 nuclides, most of which are radioactive. In nearly all cases, the radioactivity consists of the expulsion of a beta particle, frequently accompanied by gamma rays.

In a small proportion of fissions, a light nuclear particle, such as an alpha particle or a tritium nucleus, is also formed. Tritium is a radioactive isotope of hydrogen and a small quantity of this substance is always generated as a result of fission. It is thus, in a sense, a minor fission product.

Uranium Enrichment

The light-water reactors (LWRs) (p. 26), which are expected to make a significant contribution to the electrical capacity of the United States for the remainder of this century, cannot use natural uranium as the nuclear fuel. To become critical and maintain a fission chain reaction, these reactors require a fuel that contains more uranium-235 than the 0.71 percent by weight present in ordinary (natural) uranium. Such material is commonly referred to as *enriched* uranium. For LWRs, the fuel is usually enriched to the extent of 2 to 4 percent in the fissile uranium-235 (half-life = 704 million years).

Enrichment of uranium in the lighter (fissile) isotope is achieved by a process known as *gaseous diffusion*. Uranium compounds, extracted from uranium ores and purified, are converted into uranium hexaflouride, UF_6. This is a solid at room temperature, but is changed completely into vapor at 56.4 °C (133.5 °F). The vapor, like the solid, consists of a mixture of uranium-235 hexafluoride and uranium-238 hexafluoride in the same (molecular) proportion as the two isotopes occur in nature.

The uranium hexafluoride vapor is forced through a porous barrier containing a large number of very small holes. The hexafluoride molecules containing the lighter isotope, uranium-235, travel more rapidly than do those containing the heavier isotope. As a result, uranium-235 hexafluoride exhibits preferential diffusion through the porous barrier—that is, the vapor that first passes through the barrier is somewhat richer in uranium-235. The amount of enrichment attained by diffusion through a single barrier is quite small, but passage through a

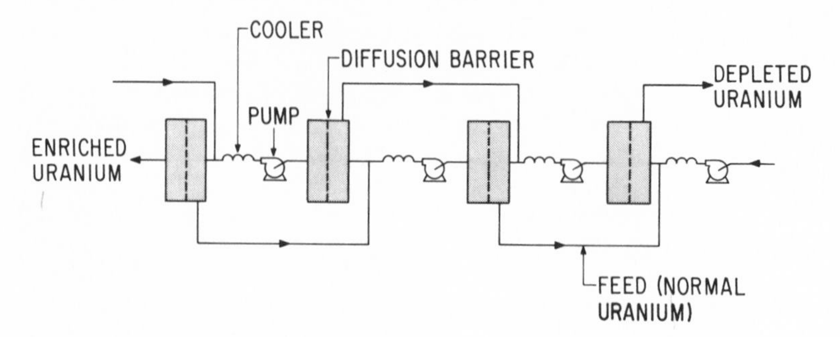

FIG. 2-4. Separation of isotopes by gaseous diffusion.

series (or cascade) of porous barriers can lead to products of almost any degree of enrichment, up to more than 90 percent of uranium-235 (Fig. 2-4). The gaseous diffusion cascade can consist of several hundred stages for enrichments of a few percent, or thousands of stages for higher enrichments. The gaseous diffusion plants thus cover large areas and consume considerable amounts of power. The amount of electricity used to operate these plants, however, is less than 5 percent of the power that can be generated from the enriched uranium that is produced.

There are three gaseous diffusion installations in the United States, located in Tennessee, Kentucky, and Ohio. With planned improvements, these three plants should be able to satisfy the projected demands of nuclear power facilities in the United States well into the early 1980s, and also to meet the commitments to provide enrichment services for some other countries.

If the use of nuclear energy for the generation of electricity develops in the manner expected, it appears that an additional capacity for uranium enrichment will be required. To meet this requirement, a new enrichment plant is to be built in Ohio. This plant will utilize the gas centrifuge method of isotope enrichment, mainly because it will need less than one-tenth the power required to operate an equivalent gaseous diffusion plant.

In a centrifuge spinning at high speed, the molecules of the heavier uranium-238 hexaflouride tend to move outward, whereas the lighter uranium-235 hexafluoride molecules move inward toward the axis of rotation. The product withdrawn from near the axis of the centrifuge is thus enriched in uranium-235.

Uranium hexafluoride always has the same chemical properties, regardless of the uranium-235 content. The product of the gaseous diffusion (or alternative enrichment) process can thus be converted into uranium metal, uranium dioxide, or other uranium compound for use as the nuclear fuel in a reactor. The uranium-235 enrichment of the fuel material is the same as that of the uranium hexafluoride from which it is made.

Production of Plutonium-239

The fissile species plutonium-239 is obtained from the heavier uranium isotope, uranium-238, as the fertile material. If uranium-238 is present in a nuclear fission reactor, it will capture some of the free neutrons and thereby be converted into a still heavier isotope, uranium-239, which is not found in nature. The uranium-239 is radioactive, with a half-life of about 23.5 minutes.

Thus, it decays fairly rapidly, and the decay product is neptunium-239, an isotope of a synthetic element that does not occur naturally on the earth except perhaps in minute traces. Neptunium-239 has a half-life of 2.3 days, and it in turn decays into plutonium-239. Like neptunium, plutonium is a synthetic element. The plutonium-239 isotope is radioactive, but has a half-life of 24 400 years; hence, it decays so slowly that it remains essentially unchanged in amount for many years after it is formed. Furthermore, when it does decay, the product is the long-lived fissile uranium-235.

The production of fissile plutonium-239 from the fertile material uranium-238 is summarized below:

Uranium-238 + Neutron ⟶ Uranium-239
Uranium-239 (half-life 23.5 min) ⟶ Neptunium-239
(+ beta particle)
Neptunium-239 (half-life 2.3 days) ⟶ Plutonium-239
(+ beta particle)
(half-life 24 400 yr).

It is evident that if uranium-238 is present in an operating reactor, it will generate fissile plutonium-239 as a result of neutron capture followed by two stages of beta decay. Since plutonium-239 has such a long half-life, it tends to accumulate in the reactor. However, the amount does not increase beyond a certain limit. The reason for this is that the accumulation of plutonium-239 is increasingly counteracted by its removal in fission and nonfission reactions.

When material containing plutonium-239 is discharged from the reactor with the remaining uranium and fission products, it could be sent to a reprocessing plant. Here plutonium-239 would be extracted, usually in the form of plutonium nitrate. Plutonium nitrate could then be converted chemically into the particular material (i.e., the metal, oxide, or carbide) that may be required for use as reactor fuel.

Production of Uranium-233

The third fissile material, uranium-233, is produced from the fertile thorium-232 by a series of reactions that are quite similar to those involved in the formation of plutonium-239 from uranium-238. When exposed to neutrons in a nuclear reactor, the thorium-232 nucleus captures a neutron and forms radioactive thorium-233, with a half-life of about 23 minutes. The decay product is protactinium-233, which has a half-life of 27.6 days. Decay of protactinium-233 leads to the formation of the fissile uranium-233 with a half-life of 158 000 years. The mode of production of uranium-233 in a reactor is then as follows:

Thorium-232 + Neutron ⟶ Thorium-233
Thorium-233 (half-life 23 min) ⟶ Protactinium-233
(+ beta particle)
Protactinium-233 (half-life 27.6 days) ⟶ Uranium-233
(+ beta particle)
(half-life 158 000 yr).

Only moderate amounts of uranium-233 have been produced so far. Processes have been developed, however, for the separation of uranium from thorium and admixed fission products. Since uranium-233 has the same chemical properties as the other isotopes of uranium, it can be converted into the desired fuel material (e.g., metal, oxide, or carbide) in the same manner.

NUCLEAR REACTORS

Thermal and Fast Reactors

There are many possible variations in the design and components of nuclear reactors in which the fission process is used for the release of energy as heat. Nevertheless, all power reactors have certain features in common. Details of the important types employed for the generation of electricity in the United States are presented later in this book, but a general description, emphasizing certain common features, is given here.

First, however, a distinction must be made between the two main reactor categories—namely, *thermal reactors* and *fast reactors*. Both of these terms are somewhat misleading, but they have come into general use and must be defined.

The neutrons released in the fission process have high (kinetic) energies and, hence, travel with very high velocities. Consequently, they are called *fast neutrons*. As a result of collisions with the nuclei of light atoms, such as hydrogen and carbon, the fast neutrons lose part of their energy and are slowed down. The relatively slow neutrons are referred to as *thermal neutrons*, because their velocities are largely determined by the temperature of the medium in which they move. In a fast reactor, the great majority of the fissions are caused by fast neutrons, whereas in a thermal reactor, most fissions result from the absorption of much slower neutrons.

Reactor Core and Fuel Elements

A schematic outline of a common type of nuclear steam supply system (p. 1) is shown in Fig. 2-5. The reactor (at left), surrounded by a shield, consists of a *core*, in which the fission chain is sustained and in which the fission energy is released as heat. The core contains nuclear fuel, usually in the form of long, thin cylindrical rods. The fuel rods or assemblies of rods are sometimes called *fuel elements*. In power reactors, the fuel consists of a fissile nuclide, which is uranium-235 in most existing reactors, together with a substantial proportion of a fertile material, usually uranium-238. In the thermal power reactors in the United States, the fuel generally contains an average of about 3 percent of uranium-235. Fast reactors, on the other hand, require fuel with roughly 15 percent or more fissile material.

The fissile (and fertile) species in a fuel element can be in the form of metal, but an oxide or carbide is commonly used because it can withstand much higher temperatures without melting. The fuel material is nearly always enclosed by a thin metal cladding. This cladding has two purposes: it prevents the escape of fission products into the coolant used to remove the fission heat, and it protects the fuel from chemical attack by the coolant.

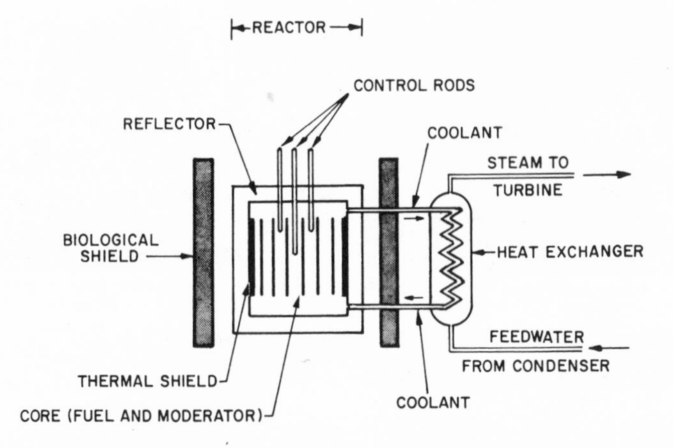

FIG. 2-5. Schematic outline of a common type of nuclear reactor system for generating steam.

Moderators in Thermal Reactors

In a thermal reactor, the core contains a material known as a *moderator*, the purpose of which is to decrease (i.e., moderate) the speed of the fast neutrons that are liberated in the fission process. Most thermal power reactors in the United States employ ordinary (light) water as the moderator (and coolant). It is the hydrogen nuclei in the water that are effective in slowing down the neutrons. In some (gas-cooled) power reactors, the moderator is the element carbon in the form of graphite. Fast reactors do not have a moderator as such; elements that might slow down the fission neutrons are avoided as far as possible, but some will be present in the oxide fuels commonly used.

Reflectors

The core is surrounded by a neutron *reflector*, the nature of which depends on whether the reactor is thermal or fast. The purpose of the reflector is to decrease the loss of neutrons that have escaped from the core. As the name implies, the reflector returns many of these neutrons to the core where they can contribute to the fission chains. Among the consequences are a decrease in the critical mass of the fuel and an improved distribution of fissions (and hence improved heat generation rate) within the core.

In thermal power reactors, the reflector usually consists of the same material that serves as the moderator. A water-moderated reactor, for example, would generally have a water reflector, whereas in a reactor with graphite as a moderator, the reflector would also be made of graphite. In fast reactors, the reflector, called a *blanket*, is usually chemically similar to the fuel (e.g., uranium oxide), but the uranium is not enriched in fissile material. Part of the breeding process, which is a characteristic of fast reactors, occurs in the blanket (p. 25).

Reactor Coolant

The heat generated in the reactor core, as a result of the fissions occurring there, is removed by circulation of a suitable *coolant*. The coolant may be either a liquid or a gas. In power reactors in the United States that use water as the moderator (and reflector), the same water acts as the coolant. Thus, water-cooled reactors have the advantage of using the same material as coolant, moderator, and reflector. On the other hand, there are drawbacks, such as the necessity for operating the reactor at high pressure to attain the high steam temperatures required to increase turbine efficiency (p. 2).

Most reactors in which graphite is the moderator utilize a gas as the coolant. In the United Kingdom power reactors have been operated for several years with carbon dioxide gas as the coolant. In the United States, however, the preferred gaseous coolant is helium. This gas is very inert chemically, and high temperatures can be attained without danger of any material in the core suffering attack from the coolant, provided impurities can be kept out.

A special type of power reactor, called the "N-reactor," near Richland, Washington, has graphite as the moderator, but water as the coolant.

Since water slows down neutrons, this coolant is not used in fast reactors. For such reactors, the preferred coolant at present is molten sodium. Although it is very reactive chemically, with both air and water, sodium has the advantage of being an excellent heat-transfer material, which can be used at very high temperature without the need for pressurization. Furthermore, sodium is desirable because it does not slow down neutrons to any great extent. Considerable experience has been accumulated over more than 20 years in procedures for the safe handling of sodium. Consideration is being given, however, to the design of a fast reactor cooled with helium gas at a moderately high pressure.

In some water-cooled reactors, fission heat converts the liquid water directly into steam in the reactor vessel. In other reactors employing either water, sodium, or helium gas as the coolant, the heated coolant produces steam in a separate heat-exchange system, as indicated at the right of Fig. 2-5. The characteristics of the more common nuclear steam supply systems are described later in this chapter.

Reactor Control

When a reactor is exactly critical (p. 14), the neutron density remains constant; for the neutron density to be increased, the system must be supercritical; for it to be decreased, the reactor must be subcritical. A convenient way of expressing the extent of departure of a reactor from criticality is by means of a quantity called the *reactivity*. In an exactly critical system, the reactivity is zero, and the neutron density does not change. The reactivity is positive in a supercritical system, and the neutron density steadily increases. Finally, a negative reactivity signifies a subcritical system with a decreasing neutron density.

The rate of heat generation (or power level) in a reactor core is determined by the rate of fission and this depends on the neutron density (i.e., the average number of neutrons per unit volume). Consequently, the power level can be changed by varying the reactivity. In a thermal reactor, variations are achieved by inserting or withdrawing control rods containing a neutron *poison*—that is, a substance that can readily absorb neutrons. Among such poisons are the elements boron and cadmium; control rods for thermal reactors usually contain one or the other of these elements in convenient form.

If a neutron-absorbing control rod is inserted into the reactor core, it competes with the fissile material for the available neutrons. Consequently, the absorber interferes with the propagation of the fission chain. This results in a decrease in the reactivity and, hence, in the power level. Insertion of the control rods beyond their normal operating position causes the reactivity to become negative. The reactor will then be subcritical and will shut down. On the other hand, withdrawal of the poison control rods is accompanied by an increase in the reactivity and in the operating power level of the reactor. Thus, the power of the reactor can be adjusted to the desired level by appropriate movement of the control rods.

In small fast reactors, the reactivity has sometimes been varied by controlling the escape of neutrons from the reactor core by moving part of the reflector. In large fast (power) reactors, however, control is based on the use of a neutron absorber, usually boron, in a manner similar to that in thermal reactors.

An important feature of the fission process that facilitates safe operation and control of reactors is that not all the fission neutrons are released instantaneously. A small fraction of the neutrons accompanying fission arise in an indirect manner, and their emission is delayed. As long as the conditions in the reactor are such that these delayed neutrons, in addition to those released promptly, are required to maintain the fission chains, the neutron density (and power level) can be increased at a readily controllable rate. Attainment of an undesirably high power level can then be prevented.

The average number of free neutrons produced in fission and the fraction of these that are delayed depend primarily on the identity of the fissile nuclide. Neutron speed does have some effect, but it is minor. Hence, there is no difference in principle in the ease of control of thermal and fast reactors using the same fissile material. The number of delayed neutrons per fission is less, however, for plutonium-239 than for uranium-235; consequently, reactors employing the former as the fissile material (e.g., fast reactors) have less control margin than those using uranium-235. Within this margin, however, they can be controlled safely.

Shielding and Containment

To ensure the safety of nuclear power operators, the reactor is surrounded by biological shielding (Fig. 2-5). The purpose of this shielding is to attenuate the neutrons and gamma rays escaping from the reactor to such an extent that the radiations do not constitute a hazard to persons in the immediate vicinity. Power reactors often have two different shields: a *thermal shield* and a *biological shield*.

The thermal shield, which is within the reactor vessel and is fairly close to the core, usually consists of a few inches of iron or steel. By absorbing much of the gamma radiation and some of the neutrons, the thermal shield protects the biological shield from possible damage due to overheating. It also protects the steel reactor vessel from embrittlement that could result from continuous bombardment by neutrons. The biological shield is generally a layer of concrete several feet thick surrounding the vessel containing the reactor core, reflector, and coolant. This shield attenuates neutrons and gamma rays that are not removed by the thermal shield.

As a precaution against the possible spread of radioactive materials (e.g., the fission products) in the unlikely event the reactor core is damaged, the whole nuclear steam supply system is enclosed in a containment vessel made of reinforced concrete with a steel liner. This vessel, which is a characteristic feature of nuclear power plants, is designed to withstand the highest pressure that could be attained as the result of the worst credible accident with the particular reactor system.

Excess Reactivity and Its Control

In the course of the operation of a nuclear reactor, the fissile material is gradually consumed. Although there may be some replacement by conversion of fertile into fissile species, there is generally a net decrease in the quantity of fissile material present.[e] If the reactor were initially critical or slightly supercritical, this reduction in the fissile species would tend to make the reactor subcritical. There is also another factor that has the same effect. Some of the fission products and heavier nuclei formed in the fuel can capture neutrons very readily; that is, they act as neutron poisons. As a result, fewer neutrons are available for maintaining the fission chains.

Both a decrease in the fissile material and an accumulation of poisons in the core tend to decrease the reactivity of the system. Unless something were done to overcome this situation, the reactor would soon become subcritical and cease to operate. Consequently, a freshly loaded reactor core contains additional fuel, over and above the minimum required for criticality, to provide what is called *excess reactivity*. Not only can this excess compensate for the tendency of the reactivity to decrease during operation, but it also allows for changes in reactivity arising from temperature and other effects.

Various methods are used in different reactor designs to overcome the initial excess reactivity and thus prevent the reactor core from becoming appreciably supercritical. The procedures are invariably based on the utilization of neutron poisons to decrease the reactivity. The methods used in water-cooled reactors are described in Chapter 4.

Treatment of Spent Fuel

In the interest of both safety and economy, the excess reactivity that is built into the reactor core at the beginning of its operation must be kept within reasonable limits. Hence, after operating for some time, there is not sufficient reactivity available to maintain criticality, and power production ceases. Part of the spent fuel must then be discharged and the reactor reloaded with fresh fuel. In some cases, the operating time of the nuclear reactor is not determined so much by reactivity considerations as by fuel deterioration—that is, by detrimental physical and mechanical changes due to the action of fission fragments and prolonged exposure to the high density of neutrons in the core.

[e]Breeder reactors are exceptional in this respect (p. 23).

In general, no more than three-fourths of the fissile material and only about 2 percent of the fertile species in a uranium fuel will have been consumed when the fuel elements are discharged from an LWR. Thus, the spent fuel still contains substantial amounts of the original fissile and fertile nuclides. In addition, some fissile material (e.g., plutonium-239) will have been formed from the fertile species (e.g., uranium-238). At the fuel reprocessing plant, the accumulated fission products would be removed, and then the uranium and plutonium would be separated and purified. (The reprocessing of commercial reactor fuel has been postponed indefinitely in the U.S. [p. 254].)

There is no change in the proportions of different isotopes during reprocessing, so uranium-235 in the recovered uranium would be the same as in the spent fuel (about 0.8 percent). The uranium, consequently, could be converted into hexafluoride and returned to the gaseous diffusion (or gas centrifuge) plant for re-enrichment. Most of the recovered plutonium would be set aside for possible use in two ways. First, it could be used to replace part of the uranium-235 in LWRs, in what is called *plutonium recycle*, thereby extending the available supply of fissile material. Second, the recovered plutonium could be used as the starting fuel for fast breeder reactors, a discussion of which follows.

Conversion and Breeding

On the average, the fuel material in a water-cooled thermal power reactor generally contains about 3 percent of uranium-235; the remaining 97 percent (by weight) is usually the fertile species uranium-238. Hence, in normal operation of the reactor, plutonium-239 is formed as a consequence of the capture of neutrons by uranium-238. In the early stages of reactor operation, uranium-235 undergoes fission and produces energy. At the same time, some of the fission neutrons in excess of those required to maintain the fission chain are taken up by uranium-238 nuclei, and plutonium-239 is formed. As plutonium-239 accumulates, it serves as an alternative to uranium-235 in the fission process. A reactor in which a fertile species is changed into a fissile material, while generating heat by fission, is called a *converter*. For example, in a water-cooled reactor, the total (or gross) quantity of plutonium-239 produced from uranium-238 is roughly half the amount of uranium-235 that has undergone fission.

In general, the plutonium-239 formed in a thermal reactor is less than the uranium-235 consumed. But in a fast reactor, the plutonium-239 generated from uranium-238 can exceed the quantity of fissile material used in operating the reactor. In other words, the remarkable situation is possible—and has, in fact, been demonstrated—in which a nuclear reactor produces energy while, at the same time, generating more fissile material than is consumed. A reactor in which such a net increase in fissile material occurs is called a *breeder reactor*. Breeding plutonium-239 from uranium-238 can be achieved only in a fast reactor, but breeding fissile uranium-233 from thorium-232 is (theoretically) possible in a thermal reactor.

Breeding in Fast and Thermal Reactors

The difference in the interactions with nuclei of fast and slow neutrons accounts for the fact that breeding plutonium-239 is feasible in a fast reactor, but not in a thermal reactor. The basic requirement for breeding to be possible is that, after allowing for neutrons lost by escape and by nonfission captures in both fissile and nonfissile (other than fertile) materials (e.g., fuel element cladding, coolant, control materials, etc.), more than two neutrons must be available for every neutron absorbed by fissile material. One of the available neutrons serves to maintain the fission chain, whereas the remainder (more than one) can be captured by the fertile species to generate new fissile material. Thus, on the average, for every fissile nucleus removed by fission or in other ways, more than one new fissile nucleus can be produced (Fig. 2-6).

In a thermal reactor, there is a significant loss of neutrons as a result of capture by uranium-235 or plutonium-239 in nonfission reactions. The essential requirement for breeding cannot be satisfied because, after allowing for losses, less than two neutrons are available for every one that is absorbed in the fissile species. Breeding plutonium-239 in a thermal reactor is thus not possible, although there is some conversion. In a fast reactor, however, the fraction of neutrons lost by nonfission capture in the fissile species is substantially less than in a thermal reactor, and breeding plutonium-239 becomes feasible.

Under appropriate conditions, breeding fissile uranium-233 from thorium-232 as the fertile species can be achieved in a thermal reactor, as well as in a fast or even moderately fast reactor. Nonfission capture of neutrons in uranium-233 is relatively small for both thermal and fast neutrons and, provided the loss of neutrons in other ways is kept to a minimum, more than two neutrons are available for every one absorbed in uranium-233. There is relatively little to be gained, as far as breeding uranium-233 is concerned, in using fast neutrons. Since it appears to be more convenient to utilize thermal reactors for breeding uranium-233 from thorium-232, it is mainly in this direction that studies are being made.

Importance of Breeding

Breeding, in the general sense of creating more fissile material than is consumed, regardless of its nature, is attainable with any of the three fissile species in the fuel and either uranium-238 or thorium-232 as the fertile material. If the fuel contains uranium-235 or plutonium-239 as the fissile substance, then a fast reactor system is required. With uranium-233 as the fissile species, either a thermal or a fast reactor could be used. Thus, in principle, many different types of breeder reactors are conceivable.

Most experimental breeder reactors have utilized uranium-235 as the fissile material to demonstrate the feasibility of breeding plutonium-239 or uranium-233. However, breeder reactors to be used for electric power generation will eventually produce the same fissile species as they consume. In these reactors, the fertile material will be either uranium-238 (fast reactor) or thorium-232 (thermal reactor), and the same fissile species, plutonium-239 or uranium-233, respectively, will be both consumed and produced. There will then be a decreased dependence on the limited supply of uranium-235. At present, the main interest in the United States and in several other countries is in the breeding of plutonium-239.

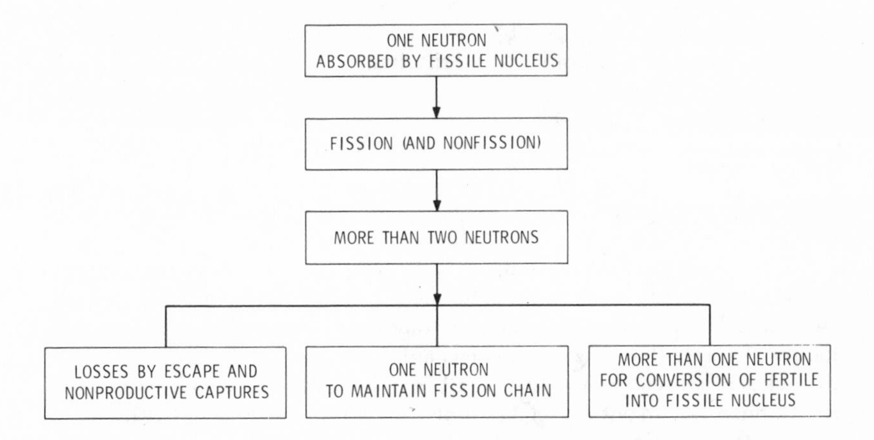

FIG. 2-6. Conditions for breeding fissile material.

With breeder reactors, it should be possible to utilize a large proportion of the available uranium-238, including the supplies of uranium depleted in uranium-235 from gaseous diffusion plants, and thorium-232 by converting them into fissile species. The large reserves of uranium and thorium that are not economically useful at present would then be available as sources of nuclear energy.

The design of a breeder reactor is essentially similar to that of any other reactor, as described earlier. One differnce is that the fertile species is included in the reflector (or blanket), as well as in the fuel elements. In these circumstances, the reflector is called a *breeding blanket*. The fertile nuclide is converted into a fissile nuclide by neutrons in the reactor core and also by many of those that escape into the surrounding blanket. If breeding occurs, the total amount of fissile species produced in the core and blanket exceeds that consumed in the release of energy.

The feasibility of breeding plutonium-239 has been established by the operation of experimental fast reactors in the United States and in other countries. But a number of technological problems remain to be solved before breeder reactors can be employed for the economic generation of electricity on a commercial scale. One of these problems is the development of fuel elements and structural materials that will not be significantly damaged by protracted exposure to high densities of fast neutrons.

To provide an alternative (or supplement) to the fast-neutron breeder reactor, studies have been conducted of thermal breeders in which thorium-232 is converted into uranium-233. In one system, the fuel is in the form of a molten salt at high temperature. This scheme avoids the problems arising from deterioration of solid fuel elements as a result of continued neutron exposure, but other problems, such as containment of the molten salt and removal of fission and other products from the spent fuel, are encountered.

The long-range future of nuclear power depends on the successful realization of one (or both) of the possible methods for breeding fissile material from the abundant supplies of uranium and thorium.

NUCLEAR STEAM SUPPLY SYSTEMS

Light-Water Reactors

Nearly all the nuclear steam supply systems that are expected to be in operation for some time to come—probably into the middle 1980s and beyond—are based on light-water reactors,—that is, thermal reactors employing ordinary (light) water as both moderator and coolant. Such reactor systems, with which this book is mostly concerned, are described in some detail in Chapter 4. For the present, however, an outline is given of the operating principles of LWRs and of two other types that might come into increasing use during the later years of this century.

Light-water reactors fall into two categories: pressurized-water reactors (PWRs) and boiling-water reactors (BWRs). In a PWR, the cylindrical core (Fig. 2-5) consists of several thousand long, thin vertical rods made of tubes packed with pellets of enriched uranium dioxide (UO_2) fuel. The enrichment ranges from about 2 to 4 percent, with an average of about 3 percent uranium-235. The core and coolant water are contained in a strong steel vessel capable of withstanding very high pressures.

At ordinary atmospheric pressure (101 kilopascals [kPa], or 14.7 pounds per square inch [psi]), water boils at a temperature of 100 °C (212 °F), but the boiling point increases as the pressure over the water is increased. The pressure in the PWR reactor vessel is maintained so high that the coolant water becomes very hot but does not boil. At the common operating pressure of 15.5 megapascals (MPa) (2250 psi), water boils at 345 °C (653 °F), but the maximum water temperature in a PWR is about 327 °C (620 °F). It is this use of high pressure to prevent boiling that led to the name pressurized-water reactor.

The pressurized water, heated by fission energy released in the fuel, passes from the reactor vessel to a separate heat exchanger (or steam generator) where the steam is actually produced. A common type of PWR steam generator is over 18.3 meters (m) (60 ft) high and contains a vertical bundle of more than 3000 tubes shaped like an inverted U, as shown in Fig. 2-7. Water from the reactor vessel circulates through these tubes and back to the reactor. The feedwater for generating steam flows at a lower (but still high) pressure around the outside of the tubes.

Heat passes through the walls of the tubes from the hot, pressurized reactor coolant water to the surrounding feedwater. Since the latter is at a lower pressure, it boils and produces steam; typically, in a modern facility, the steam has a temperature of about 293 °C (560 °F) and a pressure of around 7.6 MPa (1100 psi). An arrangement of vanes and baffles in the upper part of the steam generator serves to remove excess moisture from the steam.

Apart from dimensional (and related) differences, the core of a BWR is similar to that of a PWR, as described above. The main difference is that in the

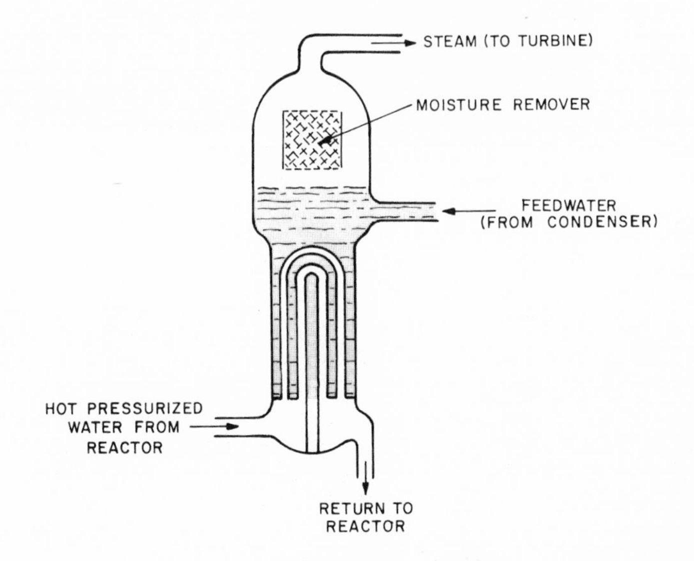

FIG. 2-7. Representation of a common type of steam generator for a PWR.

BWR the pressure is much lower, about 7.2 MPa (1040 psi) and the coolant water actually boils within the reactor vessel. Thus, steam is produced directly by fission heat within the core, and no separate steam generator is required. The wet steam rising from the boiling water enters a bank of steam separators and dryers in the upper part of the reactor vessel. The relatively dry steam leaves the vessel at much the same temperature and pressure as the steam from a PWR steam generator.

As noted earlier, the reactor fuel must be replaced periodically. Because the fuel rods have different enrichments (and for other reasons), they do not all require replacement at the same time. In the refueling operation, which is usually conducted about once a year in an LWR, approximately one-third or one-fourth of the fuel rods are removed. The remainder are rearranged and the vacant spaces are filled with fresh fuel. On the average, about 3 percent of the fuel is consumed; most of this is from the uranium-235 that was initially present, and the rest is mainly from the plutonium-239 produced from uranium-238 during reactor operation.

The first large-scale PWR plant was operated in 1957 at Shippingport, Pennsylvania; it had an electrical capacity of 60 MW, which has since been increased.[f] The first commercial BWR, the Dresden-1 plant near Morris, Illinois, commenced operation in 1960; its initial electrical generating capacity was about 170 MW. In recent years, the outputs of nuclear steam supply systems have been increased, and the larger PWR and BWR installations generate more than 1000 MW of electric power. The thermal efficiencies (p. 2) of these plants is about 32 to 34 percent.

[f]Nuclear submarines have used PWRs to generate steam for propulsion since 1954.

High-Temperature Gas-Cooled Reactors

As of 1980, only two commercial nuclear steam supply systems based on the high-temperature gas-cooled reactor (HTGR) had been constructed in the United States. The larger of these, at Fort St. Vrain, Colorado, has an electrical capacity of 330 MW, but larger plants, with capacities from 770 to 1160 MW, have been designed.

In the HTGR, the moderator is graphite, and the coolant is helium gas. Graphite is a material that has good mechanical properties at high temperatures, and helium is an inert gas. As a consequence of these factors, the coolant can attain much higher temperatures than are possible with water-cooled reactors, with the gas pressurized only to the extent required to remove heat effectively from the fuel (about 4.8 MPa [700 psi]). The temperature of the (superheated) steam generated approaches that in modern fossil-fuel plants.

The core of an HTGR is made up of a number of hexagonal graphite blocks stacked in close-packed columns. Vertical holes are drilled in the blocks for insertion of some 10 000 fuel rods and for passage of the coolant gas. The fuel rods are made of graphite containing separate particles of highly enriched uranium dioxide (or carbide or a combination of the two) and of thorium dioxide (or carbide). The combination of fissile uranium-235 and fertile thorium-232 is roughly equivalent to uranium of about 5 percent enrichment in uranium-235. During reactor operation, some of the thorium-232 is converted into fissile uranium-233 (p. 17).

Eventually, when sufficient uranium-233 has been accumulated, it will be used in place of part of the enriched uranium-235 in the fuel rods. It is expected that, as a result of the efficient conversion of thorium-232 into uranium-233 and the radiation stability of the fuel, as much as 10 percent of the fuel will be utilized in an HTGR, compared with a present maximum of about 3 percent in an LWR.

The steam generator in an HTGR system is similar to a conventional water-tube boiler. The feedwater flows under pressure through a bank of tubes that are surrounded by helium gas that has been heated in the reactor core. Because the water is heated by hot gas, rather than by hot water as in a PWR steam generator, HTGR steam can be *superheated*. It leaves the generator at a temperature of about 538 °C (1000 °F) and a pressure of 16.5 MPa (2400 psi).[g] Because of the high temperature of the steam entering the turbine, the thermal efficiency of an HTGR power plant is nearly 40 percent.

Liquid-Metal Fast Breeder Reactor

Several experimental fast reactors of low power, using liquid metal as coolant and capable of breeding plutonium-239 from uranium-238 (p. 23), have been operated in the United States. However, none of these is a prototype for an economically competitive liquid-metal (cooled) fast (neutron) breeder reactor (LMFBR) that could be used in a central-station electric power plant. In 1980, such prototype plants were in operation in France, the United Kingdom, and

[g] The temperature of saturated (i.e., not superheated) steam at this pressure would be 317 °C (603 °F).

the USSR, but the time when the first LMFBR demonstration power plant in the United States will generate electricity is uncertain.

The fuel in an LMFBR is a mixture of about 85 percent uranium dioxide and 15 percent plutonium dioxide. The uranium may be normal uranium (99.3 percent uranium-238), or it may consist of the depleted residues (99.7 to 99.8 percent uranium-238) from a gaseous diffusion plant, for which there otherwise is little use. The fuel rods consist of long, thin stainless-steel tubes packed with pellets of the mixed (uranium and plutonium) oxides. The core contains a large number of fuel rods separated by channels to permit flow of the molten sodium coolant for heat removal. Conversion of uranium-238 to plutonium-239 occurs within the core as well as in a breeding blanket (reflector) of natural (or depleted) uranium dioxide that surrounds the core. There is, of course, no moderator since fast neutrons are required to achieve breeding.

In its passage through the reactor core, sodium captures neutrons to a small extent and is partly converted into a radioactive isotope. The heated sodium coolant leaving the reactor is consequently slightly radioactive. To provide protection for plant operators, steam generation involves an intermediate stage. Heat from the radioactive coolant is first transferred to nonradioactive sodium in a well-shielded heat exchanger. The high-temperature nonradioactive sodium then produces steam from feedwater in a heat-transfer steam generator operating on the same principle as that in a PWR system.

Just as when a hot gas is used to produce steam, hot sodium can superheat the steam generated. The temperature of the steam from an LMFBR system is expected to be in the range of 482 to 538 °C (900 to 1000 °F), and the thermal efficiency for the production of electricity should be about 38 or 39 percent. Furthermore, as a consequence of breeding fissile plutonium-239 from fertile uranium-238, the LMFBR would, in the long run, be capable of making use of at least half the uranium in the fuel.

Bibliography—Chapter 2

General Electric Company. "General Description of a Boiling Water Reactor." 1974.

Glasstone, S. *Sourcebook on Atomic Energy.* New York: Van Nostrand Reinhold Company, Inc., 1967.

Hogerton, J. F. "The Arrival of Nuclear Power." *Scientific American* 218, no. 2 (1968): 21.

Nero, A. V., Jr. *A Guidebook to Nuclear Reactors.* Berkeley: University of California Press, 1979.

U.S., Atomic Energy Commission. *Current Status and Future Technical and Economic Potential of Light Water Reactors* (WASH-1082), 1968. (Contains detailed descriptions of the PWR and BWR.)

U.S., Energy Research and Development Administration. *The Nuclear Industry 1974* (WASH-1174-74). 1975.

Westinghouse Electric Corporation. "Systems Summary of a Westinghouse Pressurized Water Reactor Nuclear Power Plant." 1973.

3

Licensing of Nuclear Power Plants

INTRODUCTION

Protecting Public Safety and the Environment

By the terms of the Atomic Energy Act of 1954 (and its amendments), the U. S. Atomic Energy Commission (AEC) was responsible for licensing and regulating the facilities concerned with the production and utilization of nuclear energy, in order to assure the health and safety of the public. Early in 1975, when the AEC was abolished by the Energy Reorganization Act of 1974, these responsibilities were transferred to the newly formed U. S. Nuclear Regulatory Commission (NRC). Furthermore, by the National Environmental Policy Act (NEPA) of 1969, the NRC is required to prepare a detailed environmental statement on any proposed action, such as the licensing of a nuclear power plant, that may significantly affect the human environment. In its decision-making process, the NRC has to consider the environmental impacts of each proposed major action and the available alternative actions.

To protect the public health and safety and to examine the effects of its decisions on the environment, the NRC has developed a comprehensive program of licensing and inspecting nuclear facilities. This program is implemented through rules and regulations that specify procedures to be followed in activities relating to nuclear energy. These procedures are subject to change from time to time.

Before a nuclear power plant (or related facility, such as a fuel fabrication or reprocessing plant) can be constructed, a permit must be obtained from the NRC; and after construction, the plant may be operated only if a license to do so has been granted. Such permits and licenses are issued after thorough reviews, which include public hearings, have shown that the proposed construction and operation can be carried out with due care for the public health and safety and the quality of the environment.

As part of the application for a permit to construct a nuclear power plant, the applicant must demonstrate compliance of the proposed facility with local zoning and land-use regulations. The thermal and other water pollution and use limitations or requirements, imposed by federal, state, regional, or local agencies in accordance with the Federal Water Pollution Control Act (see Chapter 12), must also be taken into consideration.

Thus, many precautions are taken in the licensing procedures to provide reasonable assurance that the construction and operation of the nuclear plant will be acceptable from both safety and environmental standpoints. In addition, the NRC monitors the construction, testing, and operation of the plant to assure compliance with the conditions of the permit or license.

NRC Rules and Regulations

The NRC rules and regulations, which are issued as parts of the Code of Federal Regulations, constitute essential features of the regulatory process. They form the framework of standards, criteria, practices, etc., within which nuclear plants must be designed, constructed, and operated. Although they do not specify detailed designs, they do indicate the requirements that must be satisfied by a proposed design. The rules and regulations have evolved with time; new ones are proposed, and existing ones are modified as experience is gained in the design, construction, and operation of nuclear power (and related) plants and as new information is obtained from research and development programs. In particular, changes have followed from the accident in March 1979 at the Three Mile Island Unit 2 reactor near Harrisburg, Pennsylvania, when a combination of operator errors and equipment failures led to a potentially serious situation (p. 105). It is unlikely, however, that the NRC's basic program for licensing nuclear power plants, as outlined in this chapter, will be affected, although greater emphasis will be placed on the clarification and enforcement of regulations.

Proposed changes in or additions to rules and regulations are published in the *Federal Register* with an explanation of their purpose. Comments from the public are invited and considered prior to adoption of the rule. If a contemplated change affects a wide variety of interests, the NRC may hold public rulemaking hearings. Such hearings provide an opportunity for the presentation of a wide range of views. All comments are given careful consideration before the NRC issues the final form of the regulation.

Code of Federal Regulations, Title 10

Regulatory directions incorporated in Title 10 (Energy) of the Code of Federal Regulations (10CFR) include both standards and criteria. In general, *standards* specify limits for or definitive ways of accomplishing an objective, whereas *criteria* provide a basis of comparison for judging the acceptability of a particular action or procedure. Standards and criteria that are incorporated in Title 10 have the force and effect of law, and, where appropriate, compliance with them is required. If specific standards or criteria are not mentioned in the NRC regulations, compliance is required if such standards or criteria are specifically included in the license conditions.

The most important parts of 10CFR, insofar as the standards and criteria of nuclear power plants are concerned, are Part 20 (10CFR20), Part 50 (10CFR50), Part 51 (10CFR51), and Part 100 (10CFR100). The main features of these regulations are given below.

Entitled "Standards for Protection Against Radiation," 10CFR20 governs the release of radioactive materials to the environment and places limits on the radiation exposures of plant workers and members of the public from nuclear power facility (and related) operations. Some of the requirements of 10CFR20 are described in Chapter 5.

Entitled "Licensing of Production and Utilization Facilities," 10CFR50 describes the procedures for preparing, filing, and processing applications for construction permits and operating licenses; detailed technical guidance is provided in appendices. The major topics in 10CFR50 are considered in this chapter.

Concerned with "Licensing and Regulatory Policy and Procedures for Environmental Protection," 10CFR51 deals with, among other matters, the preparation of Environmental Reports, which must be submitted with applications for construction permits and operating licenses, as described later in this chapter.

Entitled "Reactor Site Criteria," 10CFR100 outlines some of the criteria for determining the suitability of a proposed site for a nuclear power (or related) plant, taking into consideration protection of the public in the event of an accident. These criteria are reviewed in one of the following sections.

Two parts of Title 10 are of interest since they indicate the means whereby the public is kept informed of deliberations relating to the applications for a construction permit and an operating license. First, 10CFR2, "Rules of Practice," describes how hearings are initiated and conducted and how interested parties may participate in the hearings (p. 46). In addition, 10CFR9, "Public Records," explains the conditions under which records in the possession of the NRC pertaining to a specific application are (or can be made) available to members of the public. Finally, mention should be made of 10CFR73, "Physical Protection of Plants and Materials," and 10CFR140, "Financial Protection Requirements and Indemnity Agreements."

Industry Codes and Standards

Codes and standards represent accepted practice in the engineering and construction industries for assuring an acceptable (minimum) level of quality and performance in any product. The NRC defines the codes and standards for nuclear power plants; these are often more stringent than those accepted by industry in general. In the absence of an appropriate code or standard, the organization responsible for constructing the plant must develop its own specifications for particular components and obtain approval for their use. Many nationally accepted codes and standards are already available, and more are being developed to meet the stringent safety objectives required for nuclear plants.

Preexisting codes have been improved, supplemented, or rewritten for the components of nuclear power facilities. An example is the widely used Boiler and Pressure Vessel Code of the American Society of Mechanical Engineers (ASME). A completely new section (Section III) has been added to provide specifications for reactor vessels, which must meet the special conditions of nuclear service. The code has been further amended and extended to include piping, pumps, and valves used in nuclear high-pressure systems. Moreover, sections have been added on quality assurance and nondestructive testing.

The principal organization for the establishment of comprehensive standards in the United States is the American National Standards Institute (ANSI), which serves as a central clearinghouse for all standards in this country. In an effort to respond to the needs of the nuclear power industry, ANSI has set up several committees to draft nuclear standards, which are subsequently reviewed by a Nuclear Technical Advisory Board of experts in various areas of science and engineering.

Through the Advisory Board, ANSI defines the scope of its committees and assigns them for sponsorship to an appropriate technical society. Thus, nuclear aspects are sponsored by the American Nuclear Society, mechanical aspects by ASME and electrical aspects by the Institute of Electrical and Electronics Engineers. In addition, technical advice is provided by the NRC and other organizations.

Quality Assurance Program

In general, the term *quality assurance* refers to all actions necessary to provide adequate confidence that a component, structure, or system will perform satisfactorily in service. According to Appendix B of 10CFR50, the NRC quality assurance requirements are applicable to "the design, construction, and operation of those structures, systems, and components . . . [intended] to prevent or mitigate the consequences of postulated accidents that could cause undue risk to the health and safety of the public." The quality assurance program covers "all activities affecting the safety-related features [of the facility] . . . including engineering, purchasing, fabricating, handling, shipping, storing, cleaning, erecting, installing, inspecting, testing, operating, maintaining, repairing, refueling, and modifying."

An applicant for a permit to construct a nuclear power plant must submit for NRC approval its proposed Quality Assurance Program. The purpose is to provide assurance that the design of the plant can meet the requirements as defined and agreed upon by the NRC and the plant owner, and that construction will be carried out in accordance with the design and with the materials and components that satisfy the accepted codes and standards. Materials and components must be inspected and tested during fabrication and construction. Procedures and results must be documented throughout all stages. The qualifications of certain craftsmen, inspectors, and others concerned must be recorded, and the individuals and management responsible for the quality assurance and control programs must be identified.

Regulatory Guides

As stated earlier, the NRC rules and regulations (e.g., 10CFR20, 10CFR50, 10CFR51, and 10CFR100) provide the framework of standards, criteria, practice, etc., within which nuclear plants must be designed, constructed, and operated. The rules and regulations are supplemented by Regulatory Guides.

In 1970, the AEC (as it then was) began to issue a series of "Safety Guides for Water-Cooled Nuclear Power Plants," the primary purpose of which was to explain the positions that had been developed by the AEC Regulatory Staff and the Advisory Committee on Reactor Safeguards (ACRS) (p. 37) on safety issues. The guides indicate various matters relating to reactor safety and quality assurance that should be considered in the design and evaluation of water-cooled reactors, and they describe the principles and specifications that would represent acceptable solutions to these matters. Near the end of 1972, the Safety Guides were incorporated in a more extensive series of Regulatory Guides, covering many aspects of the regulatory activities of the AEC (now of the NRC) in addition to nuclear power plants. The guides do not have the force of CFR rules and regulations, but rather describe methods acceptable to the NRC for achieving compliance with certain regulations. The guides, like the regulations, are modified by experience, such as the Three Mile Island plant accident referred to earlier. The Regulatory Guides of special interest for safety and quality assurance are included in "Division 1: Power Reactor Guides"; those concerned with environmental matters are in "Division 4: Environmental and Siting Guides." Reference to some of these guides will be made in due course.

Population Distribution in Plant Siting

Many factors must be taken into consideration in selecting a site that would be acceptable for a nuclear power plant (see Chapter 14). Some of these factors, especially those related to the population distribution in the vicinity of the proposed site, are described in 10CFR100.[a] Three more or less distinct areas (or distances) are defined; they are (a) an exclusion area, (b) a low-population zone, and (c) a population-center distance. The general characteristics of these areas are discussed below.

The nuclear power facility must be located within an *exclusion area,* which is defined as the area surrounding the installation in which the plant licensee has the authority to determine all activities—in particular, the exclusion or removal of people and property. Residence within the exclusion area would normally be prohibited, but if there are any residents, they must be subject to ready removal, if necessary. The area may be traversed by a highway, railroad, or waterway, provided it is not so close to the facility as to interfere with or present a potential hazard to its normal operation. Furthermore, even before the plant starts to operate, appropriate plans must be ready to control traffic in the event of an emergency.

The main purpose of the exclusion area is to limit the radiation doses that might be received by offsite individuals in the event of a hypothetical major, but improbable accident. Such an accident is generally assumed to result in substantial reactor core meltdown with subsequent release of appreciable quantities of fission products to the environment. The radius of the exclusion area should be such that a person located (and remaining) at any point on its boundary for 2 hours immediately following the postulated release would not receive, even under adverse meteorological conditions, a whole-body dose of radiation exceeding 25 rems or a thyroid dose exceeding 300 rems from exposure to radioisotopes of iodine.[b] (The rem is a unit of radiation dose, as explained in Chapter 5).

[a] Appendix A sets forth the principal seismic and geological considerations to be used in evaluating the suitability of a plant site.

[b] The whole-body dose of 25 rems is numerically equal to the limit imposed by the National Council on Radiation Protection and Measurements for persons performing planned actions in an emergency situation.

It should be emphasized that the maximum radiation doses given here (and below for the low-population zone) are not meant to imply acceptable limits for radiation exposure of the general public under accident conditions. They are intended, rather, as reference values only, to be used in evaluating reactor sites with respect to potential accidents of low probability.

Surrounding the exclusion area, there must be a *low-population zone;* the total number and density of residents in this zone must be such that there is a reasonable probability that appropriate protective measures can be taken on their behalf (e.g., evacuation if deemed necessary) in case of a major accident. As a guide to the dimensions of the low-population zone, 10CFR100 states that an individual located at any point on the outer boundary of the zone and remaining there during the *entire period* of passage of the radioactive cloud resulting from the postulated accident would not receive a whole-body dose of radiation exceeding 25 rems or a thyroid dose exceeding 300 rems. The purpose of the protective measures referred to above is to reduce actual radiation dose to well below these reference values.

The *population-center distance* is defined as the distance from the reactor to the nearest boundary of a densely populated center containing more than 25 000 residents. The population-center distance shall be at least one and one-third times the distance from the reactor to the outer boundary of the low-population zone defined above. Where the population center is a large city, a greater distance may be necessary. In applying this guide, the boundary of the population center shall be determined upon consideration of the population distribution, regardless of political (municipal) boundaries.

An acceptable population-center distance is not dependent on the radiation dose that might be received by an individual, as is the case for the exclusion area and low-population zone, but rather on the *population dose.* This is the product of the number of individuals in the exposed population and the average dose per individual, and is expressed in person-rem (man-rem) units. It is also equal to the sum of the doses to all the members of the population group.

The population dose is important in several respects. In the first place, it relates to real people rather than to the hypothetical individuals who are supposed to spend their lives at the boundary of the exclusion area or of the low-population zone. Furthermore, as explained in Chapter 7, the long-range (or eventual) effects of small doses of radiation are believed to be statistical in nature. The overall incidence of such effects will then be dependent on the population dose in person-rems.

In determining the population-center distance, therefore, it is necessary to take into consideration the number and distribution of people in populated areas in the vicinity of the proposed plant site. If the population is large, the average radiation dose per individual (in rems) must be less than for a smaller population so that the population dose (in person-rems) will be the same. This is achieved by making the population-center distance greater in the former case; that is, the plant would have to be located farther from the larger population center.

The AEC (now the NRC) has provided guidance concerning acceptable techniques and assumptions for calculating the radiation doses that might be received by individuals at various distances from a reactor as a result of certain postulated accidents. For example, Regulatory Guides 1.3 and 1.4 deal with "Assumptions Used for Evaluating the Potential Radiological Consequences of a Loss of Coolant Accident" for BWRs and for PWRs, respectively. The assumptions made concerning the amount of radioactivity released and the existing meteorological conditions are very conservative (i.e., unfavorable). The maximum radiation doses at the various distances defined in 10CFR100 are thus much greater than would actually be realized if any of these postulated accidents should occur (p. 46).

In 1978, the NRC directed its staff to develop a policy statement on nuclear power plant siting for the purpose of updating the criteria of 10CFR100. In its report, issued in August 1979, the Siting Policy Task Force suggested that calculated radiation doses no longer be used to establish minimum exclusion distances and low-population zone boundaries. The Task Force recommended that more emphasis be placed instead on isolation and emergency planning for evacuation. According to 10CFR100, the dimensions of the exclusion area and the low-population zone can vary, depending on features designed to reduce the escape of radioactivity if a serious reactor accident should occur. However, the Task Force suggested that these variable distances be replaced by fixed minimum distances applicable in all cases.

The following modifications to the population distribution aspects of 10CFR100 were recommended:

1. Establishment of a fixed minimum exclusion area radius, possibly 0.8 km (0.5 mile).

2. Replacement of low-population zone distance by a fixed minimum emergency planning zone distance of about 16 km (10 miles); this zone would correspond with that for which local and state agencies have made plans for emergency evacuation of the population, as described in the next section.

3. Elimination of the population-zone distance and replacement by specific limits on population density and population distribution outside the exclusion area to a distance of 32 km (20 miles) or so.

As of March 1980, these recommendations were still under consideration by the NRC.

Emergency Planning

Before the Three Mile Island accident, it was generally assumed that compliance with 10CFR100, coupled with the defense-in-depth approach to reactor safety (p. 60), would be adequate to protect the public from the effects of a major reactor accident. Emergency planning was considered as a secondary measure to be exercised in the event of a substantial release of radioactivity to the surroundings. The experience at Three Mile Island, however, showed that additional attention should be given to emergency response planning in the course of reactor licensing. Consequently, the NRC now requires that, in addition to dealing with radiological emergencies within the nuclear plant site, the licensee make arrangements with state and local organizations to respond to accidents that might involve the release of radioactivity outside the plant site.

A Joint Task Force of the U.S. Environmental Protection Agency and the NRC recommended in 1978 that state and local agencies establish two *emergency planning zones* (EPZs) around the plant site. In the first zone, with a radius of roughly 16 km (10 miles), the population should be protected primarily against exposure to airborne radioactivity (e.g., by evacuation), whereas in the second zone, with a radius of some 80 km (50 miles), protection should be against the consumption of contaminated food and water (e.g., by impounding or proscription). The exact size and shape of the EPZs will depend on the specific conditions around the plant site.

Responsibility for evaluating state and local emergency planning and for assuring that the plans are adequate and capable of implementation by trained persons is shared by the Federal Emergency Agency (FEMA) and the NRC. The FEMA has the lead responsibility for off-site planing in cooperation with the NRC, which still has the statutory mandate to protect the public from the effects of radiation. In addition to reviewing the FEMA findings concerning state and local emergency plans, the NRC is responsible for evaluating the plant licensee's plans for on-site protection and for integrating them with state and local plans for the off-site population.

Outline of Licensing Procedures

The granting of a license to an applicant for the construction and operation of a nuclear power plant at its full design power is a multi-step process, involving a sequence of events that may continue over several years. These stages are first summarized below to provide an overall picture of the procedures; then they are discussed in some detail in later parts of this chapter.

At the outset, the utility desiring to construct and operate a nuclear power plant makes a formal application for a *construction permit*. This is reviewed by the staff of the NRC, and the safety aspects are considered independently by the ACRS. This committee is a statutory body established by the Atomic Energy Act. It consists of a maximum of 15 members appointed by the NRC for overlapping terms of four years each. Members of the ACRS are individuals experienced in various disciplines, such as engineering, metallurgy, reactor operation, reactor physics, and instrumentation. They are not NRC employees, but are selected from industry, private practice, universities, and the national laboratories.

After the application has been evaluated by the NRC staff and the ACRS, a public hearing is held to consider the issuance of a construction permit. The hearing is conducted by members of the Atomic Safety and Licensing Board (ASLB) panel. At present, the panel consists of approximately 70 persons appointed by the NRC; the members include lawyers, economists, engineers, and public health, environmental, and physical scientists. Three members of the panel, with a lawyer as chairman, are chosen by the NRC to constitute the ASLB that conducts the hearings on a particular application for a construction permit. The board reviews both safety and environmental aspects of the proposed plant and decides whether a construction permit should be granted or not. The decision is reviewed by the Atomic Safety and Licensing Appeal Board, which is selected from a separate panel of lawyers, engineers, and scientists. At their own discretion, the commissioners of the NRC may review the decision of the appeal board.

If a construction permit is granted, the plant is built in accord with plans, procedures, and criteria that have been approved by the NRC. Frequent inspections of the plant during construction are made by NRC staff to ensure conformance with approved quality assurance procedures and NRC criteria.

Prior to completion of construction, an application is filed for an *operating license*. This application, like that for a construction permit, is reviewed by the NRC staff and by the ACRS. If safety and environmental requirements have been met, an operating license will be issued. A public hearing will be held only if there is a showing of good cause.

Procedures similar to those outlined above, covering both safety and environmental aspects, are also followed in licensing nuclear fuel fabrication and spent-fuel reprocessing plants. Since the number of fabrication and reprocessing facilities is (and will continue to be) quite small, the discussion in the remainder of this section is concerned more specifically with nuclear power plants.

APPLICATION FOR A CONSTRUCTION PERMIT

Introduction

Although it is not presently mandatory, a prospective applicant for a permit to construct a nuclear power plant is encouraged to discuss with the NRC staff the suitability of possible sites before filing an application. If a satisfactory site has been identified in this or any other way, the applicant must submit two major documents: a Preliminary Safety Analysis Report (PSAR) and an Environmental Report. The essential contents of these reports are described in 10CFR50 and 10CFR51. To help applicants prepare the reports, comprehensive guides entitled "Standard Format and Content of Safety Analysis Reports for Nuclear Power Plants" (NRC Regulatory Guide 1.70) and "The Preparation of Environmental Reports for Nuclear Power Plants" (Regulatory Guide 4.2) have been prepared. If the application is for a Limited Work Authorization (LWA) (p. 48), then only the Environmental Report need be submitted at this stage.

If an initial examination indicates that the application is sufficiently complete to permit an NRC review, notice of its receipt is published in the *Federal Register* as a means of informing the public. Copies of the PSAR and Environmental Reports are accessible to interested parties at the NRC Public Document Room in Washington, D.C., and at a location, commonly a public library, in the vicinity of the proposed plant site. Copies are also available to the public at appropriate state, regional, and metropolitan clearinghouses.

Preliminary Safety Analysis Report

In accordance with the requirements of 10CFR50, the PSAR must cover the topics outlined in the following paragraphs. They are concerned mainly with safety considerations in the design, construction, and operation of the plant.

1. A description and safety assessment of the site on which the plant is to be located with appropriate attention to features that affect design of the plant. The assessment should contain an analysis and evaluation of the major structures, components, and systems of the facility that bear on the acceptability of the site as defined by the criteria in 10CFR100 (p. 34).

2. A summary description and discussion of the facility, with special attention to design and operating characteristics, unusual or novel features, and the principal safety considerations.

3. The principal design of the plant must be shown to be in conformance with Appendix A to 10CFR50, "General Design Criteria for Nuclear Power Plants." The criteria in this appendix establish minimum requirements for nuclear power plants similar in design and location to the plants for which construction permits have been issued previously. The preliminary design information includes the *design bases* and their relation to the principal design criteria.[c] Information is required relating to the materials of construction, general arrangement, and approximate dimensions sufficient to provide reasonable assurance that the final design will conform to the design bases with adequate margins for safety.

Appendix A contains 64 criteria in the following five categories: (a) overall requirements (including design bases for protection against floods, tornadoes, hurricanes, and earthquakes); (b) protection by multiple fission-product barriers (including design of the reactor core and associated coolant, control, and protection systems); (c) protection and reactivity control systems (including the reliability, testability, and capability of the systems); (d) fluid systems (including emergency core-cooling and residual heat removal systems); (e) reactor containment (for minimizing release of radioactivity to the environment in case of an accident).

4. A preliminary analysis and evaluation of the design and performance of structures, systems, and components of the plant with the objective of assessing the risk to public health and safety, both during normal operation and as a result of transient conditions that may be anticipated. The adequacy of structures, systems, and components provided for the prevention of accidents and the mitigation of the consequences of accidents, should they occur, must be examined.

5. An identification and justification for the selection of those variables, conditions, or other items that are determined to be probable subjects of Technical Specifications (p. 57) for the facility. Special attention is required to those items that may significantly influence the final design.

6. A preliminary plan for the applicant's organization, training of personnel, and conduct of operations.

[c] *Design bases* are defined as "information which indentifies the specific functions to be performed by a structure, system, or component of a facility and the specific values (or ranges of values) chosen for controlling parameters as reference bounds for design" (10CFR50, Section 50.2 [u]).

7. A description of the quality assurance program (p. 33) to be applied to the design, fabrication, and testing of the structures, systems, and components of the facility. The requirements of this program are set forth in Appendix B of 10CFR50, "Quality Assurance Criteria for Nuclear Power Plants." The manner in which the requirements of Appendix B will be satisfied must be discussed.

8. An identification of those structures, systems, or components, if any, of the facility that may require research and development to confirm the adequacy of their design. The research and development program must be described, together with a schedule showing that outstanding problems relating to safety will be resolved before construction of the facility is complete.

9. A description of the technical qualifications of the applicant to engage in the proposed activities.

10. A discussion of the applicant's preliminary plans for coping with emergencies. The items to be included are described in Appendix E to 10CFR50, "Emergency Plans for Production and Utilization Facilities." They include such matters as measures to be taken, within and outside the site boundary, to protect health and safety in the event of an accident. A training program plan is required for individuals who may be assigned to cope with an emergency. The PSAR must describe the arrangements made with the state and local agencies for establishing the EPZs, referred to earlier. Consideration must be given to access routes, population distribution, and land use.

In addition to the foregoing items, which are specifically mentioned in 10CFR50 (Section 50.34), the PSAR must include a discussion of a range of conceivable abnormal events and show that the plant is designed in such a manner that the various situations can be controlled or accommodated safely. The abnormal events considered must include what are called *design-basis accidents*. These are events of low probability in terms of which the safety systems (e.g., the "engineered safety features" described in Chapter 4) of the plant and the suitability of the site have been evaluated. The worst design-basis accident is generally regarded as a major break in the primary coolant system, leading to a *loss-of-coolant accident* (LOCA) (p. 77) accompanied by damage to some fuel rods. Conservative assumptions are used in calculating the potential doses of radiation that might result from a design-basis accident. These calculated doses must be shown to be less than the guidance values described earlier, as given by the site criteria in 10CFR100.

The application for a construction permit for a nuclear power plant must include a description of the preliminary design of equipment to be installed to maintain control over radioactive materials in gaseous and liquid effluents released during normal operations (see Chapter 9). The design objectives and means to be employed for keeping radioactivity levels in effluents to unrestricted areas below the limits imposed by the regulations in 10CFR20 and 10CFR50 (Appendix I) must be identified. An estimate is required of the quantity of each of the principal radioactive species expected to be discharged annually in the effluents. A general account must be given of the provisions for packaging, storing, and offsite shipment of solid radioactive wastes resulting from the treatment of gaseous and liquid effluents and from other sources.

SAFETY REVIEW BY THE NRC STAFF

When an application for a construction permit is received by the NRC, the PSAR is examined by the NRC staff to determine if it meets minimal requirements for completeness. If the application appears to be adequate, the PSAR is subjected to a thorough technical review to assure that the proposed plant will satisfy all appropriate NRC criteria and other requirements. The review is performed in accordance with the NRC's "Standard Review Plan for the Review of Safety Analysis Reports for Nuclear Power Plants" (NUREG-75/087). This plan gives the acceptance criteria to be used in evaluating the plant systems, identifies components and structures important to safety, and describes the procedures to be used in performing the safety review.

The review process generally involves a number of meetings between the NRC staff and representatives of the applicant. Consultants from other agencies (e.g., the U.S. Geological Survey) are frequently called upon for expert advice. Additional information requested by the staff is supplied by the applicant in the form of amendments to the application. When the NRC staff is finally convinced that the plant will be built and operated with minimum risk to the public, the staff prepares a Safety Evaluation Report. The principal features of this report are outlined in the following paragraphs.

The physical characteristics of the site, including its seismology, meteorology, geology, and hydrology are examined to determine that these characteristics have been assessed adequately in the applicant's PSAR. Other essential factors relating to the suitability of the site include a description of the area under the applicant's control, the population density in the locality of the proposed site, and the present and projected uses of the area, such as industry, farming, recreation, etc. The engineered safety features to be included in the plant are considered in assessing whether the site meets the requirements of 10CFR100.

The design, fabrication, construction, testing, and expected performance of the plant structures, systems, and components that are important to safety are reviewed to determine that they satisfy the General Design Criteria (10CFR50, Appendix A), the Quality Assurance Criteria (10CFR50, Appendix B), and the appropriate codes and standards. Checks are made of calculations and design procedures to establish the validity and adequacy of the applicant's analyses and evaluations as related to reactor safety.

In some circumstances, further research and development work may be required to confirm the adequacy of the safety features or to resolve questions associated with a particular design. These matters are identified in the Safety Evaluation Report, and the applicant for a construction permit must provide a time schedule for the resolution of outstanding problems.

Evaluations are made of the design of systems provided for the control of radioactive effluents from the plant (i.e., the *radwaste* system) to determine if the facility can be operated in such a manner as to comply with the NRC standards (10CFR20). Radiation exposures of the general public, from plant effluents beyond the plant boundaries, must satisfy the requirement that they be "as low as is reasonably achievable," as defined in 10CFR50, Appendix I (see Chapter 6).

The expected responses of the plant systems to various anticipated abnormal events and to a wide range of hypothetical accidents are considered. The potential consequences of such incidents, particularly regarding radiation exposures beyond plant boundaries, are evaluated independently on the basis of conservative assumptions. The calculated radiation doses must be within the guidelines for site acceptability.

The applicant's plans for the conduct of plant operations are evaluated, as are the organizational structure, the technical qualifications of operating and technical support personnel, and the planning for emergency actions. This review serves to determine whether the applicant's staff is technically qualified to operate the plant and whether effective organization and planning have been established for continuing safe operation.

Review by the ACRS

Within a few days after an application is filed for a permit to construct a nuclear power plant, copies of the PSAR are provided to the ACRS. A subcommittee is then appointed to review the application. The subcommittee is normally composed of four or five members of the ACRS, representing each of the major disciplines, such as reactor physics, engineering, and materials science, concerned with safety issues. If necessary, the services of expert consultants are called upon.

At the meetings of the subcommittee, attention is first devoted to matters relating to the suitability of the proposed site; at subsequent meetings, special consideration is given to new or unusual features in the plant design. Members of the NRC staff and representatives of the applicant for a construction permit attend the meetings, which are open to the public. Written questions may be submitted to the applicant in advance of a subcommittee meeting if an extensive amount of information must be developed in response to these questions. In addition, the staff keeps the subcommittee informed of the staff's requests for information and of meetings held with the applicant's representatives to resolve any points that may be at issue.

When the necessary information has been developed by the subcommittee, the project is scheduled for consideration by the full ACRS. Prior to this time, the NRC staff will have published its Safety Evaluation Report, stating its judgments and identifying areas that, in its opinion, warrant special attention or reservations. Following discussion with the staff, the full committee continues its review of the application by questioning the representatives of the applicant. Further meetings may be scheduled before the committee's concerns have been completely satisfied.

When the ACRS has completed its review and has determined that the plant can be constructed in such a manner as to assure safe operation, it submits its report in the form of a letter to the chairman of the NRC. This report, which is made available to the public, contains a brief description of the main characteristics of the proposed plant and its location. Attention is called to problems of special interest in relation to safety and to the design of components and systems that will require resolution during construction of the plant.

Evaluation of Environmental Effects

To provide a basis for the NRC's evaluation of the environmental impact of a proposed nuclear power plant, in accordance with NEPA (p. 30), an applicant for a construction permit is required to submit an "Applicant's Environmental Report—Construction Permit Stage." In the report, the following items must be considered, as explained in 10CFR51:

1. The environmental impact of the proposed action.
2. Any adverse environmental effects that cannot be avoided should the proposal be implemented.
3. Alternatives to the proposed action.
4. The relationship between local short-term uses by man of the environment and the maintenance and enhancement of long-term productivity.
5. Any irreversible and irretrievable commitments or resources that would be involved if the proposed action should be implemented.

The summary given below, based on NRC Regulatory Guide 4.2, indicates the extent and scope of the Environmental Report:

1. Purpose of the Proposed Facility: The applicant should demonstrate the need for the proposed nuclear plant in terms of power requirements to be satisfied and other primary objectives of the facility. The consequence of delay in the scheduled operation of the plant on these primary objectives should be discussed.

2. The Site and Environmental Interfaces: Basic relevant information should be presented concerning those physical, biological, and human (i.e., historic, scenic, and cultural) characteristics of the environment that might be affected by the construction and operation of the plant at the proposed site.

3. The Station: The essential features of the station (i.e., the power plant) and the transmission system should be described. Effluents (heat, chemicals, and radioactive materials) from the plant and plant-related systems that interact with the environment must be considered in detail.

4. Environmental Effects of Site Preparation, Station Construction, and Transmission Facilities Construction: The expected effects of site preparation and construction of the plant and transmission facilities should be discussed. The effects should be considered in terms of their physical and ecological impact on resources and populations. (Social and economic impacts are considered in Section 8.)

5. Environmental Impact from Routine Operation: The interaction of the operating plant and transmission system on the environment should be discussed. Measures planned to reduce any undesirable effects of plant operation on the environment should be described. A distinction should be made between unavoidable environmental effects that are temporary or subject to amelioration and those that are regarded as permanent. In this section, detailed consideration must be given to the possible effects of waste heat (in condenser water) and radioactive and chemical effluents on all biota in the environment. Estimates must be made of the radiological impact on man by way of various exposure pathways, including internal exposure from food (e.g., fish, meat, milk, vegetables, etc.) and external exposure from radioactive gases.

6. Effluent and Environmental Measurements and Monitoring Programs: The procedures used in collecting the baseline data presented in other sections of the report should be explained. Plans and programs for monitoring the environmental impacts of site preparation, plant construction, and operation should be described.

7. Environmental Effects of Accidents: The contents of this section are explained below.

8. Economic and Social Effects of Station Construction and Operation: The social and economic benefits and costs of the proposed facility should be examined. Apart from the availability of electric power, benefits might include increased local employment and tax revenues, enhanced recreational values (e.g., parks, artificial lakes), improvement of wildlife habitats in restricted areas, etc. Costs are both internal (i.e., capital and operating costs of the plant) and external (e.g., housing for construction workers, disturbance of local municipal, educational, and social services, and changes in recreational, historic, and natural features, adverse meteorological effects, and effects on sport and commercial fishing).

9. Alternative Energy Sources and Sites: The basis for the proposed choice of site and nuclear fuel among the available alternative sites and energy sources should be presented. The range of practicable alternatives should be discussed, bearing in mind the environmental costs and benefits, and it should be demonstrated that none of these is clearly to be preferred over the proposed site and type of plant.

10. Station Design Alternatives: The proposed plant design should be evaluated by a comparison of the internal and external costs, as described in Section 8, of alternative subsystems (e.g., for condenser water, radwaste, chemical waste, etc.).

11. Summary Cost-Benefit Analysis: By means of a cost-benefit analysis, it should be demonstrated that the aggregate benefits of the plant will outweigh the aggregate costs. Although some costs and benefits can be expressed in monetary terms, others must be assessed on a judgmental basis consistent with the underlying concept of cost-benefit analysis.

12. Enviromental Approval and Consultation: The status of all licenses, permits, and other approvals of plant construction and operations required by federal, state, local, and regional authorities for the protection of the environment should be listed.

13. References: A bibliography of sources used in preparation of the Environmental Report should be given, keyed to specific areas or topics in the individual Sections 1 through 12.

Postulated Accidents for Environmental Report

Section 7 of the Applicant's Environmental Report deals with the environmental release of radioactivity arising from a variety of accidental situations. Since it is impossible to analyze all conceivable accidents in detail, the NRC staff has divided accidents that might occur in nuclear power plant operation into nine classes ranging from minor to the most severe. Qualitative descriptions of these classes with typical examples in each class are listed in Table 3-I.

TABLE 3-I
CLASSIFICATION OF POSTULATED ACCIDENTS FOR ENVIRONMENTAL REPORT

Class	Description	Examples
1	Trivial incidents	Small spills and leaks inside the containment.
2	Small releases outside the containment	Small spills and leaks outside the containment.
3	Radwaste system failures	Equipment leakage or malfunction. Release from gas or liquid waste storage tank.
4	Fission products to primary system (BWR)	Fuel cladding defects. Off-design transients inducing fuel failures.
5	Fission products to primary and secondary systems (PWR)	Fuel cladding defects and steam generator leaks. Off-design transients. Steam generator tube rupture.
6	Refueling accidents	Fuel assembly drop. Heavy object drop onto fuel in core.
7	Spent-fuel handling accident	Fuel assembly drop in storage pool. Heavy object drop onto fuel rack. Fuel cask drop.
8	Accident-initiation events considered in design-basis evaluation in the Safety Analysis Reports	Loss-of-coolant accidents (small and large). Control rod ejection (PWR) or drop (BWR). Steam line breaks (BWR), outside containment (PWR).
9	Hypothetical sequence of successive failures more severe than those postulated for establishing the design basis.	

The applicant for a construction permit must define accidents in each class that are appropriate to the operation of the proposed plant. Class 1 events need not be considered because of their minor consequences; similarly, the environmental effects of Class 9 accidents need not be discussed because the probability of the occurrence of such accidents is considered to be so small that their environmental risk is very low (see Chapter 4). The classes and types of accidents and the methods suggested for calculating the release of radioactivity in each case are described in a proposed Annex to 10CFR50.

Class 8 events are the same as the most serious category of accidents (e.g., design-basis accidents) considered in the Safety Analysis Report (p. 40). In the latter case, highly conservative assumptions are made in calculating the potential release of radioactivity to assure adequacy of the engineered safety features. For environmental risk estimates, however, more realistic assumptions (e.g., average rather than worst meteorological conditions) are regarded as being more appropriate. The radiation doses predicted in this manner are thus less than calculated in the safety report for similar accidents.

Environmental Review by NRC Staff

When an applicant's Environmental Report is accepted for review, the NRC staff makes an independent evaluation of the expected environmental impact of the plant. A Standard Review Plan for Environmental Reviews, analogous to the standard plan for safety reviews, referred to on page 41, has been issued by the NRC (NUREG-0555, 1979).

As a first step, a Draft Environmental Statement is prepared by the NRC staff with the assistance of specialists in the national laboratories. This statement, together with the applicant's Environmental Report, is transmitted for comment to the Council on Environmental Quality, to certain federal agencies, to the governor of the state where the plant is to be located, and to appropriate local officials connected with matters affecting the environment. The Draft Statement is also announced and made available to the public in the same manner as the applicant's Environmental Report (p. 43). Comments on the Draft Statement are requested within a specified period, usually 45 days.

After receipt of the comments, the NRC issues the Final Environmental Statement, required by NEPA, relating to the planned nuclear facility. The Final Statement includes a discussion of problems and objections raised by federal, state, and local agencies, and by private organizations and individuals. The manner in which these matters have been disposed of must be explained.

In the Final Environmental Statement, the NRC staff, after weighing the environmental, economic, technical, and other benefits against environmental costs and considering available alternatives, recommends the action to be taken. The recommendation may call for issuance of the construction permit or its denial. Frequently, the statement advises issuance of the permit subject to conditions that would offer better protection of the environment. Copies of the Final Environmental Statement are transmitted to the Council on Environmental Quality and other federal and state agencies and are made available to the public through the usual channels.

Public Hearings on Construction Permit

The public hearing on a construction permit is conducted by an Atomic Safety and Licensing Board of three members selected from the Atomic Safety and Licensing Board Panel, as described earlier. Since the hearing is a legal proceeding, the chairman of the board is a lawyer with experience in administrative law and procedures; the other two members are technical experts, one of whom is usually a specialist in some aspect of environmental science. Hearings are generally held near the proposed plant site, unless the parties agree to another location.

Members of the public who wish to participate in the proceedings may request to do so either by a petition to intervene or by making a limited appearance. A person who wishes to intervene must file a written petition for leave to do so; in this petition, the individual must explain the facts relating to his special interest in the case. If he is permitted to intervene, he becomes a full party to the proceedings and has the right to participate fully in the conduct of the hearing, including presentation of evidence and examination and cross-examination of witnesses. A person making a limited appearance does not become a party to the proceedings. He may state his position and raise questions; these questions may serve to focus on information that should be clarified for the public benefit or on issues that should be considered by the board.

Prior to the hearing, the ASLB holds a prehearing conference to consider intervenors' contentions, to sharpen issues, to resolve procedural questions, and to prepare an agenda for the hearing. The ASLB listens to arguments of the NRC staff, the applicant, and the proposed intervenors. The ASLB will then admit to the hearings such parties as have established a valid interest and specific contention, as defined in 10CFR2, Sec. 2.714. These intervenors will receive all of the pertinent documents, including those generated by the staff and the applicant during the course of the staff review. This enables them to become familiar with the design of the proposed plant and the potential safety and environmental effects.

The start of the public hearing is scheduled by the ASLB after the NRC staff has completed its review of the construction permit application and published its Safety Evaluation Report and/or its Final Environmental Statement. Thus, the hearing may be in two parts, to consider safety and environmental aspects separately, or they may be combined into a single hearing.

At the hearing (or hearings), testimony is given by the applicant and by the NRC staff on the safety and environmental aspects of the proposed nuclear plant, and also on the applicant's technical and financial qualifications to construct it. Statements are offered by interested parties to the proceedings, either as intervenors or by making a limited appearance, as explained above. The recommendations of the staff are presented, including the conditions under which a construction permit might be granted.

In a hearing for an uncontested application—that is, if there is no disagreement between the applicant and the NRC staff concerning the granting of a permit or any of its conditions, and if the application is not opposed by an interested party—the ASLB will make its determination after considering the evidence. The ASLB is required to make certain that the construction permit application, the staff's Safety Evaluation Report and Final Environmental Statement, and the hearing record are sufficient to support the finding that the plant can be built safely, the applicant is qualified, and other requirements can be met. The ASLB need not make these findings, but must determine that the record and review by the staff are adequate. In contested proceedings, the ASLB is also required to decide matters in controversy and may be called upon to make technical judgments in these matters.

Regardless of whether an application is contested or not, the ASLB must determine if the appropriate requirements of NEPA, regarding the environmental impact of the proposed plant, have been complied with. Furthermore, an independent balance must be made of environmental risks and economic (and other) benefits of the plant in order to determine the appropriate action to be taken. The ASLB then decides whether a construction permit should be issued, denied, or appropriately conditioned to protect the environment.

Review of Construction Permit Decision

The initial decision of the ASLB is reviewed by an Atomic Safety and Licensing Appeal Board. The Appeal Board consists of three members chosen from the Appeal Board Panel; a chairman and two technically qualified individuals are designated for each proceeding. If any of the affected parties appeals the decision of the ASLB, the Appeal Board makes a detailed review of the matters in controversy and decides on their merits. In any event, the Appeal Board examines the record of the ASLB hearings and determines whether this board has ruled properly on the testimony presented at the hearings.

The Appeal Board may approve the ASLB decision, it may reverse it, or it may instruct the ASLB to reopen the hearing to receive further evidence on controversial matters. The commissioners of the NRC may then, on their own initiative, accept or reverse the decision of the Appeal Board. Furthermore, an interested party may appeal the decision in a federal court of law.

The procedures involved in connection with an application for a permit to construct a nuclear power plant are summarized in Table 3-II.

Limited Work Authorization

At one time, an applicant for a permit to construct a nuclear plant could perform preliminary work on the site before the permit was granted. With the passage of NEPA in 1969, this was no longer allowed until 1974, when the AEC/NRC issued new regulations. The NRC staff may now complete its review of the applicant's Environmental Report and issue its Draft Environmental Statement and Final Environmental Statement before the review of the PSAR has been completed. In these circumstances, the applicant may apply for a Limited Work Authorization, which would allow work to proceed on site preparation to the extent that safety options would not be compromised; this could include clearing and grading, excavating for foundations, and pouring of basement concrete.

Before the LWA is granted, a public hearing must be held before an ASLB to provide assurance that all NEPA requirements have been met. The matters taken up are primarily the environmental impact of the construction and operation of the proposed plant, the benefits that would result from the plant, and the balancing of risks and benefits. In addition, the hearing must cover the suitabili-

TABLE 3-II

OUTLINE OF PROCEDURE FOR A CONSTRUCTION PERMIT

Health and Safety Aspects	Environmental Impact
PSAR submitted by applicant	Environmental Report—Construction Permit Stage is submitted by applicant
PSAR reviewed by NRC staff	NRC staff prepares a Draft Environmental Statement
PSAR reviewed by ACRS	Draft reviewed by various agencies
Letter of recommendations sent by ACRS to NRC; NRC staff prepares Safety Evaluation Report	NRC staff prepares a Final Environmental Statement (sent to Council on Environmental Quality, etc.)
Public hearings on safety and environmental aspects held by the ASLB	
The ASLB determines whether the NRC staff review is adequate and safety issues have been (or will be) resolved	The ASLB determines whether the requirements of the National Environmental Protection Act of 1969 have been met. The ASLB balances environmental costs against benefits
If satisfied, the ASLB authorizes the NRC to issue a construction permit, subject to such conditions as may be considered necessary	
The ASLB decision is reviewed by the Appeal Board	
The commissioners of the NRC may make the final determination	

ty of the site for a nuclear power plant of the general type and power level proposed by the applicant. The site review (see Chapter 14) will include such topics as meteorology, hydrology (flooding), geology (faults, soil and rock characteristics, etc.), population distribution, and the probability of compliance with the site criteria discussed earlier.

After the hearing, the ASLB considers the testimony and writes an initial decision concerning environmental matters and site suitability. This decision may authorize the NRC to issue the LWA for land clearing, etc., limited in scope by 10CFR50, Section 50.10(e)(1). When the NRC staff has reviewed the applicant's PSAR and prepared its Safety Evaluation Report, the hearing will be resumed to consider health and safety questions. The second part of the initial decision then authorizes the NRC to issue a construction permit. The initial decision is then reviewed in the manner described earlier.

Standardized Designs

Another modification of the procedure for obtaining a construction permit is based on the concept of standardized designs of nuclear power reactors. Since nuclear power technology is a relatively recent development, improvements in design have been made continuously on the basis of experience. Consequently, each reactor design has differed in some respects from others. Furthermore, in the course of time, NRC safety criteria and requirements have become more stringent. Hence, the NRC staff has had to consider in detail each application for a construction permit.

Changes in the regulations allow for issuance of a single permit to cover a class of identical systems. This can be accomplished in two alternative ways:

1. In the *standard design* option (10CFR50, Appendix O) a manufacturer can apply for a license to build a number of identical nuclear power reactors (or major portions thereof). The application includes a sufficiently detailed description of the proposed system to permit the NRC staff to review it for conformity with all the appropriate NRC criteria. After the review is completed, the staff prepares a Safety Evaluation Report for the standard (or reference) design and, if all requirements are met, may recommend approval of the design.

When a utility applies for a permit to construct a nuclear plant at a selected site using such an approved design, the detailed safety evaluation is not repeated. The NRC review and the subsequent ASLB hearings will then refer only to site suitability and environmental impact, just as in the application for an LWA.

2. If a utility (or group of utilities) wishes to order two or more nuclear power plants of the same design for construction at different sites, a single review and hearing will be held on the safety of the proposed design. This is referred to as the *duplicate plant concept* (10CFR50, Appendix N). Each site is then considered for suitability and environmental impact, and a hearing is held at each site, as for an LWA.

In either of the two situations described above, after all the required hearings have been completed, the ASLB will make its decision regarding the issuance of a construction permit at the approved site. This initial decision is then subject to review by an Atomic Safety and Licensing Appeal Board in the usual manner.

License to Manufacture

As a general rule, a nuclear reactor facility is constructed at the site at which it will be operated. However, a scheme has been proposed for manufacturing floating nuclear power plants at a given location and then towing them to another location at which they are to be operated. This situation is covered in 10CFR50, Appendix M, which deals with license to manufacture nuclear power plants to be operated at sites not identified in the license.

The applicant for a manufacturing license must submit a Design Report, which takes the place of the PSAR, and an Environmental Report. The latter is directed mainly at the manufacturing site, but it must also deal in general terms with potential operating sites. Apart from these modifications, the licensing process is the same as that for a construction permit. Before the reactor can be operated at an approved site, an operating license must be obtained in the usual manner, as described below.

APPLICATION FOR AN OPERATING LICENSE

Final Safety Analysis Report

As construction proceeds on a nuclear power plant, it is inspected periodically by representatives of the NRC to assure that the requirements of the construction permit are being met. Before construction is complete, the owner-operator of the plant submits to the NRC an application for an operating license. This application must include a Final Safety Analysis Report (FSAR) and the Applicant's Environmental Report—Operating License Stage.

The FSAR covers much the same ground as the preliminary report, but the topics are treated in greater detail.[d] The final report is amended continuously to provide additional information required by the NRC staff and to reflect design changes as construction proceeds. Many items that are not applicable to the preliminary report are treated in the final report. For example, detailed analysis and safety evaluation are required for the reactor core, the coolant system, instrumentation and controls, containment and other engineered safety features, the radwaste system, and the means for handling fresh and spent fuel assemblies.

The report must contain a final analysis and evaluation of the design and performance of structures, systems, and components, taking into account any new information that has been developed since submission of the PSAR. Evidence must be provided that safety questions identified at the construction permit stage have been resolved. New information must be incorporated relative to site evaluation, such as the results, obtained during the construction phase, of environmental and meterological monitoring programs.

The FSAR must also include a description of plans for preoperational testing and startup programs (p. 54). Furthermore, plans for conducting normal opera-

[d]As nuclear power reactor designs become more standardized, the PSAR tends to cover more of the information required in the final report.

tions, including maintenance, surveillance, and periodic testing of structures, systems, and components, must be explained.

Plans for coping with emergencies and for cooperation with state and local government agencies must be described in detail, in accordance with the requirements of 10CFR50, Appendix E. This appendix is being revised to take into account the emergency planning procedures outlined on page 36. The state and local plans, as assessed by the FEMA, must be found acceptable by the NRC before an operating license can be granted.

Environmental Reports and Statements

The Applicant's Environmental Report—Operating Stage is, in effect, an updating of the Environmental Report submitted at the construction permit stage. The report should discuss differences between currently projected environmental effects of the nuclear power plant, both beneficial and adverse, and those projected in the earlier Environmental Report. Such differences may arise from changes in plans, in plant design, or in the surrounding land-use or zoning classifications, or from new or more detailed information that has become available.

The results should be given of all studies that were not completed when the construction permit was issued, but were specified to be completed before the operating stage. The way in which the results have been accommodated in the design and the proposed manner of operation of the plant should be indicated. A description should be included of planned studies and those not yet completed that may yield results relevant to the environmental impact of the plant.

The detailed monitoring programs that have been (or will be) undertaken to determine the environmental effects of the operating plant must be described. The results of preoperational monitoring activities should be included. A listing is required of types of measurements, number of samples, frequency of sampling, etc., and the locations of the monitoring stations should be described and indicated on a map of the construction site area.

The NRC staff prepares a Draft Environmental Statement followed by a Final Environmental Statement, as required by NEPA, in the same manner as described earlier in connection with the application for a constructon permit. As before, the Final Environmental Statement is transmitted to the Council on Environmental Quality and to various agencies and is made available to the public.

Technical Specifications

Together with his FSAR and Environmental Report, the applicant for an operating license must submit the proposed Technical Specifications for the nuclear power plant (10CFR50, Sections 50.36 and 50.36a). These specifications, which may vary from one installation to another, delineate the technical operating limits, conditions, and requirements imposed on operation of the plant in the interest of the health and safety of the public and the environment. After review and possible modification by the NRC staff, the Technical Specifications will become part of the operating license for the plant. To simplify the licensing process and to promote uniformity in the application and interpretation of its requirements, the NRC has prepared Standard Technical Specifications for the guidance of license applicants.

The Technical Specifications concerned with reactor safety include the following:

1. The limits on important process variables, such as pressure in the reactor vessel, coolant flow rate, water level in the steam generators (in PWRs), rate of heat generation, etc. These limits serve to protect the integrity of the fuel element cladding (and other physical barriers) that minimize the release of radioactivity to the environment.

2. The instrument settings that would cause various engineered safety feature systems to start operating automatically. (Engineered safety features [see Chapter 4] consist of several elements such as an emergency core-cooling system and a containment vessel, which encloses the nuclear steam supply system, and their various components, designed to minimize the escape of radioactivity to the environment in the event of a severe accident.)

3. The maximum rate of change in the reactor temperature (e.g., 100 °F [56 °C] per hour) during heatup (startup) or cooldown (shutdown) to avoid unnecessary stress (and accompanying strain) in the reactor vessel.

4. The primary coolant leakage rate at which inspection and evaluation of the situation must be initiated and at which the reactor must be shut down to allow necesary repairs to be made.

5. Limits on the release of radioactive gaseous and liquid effluents to the environment over long periods of time.

6. The frequency of inspection, testing, and calibration of all instruments important to safety. (As noted in Chapter 5, additional instrument channels are always available to permit testing without shutting down the reactor).

7. Details of the in-service inspection of primary coolant system components to provide assurance of continuous integrity of the system.

8. Requirements for testing components of the emergency core-cooling system, the leakage rate of the containment vessel (see Item 2) and the diesel engines that would provide electric power if all other sources were to fail (p. 74).

9. Details of the operating organization of the facility, including duties and responsibilities of each position. These details should include provisions for record keeping, review, and reporting on matters relating to safe operation.

To provide assurance that operation of the plant will have a minimum impact on the environment, the Technical Specifications are concerned with the following matters:

1. Fish and other aquatic organisms can be adversely affected by high (or low) water temperature and by the rate of increase or decrease of temperature. Hence, limits are placed on the maximum temperature, the temperature increase, and the rate of temperature change (increase or decrease) in the water discharged to an adjacent water body from the plant's turbine condenser (see Chapter 12). Chemicals, which may be harmful to aquatic life forms, may be added to the condenser water to control corrosion and the deposition of slimes, etc., on the tubes; maximum permissible concentrations of these chemicals in the effluent are specified.

2. The Technical Specifications contain details of an Environmental Ecological Survey, which must be conducted before the plant starts to operate and when it is in operation. The survey includes observations of the temperature pattern (isotherms) in the thermal "plume" formed when the condenser water is discharged into an adjacent water body (p. 236). Samples of the water must be taken at specified intervals both for chemical analysis and for counting the numbers of various aquatic organisms. The larger life forms (e.g., fish) can be injured by impingement on intake screens, which are designed to prevent entry of debris; smaller life forms, by entrainment with the water and passage through the condenser. The effects of impingement and entrainment must be included in the ecological survey.

3. A Radiological Monitoring Program is specified to assure that the radiation dose received by members of the public will not exceed certain prescribed low limits (see Chapter 6). This radiation dose may arise from the consumption of food (including milk) and water and breathing air containing minute concentrations of radioactive materials, and from exposure to external radiation sources. The monitoring program thus includes sampling at frequent intervals of air, water, milk, and other foods for radioanalysis. In addition, regular measurements are required of the external radiation by instruments located at many specified points around the plant site.

4. Complete records must be maintained of the ecological and radiological surveys and summaries must be submitted to the NRC semiannually. Any significant environmental impact must be reported, as also must deviations from the approved specifications. Urgent remedial action is required when certain limits are exceeded.

Operating Licensing Procedures

The NRC staff evaluates the operating license applicant's FSAR and Environmental Report. A review is also made of the safety aspects of the proposed operation by the ACRS. The procedures are much the same as those followed in connection with the application for a construction permit. Both the staff's evaluations and the ACRS's report to the NRC are made available to the public.

A public hearing is not required by law in an application for an operating license. Normally, the NRC will not direct that a hearing be held at this stage unless there is a special safety problem of unusual public importance, unless substantial public interest warrants a hearing, or unless a hearing is requested by a party whose interest would be adversely affected. In requesting a hearing, the reason for opposing issuance of an operating license must be stated specifically.

Should a hearing be required, it is held by an ASLB; the members of this board may be, but are not necessarily, the same as those who considered the application for the construction permit. The hearings are conducted in the same manner as at that stage. The ASLB must now decide only on matters of controversy.

Before the NRC issues a full-term operating license, the Director of the agency's Office of Nuclear Reactor Regulation must be satisfied that, in addition to compliance with the environmental requirements in 10CFR51, the following conditions are met:

1. Construction of the facility has been substantially completed in conformity with the construction permit and the applicable rules and regulations of the NRC.
2. The facility will operate in conformity with the application for an operating license and the rules and regulations of the NRC.
3. There is reasonable assurance that (a) the activities authorized by the operating license can be conducted without endangering the health and safety of the public and (b) such activities will be conducted in accordance with the NRC regulations.
4. The applicant is technically and financially qualified to engage in the activities authorized by the operating license in accordance with NRC regulations.
5. The applicable provisions of 10CFR140, "Financial Protection Requirements and Indemnity Agreements," which deal with liability insurance, have been satisfied.
6. The issuance of the license will not be inimical to the common defense and security or to the health and safety of the public.

The stages in the procedure for obtaining a license to operate a nuclear power plant are outlined in Table 3-III.

REGULATION OF PLANT OPERATIONS

Preoperational and Startup Testing

Before the nuclear plant can be operated at full power, comprehensive preoperational and startup test programs must be carried out. Guidance for planning these programs is provided in NRC Regulatory Guide 1.68, "Preoperational and Initial Startup Test Programs for Water-Cooled Power Reactors." The major purposes of the preoperational program are: (a) to assure that construction has been accomplished properly, (b) to verify that all equipment and systems meet their design objectives, (c) to verify that operating and emergency procedures are correct and reasonable, (d) to allow plant operators to familiarize themselves with equipment and procedures prior to actual operation, and (e) to assist in qualifying and training operators. The program is conducted in accordance with written procedures, and it must be approved at appropriate stages by management and quality assurance personnel.

Upon completion of the preoperational test program, an extensive reactor startup program is conducted. This program involves a deliberate step-by-step ascent to full-power operation. In the course of the program, the performance of crucial safety equipment is confirmed. The operating characteristics of the plant and its equipment are checked at increasing power levels. Operating procedures are verified, and operators develop increasing familiarity with the plant.

Operator Training and Licensing

The controls of a commercial nuclear reactor may be manipulated only by an operator licensed by the NRC in accordance with 10CFR55, "Operator's

TABLE 3-III

OUTLINE OF PROCEDURE FOR AN OPERATING LICENSE

Health and Safety Aspects	Environmental Impact
FSAR including Technical Specifications submitted by applicant	Environmental Report—Operating License Stage submitted by applicant
FSAR reviewed by NRC staff and independently by ACRS	NRC staff prepares a Draft Environmental Statement and, after review, a Final Environmental Statement, as at the construction permit stage (sent to Council on Environmental Quality, etc.)
NRC staff prepares Safety Evaluation Report	
Public hearing is not mandatory, and the NRC can grant an operating license, subject to such conditions as may be considered necessary	
A hearing may be requested by a party who would be adversely affected in a specified manner. An ASLB will then consider only matters of controversy and transmit their findings to NRC	
NRC issues an operating license	

Licenses." The licensed operator must complete a comprehensive program of classroom training, which includes general courses on nuclear reactor theory and related topics, and specific courses on the detailed systems, operating procedures, and emergency actions in the plant for which he is to be licensed. In addition, a senior operator must demonstrate a knowledge of the limits and conditions of the facility license, design modifications, Technical Specifications, and fuel and waste handling.

Besides classroom training, an operator must complete a program of on-the-job training. This commonly includes: (a) work at a plant of the same type as the one in which he is to be employed, (b) the use of training simulators programmed to simulate normal, abnormal, and emergency operating conditions that might be encountered in an actual plant, and (c) familiarization with equipment and procedures during the preoperational test program. Before receiving a license, an operator candidate must pass written and operational proficiency examinations. When issued, a license is generally restricted to a specific unit (or units in the case of an installation that includes two or more units of the same type) and must be renewed every two years.

The accident at Three Mile Island was due in part to operator errors. It appears, therefore, that more extensive training (and retraining) is necessary for operators to learn to deal with a wide spectrum of abnormal situations. The NRC now requires plant licensees to establish programs for retraining of reactor operators and shift supervisors, improved shift turnover procedures, and the presence at all times of a shift technical adviser with at least a bachelor's degree (or its equivalent) in science or engineering.

The NRC Inspection Program

The NRC program for the inspection of nuclear power plants covers all phases, from construction through preoperational testing and initial startup to normal operation. Typically, the facility is inspected 25 to 30 times while under construction. About 10 or 12 inspections are generally made during the initial operation stage, and, in the past, there have been about four or five inspections annually when the plant is in full operation. However, the NRC plans to place resident inspectors in all operating nuclear power plants.

The on-site inspection program is intended to achieve the following three principal objectives:

1. To ascertain if the plant licensee has implemented an effective quality assurance program in accordance with his application for a construction permit and with NRC rules and regulations.
2. To ascertain if the facility is constructed in accordance with the provisions of the construction permit and with sound and proven engineering principles and construction practices, and that the plant has been tested adequately.
3. To ascertain that the licensee is operating the plant and conducting the activities in compliance with the provisions of the license, with the NRC rules and regulations, and with the Atomic Energy Act (as amended).

If inspections indicate that the plant is not being operated properly, enforcement action may be taken with respect to matters relating to safety or to environmental impact or failure to comply with the requirements of the operating license. The enforcement may be routine notice of violation of the license, a monetary fine for repeated violations and/or negligence, or, in extreme cases, issuance of an order suspending or terminating the operating license.

Plant Operation

Instructions for operating a nuclear power plant are embodied in some 500 to 1000 approved written procedures covering all phases of operation. The operations must be conducted at all times in accordance with the Technical Specifications, as approved by the NRC staff, based on those submitted by the plant licensee.

In addition to procedures for normal operation, procedures must be written for coping with all expected abnormal events, such as loss of turbine load and automatic shutdown (p. 71). The written procedures also include actions to be taken in the event of an unexpected release of radioactivity to the environment. Reactor operators should be trained to place the plant in its safest condition following any questionable or unexplained operating occurrence.

Operators must maintain logs of each day's operations, and any significant abnormal event must be recorded. Any such event having safety implications must be reported to the NRC for appropriate action, as described shortly.

During the operation of a nuclear plant, the licensee may find that, as a result of experience, changes are desirable in operating procedures or equipment. Significant changes affecting safety will require amendments to the license or changes in Technical Specifications. Proposed changes are reviewed by the NRC staff, which then authorizes or denies the changes.

The Technical Specifications include the requirement that the plant licensee submit periodic reports to the NRC describing the operation of the facility and the results of specified periodic tests. In particular, twice a year the licensee must report the quantity of each of the principal radioactive substances released outside the site boundary in liquid and gaseous effluents during the previous six months of operation. The information and such other information as may be required by the NRC permits an estimate to be made of the maximum potential radiation dose to the public from effluent releases. If the dose exceeds that permitted by NRC regulations, the plant licensee must take appropriate action.

Use of Operational Experience

The occurrence of any abnormal (safety-related) event at a nuclear power plant or any violation of the Technical Specifications must be reported immediately to an NRC representative. The NRC is required to report to the U.S. Congress each quarter on abnormal occurrences at all licensed facilities.[e]

The reports submitted by licensees indicate to the NRC if operations are being conducted safely and within the license conditions, if appropriate corrective action has been taken to deal with malfunctions or unanticipated events, and if management maintains an effective program of review and audit of its operations. Furthermore, the abnormal incidents are studied to determine if they represent possible modes of failure or design deficiences that may be common to all plants of the same type. If this is the case, appropriate corrective actions or modifications will be required in all plants of that type. Several improvements have already been made in the design, construction, and operation of nuclear power plants as a result of reported experiences. An important contribution to overall safety is made by carefully reviewing abnormal occurrences, such as the incident at Three Mile Island, and requiring remedial action in all nuclear plants where the same or similar conditions could arise.

The Licensee's Maintenance Program

The licensee (or owner) of a nuclear power plant has a substantial economic incentive to maintain continuous, safe operation of the plant. The quality assurance program for design and construction, mentioned earlier, makes an important contribution in this respect. Another aspect of this program is to keep the facility in good condition while it is operating. Regulatory Guide 1.33, "Quality Assurance Program Requirements (Operation)," describes quality assurance criteria for plant operation. Plans must be prepared in writing for testing various plant subsystems that are called upon only in the event of an emergency to assure proper operation when required. Provisions must also be made for inspection and for preventive and corrective maintenance of operational components. A major problem in a nuclear plant is that many components become radioactive during operation and cannot be manipulated by normal procedures. Careful attention to design can eliminate (or minimize) the problems associated with radioactivity and also make components more readily accessible for inspection and maintenance.

Inspection of piping and other equipment is performed periodically by the licensee during the life of the plant. Nondestructive techniques used in the in-

[e]Safety-related occurrences are reported every two months in *Nuclear Safety*, a publication prepared by the Nuclear Safety Information Center, Oak Ridge National Laboratory, Oak Ridge, Tennessee.

spections include radiographic, ultrasonic, and dye-penetrant procedures. If significant deterioration is detected, suitable maintenance or possible replacement of a component is indicated.

One of the most important contributions to the safety of a nuclear power plant is that it should operate properly at all times. Adequate testing, inspection, maintenance, and effective operator training are essential to the achievement of this objective.

Bibliography—Chapter 3

Bernsen, S. A. "Quality Assurance in the Construction of Nuclear Power Plants." *Nuclear Safety* 16 (1975): 127.

Bush, S. H. "The Role of the Advisory Committee on Reactor Safeguards on the Reactor Licensing Process." *Nuclear Safety* 13 (1972): 1.

Davis, J. P. "Regulation of the Environmental Effects of Nuclear Power Plants." *Nuclear Safety* 14 (1973): 165.

Eichholz, G. G. "Planning and Validation of Environmental Programs at Operating Nuclear Power Plants." *Nuclear Safety* 19 (1978): 486.

Goodrich, N. H. "The Role of the Atomic Safety and Licensing Board Panel." *Nuclear Safety* 15 (1974): 383.

International Atomic Energy Agency. *Proceedings of the Symposium on the Application of Reliability Technology to Nuclear Power Plants.* Vienna: International Atomic Energy Agency, 1978.

______. *Proceedings of the Symposium on the Reliability of Nuclear Power Plants.* Vienna: International Atomic Energy Agency, 1975.

Joslin, W. M.; Moore, J. S.; and Russ, J. C. "Review of Engineering Standards Development for Nuclear Power Stations." *Nuclear Safety* 12 (1971): 212.

Langston, M. E. "Quality Assurance Standards and Practices." *Nuclear Safety* 12 (1971): 549.

Lawroski, S., and D. W. Moeller. "Advisory Committee on Reactor Safeguards: Its Role in Nuclear Safety." *Nuclear Safety* 20 (1979): 387.

Long, J. A. "The Reactor Licensing Process—A Status Report." *Nuclear Safety* 18 (1977): 281.

"Regulation of Nuclear Power Reactors and Related Facilities." *Nuclear Safety* 15 (1974): 1.

Thro, Ellen. "Nuclear Standards Today." *Nuclear News* 19, no. 11 (1976): 88.

U.S., Atomic Energy Commission. *Population Distribution Around Nuclear Power Plant Sites* (WASH-1308). Washington, D.C., 1973. See also, *Nuclear Safety* 16 (1975): 1.

______. *The Safety of Nuclear Power Reactors and Related Facilities* (WASH-1250), Chapters 3 and 5. Washington, D.C., 1973.

U.S., *Code of Federal Regulations*, Title 10 (Energy), Washington, D.C.,
Part 2. Rules of Practice.
Part 7. Advisory Committees.
Part 9. Public Records.
Part 20. Standards for Protection Against Radiation.
Part 50. Licensing of Production and Utilization Facilities,
Part 51. Licensing and Regulatory Policy and Procedures for Environmental Protection.
Part 55. Operator's Licenses.
Part 100. Reactor Site Criteria.
Part 140. Financial Protection Requirements and Indemnity Agreements.

U.S., Nuclear Regulatory Commission. "Assumptions Used for Evaluating the Potential Radiological Consequences of a Loss of Coolant Accident for Boiling Water Reactors" (Regulatory Guide 1.3). Washington, D.C.

______. "Assumptions Used for Evaluating the Potential Radiological Consequences of a Loss of Coolant Accident for Pressurized Water Reactors" (Regulatory Guide 1.4). Washington, D.C.

______. *Environmental Standard Review Plan for the Environmental Review of Construction Permit Application for Nuclear Plants* (NUREG-0555). Washington, D.C., 1979.

______. "The Preparation of Environmental Reports for Nuclear Power Plants" (Regulatory Guide 4.2). Washington, D.C.

______. "Quality Assurance Program Requirements (Operation)" (Regulatory Guide 1.33). Washington, D.C.

______. "Quality Assurance Requirements for the Design of Nuclear Power Plants" (Regulatory Guide 1.64). Washington, D.C.

______. *Report of the Siting Policy Task Force* (NUREG-0625). Washington, D.C., 1979.

______. *Review of the Commission Program for Standardization of Nuclear Power Plants* (NUREG-0427). Washington, D.C., 1978. See also, W. F. Kane, *Nuclear Safety* 20 (1979): 258.

______. "Standard Format and Content of Safety Analysis Reports for Nuclear Power Plants" (Regulatory Guide 1.70), Washington, D.C.

______. *Standard Review Plan for the Review of Draft Environmental Reports for Nuclear Power Plants* (NUREG-0158). Washington, D.C., 1978.

______. *Standard Review Plan for the Review of Safety Analysis Reports for Nuclear Power Plants* (NUREG-75/087). Washington, D.C., 1975.

U.S., Nuclear Regulatory Commission, and U.S. Environmental Protection Agency. *Planning Basis for the Development of State and Local Government Emergency Response Plans* (NUREG-0396; EPA 520/1-78-016). Washington, D.C., 1978.

4

Nuclear Reactor Safety

INTRODUCTION

General Considerations

The safe design, construction, and operation of a commercial nuclear power plant are the responsibility of the owner-operator (or licensee) of the facility. The U.S. Nuclear Regulatory Commission (NRC), however, has the statutory responsibility for protecting the health and safety of the public (and the quality of the environment) in connection with the development and use of nuclear energy. This responsibility is implemented through the NRC's licensing and inspection procedures, as discussed in some detail in Chapter 3. These procedures are aimed at providing assurance that nuclear power plants can be operated in a reliable manner without endangering the health and safety of the public.

Nuclear power plants are unique in the respect that they generate large amounts of radioactive materials that could be potentially hazardous to all forms of life if substantial quantities should be released to the environment. However, NRC regulations and the Technical Specifications that form part of the plant's operating license require that the plant be designed and operated in such a manner as to limit the amounts of radioactive material that may be released. Consequently, the radiation doses received by people, animals, and plants during normal operation of a nuclear plant are smaller than the variations in the dose received from the natural radiation background (Chapter 6).

In the operation of a nuclear power plant, there is always a possibility that abnormal situations may arise that, if not controlled, could result in the escape of significant amounts of radioactive materials to the environment. The main purpose of this chapter is to examine the safety features used to control a variety of abnormal events or to mitigate their consequences if complete control is not possible. In addition, consideration is given to the probability and consequences of accidents resulting from failure of the plant's safety features.

Defense in Depth

The basic philosophy of the NRC in assuring safety in the design, construction, and operation of nuclear plants has been called "defense in depth," as expressed in terms of three levels of safety. These levels concern a variety of considerations that frequently intermesh, and assignment of certain design features to one level or another is somewhat arbitrary. Nevertheless, the three levels do serve to indicate in a useful, if general, manner successive levels in the safety design of a nuclear power plant.

The first level of safety is to design the reactor and its associated components so that they will operate with a high degree of reliability with only a small chance of a malfunction occurring. To this end, the NRC's quality assurance program and the use of strict codes and standards place special emphasis on the quality of materials and workmanship in plant construction. The plant must be designed and constructed to withstand hurricanes, tornadoes, and other natural phenomena, and to be capable of a safe shutdown in the event of the worst earthquake that might be expected at the plant site. Furthermore, the design must permit continuous or periodic monitoring of components and systems to detect signs of wear or incipient failure. In summary, the overall aim is to assure that the facility is designed in a sound and conservative manner and that it can be built, tested, operated, and maintained in accordance with stringent quality standards and engineering practices.

Despite the assurance offered by careful plant design and construction, it is necessary to anticipate that some incidents or malfunctions will occur during the service life of the installation. The purpose of the second level of safety is to provide means that will forestall or cope with such events. The nuclear reactor is consequently equipped with a *protection system* designed to prevent, arrest, or accommodate safely a range of conceivable abnormal situations.

Finally, the third level of safety is based on the view that it is prudent to go beyond the levels of safety described above and to incorporate additional systems and barriers to the escape of radioactivity to protect the public even if certain highly unlikely accidents should occur. To establish these additional features, major potential failures of components and systems are postulated and their conceivable consequences are analyzed. The analysis of such hypothetical events determines the set of design-basis accidents (referred to on page 40), and safety systems (called the *engineered safety features*) are designed to control or accommodate them.

In brief, the ultimate goal of reactor safety is to reduce to an acceptably small level the risk of injury to members of the public, as well as to those working in nuclear plants. In the design, building, and operation of these plants, safety is of overriding importance, very much more so than in any other type of industrial construction.

Descriptions of Light-Water Reactors

During the 1980s and probably into the 1990s, nearly all the commercial power reactors in operation in the United States will be either pressurized water reactors (PWRs) or boiling water reactors (BWRs). Consequently, the subsequent discussion of reactor safety is restricted to reactors of these two types. As stated in Chapter 2, both PWRs and BWRs employ ordinary (i.e., light) water as the coolant and moderator. They are consequently referred to by the general name of light water reactors (LWRs). To understand the features that are important to the safety of LWRs, somewhat more detailed descriptions are given of PWR and BWR design.

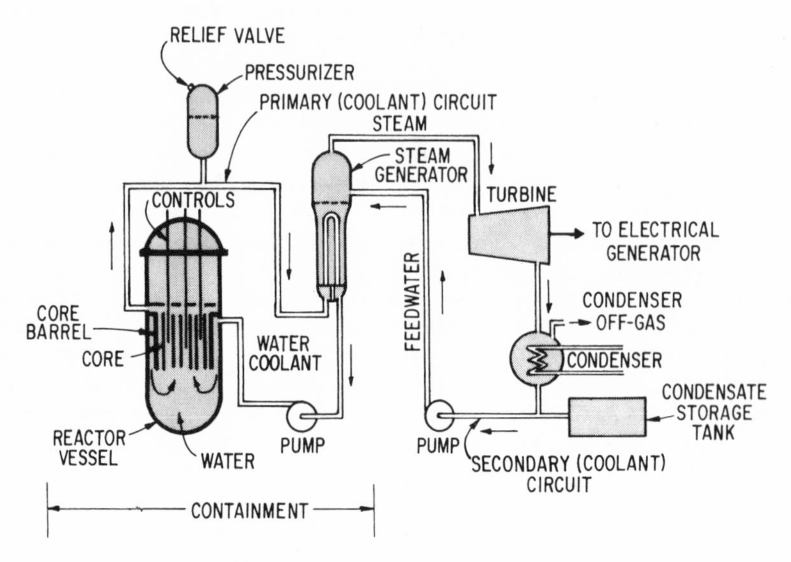

FIG. 4-1. Outline of a PWR power plant.

Pressurized Water Reactors. In a PWR (Fig. 4-1), the pressure is maintained at about 15.5 megapascals (MPa) (2250 psi) so that the water does not boil even at the maximum temperature of 326 °C (620 °F) reached during normal operation. The reactor vessel (or pressure vessel) containing the core of nuclear fuel and water under pressure is a tall cylinder, made of high-strength steel, with a hemispherical bottom and a domed head. The head is bolted to the top of the cylinder to permit removal when it is necessary to withdraw and replace spent fuel in the core, usually about once a year. The neutron-absorbing (control) rods (p. 20) pass through the head of the reactor vessel. A typical PWR vessel in a modern nuclear power plant is up to 13.7 meters (m) (45 ft) high and has an inside diameter of about 4.6 m (15 ft). The walls of the vessel are 20 to 23 centimeters (cm) (8 or 9 in.) thick.

The reactor core is approximately cylindrical in form and is about 4.0 m (13 ft) high and 3.4 m (11 ft) across. It consists of roughly 40 000 vertical fuel rods, about 1 cm (0.4 in.) in diameter, arranged in clusters (or assemblies), each containing some 200 or more rods. The fuel material is uranium dioxide (UO_2), enriched to the extent of 2 to 4 percent in the fissile isotope uranium-235. A cladding, usually made of a zirconium alloy, prevents access of the high-temperature water to the fuel and also aids in retaining the products of the fission reaction. More is said about the fuel rods later. The total weight of uranium dioxide in a PWR core is generally about 110 000 kilograms (kg) (240 000 lb).

Coarse (or shim) control of the fission rate, and, hence, of the thermal power output, is achieved by means of the neutron absorber boron, in the form of boric acid, disssolved in the reactor water. The concentration of boron is changed during the course of reactor operation to balance the gradual decrease in reactivity resulting from fuel depletion and the accumulation of fission-product poisons. For fine control, use is made of rods containing a neutron-absorbing material, such as an alloy of silver, indium, and cadmium. The rods enter from the top of the core and can be moved up or down, as required, within guide tubes located in some fuel assemblies. The control rods are combined in clusters of up to about 20 rods, which are moved together, and there may be as many as 70 clusters distributed throughout the core.

The drive mechanisms for the individual control rod clusters are located above the top of the reactor vessel. The mode of operation of the mechanisms varies to some extent from one PWR design to another, but it always depends on the use of electricity. In normal operation, the rods are moved automatically to maintain reactor power within prescribed safe limits. When it is required to shut down the reactor, the control rods are usually inserted gradually into the core. In the event of an emergency requiring the reactor to be shut down quickly, the electrical supply to the electromagnetic clutches holding the control rods is cut off, and the rods drop into the core under the influence of gravity.

The water under pressure is heated as it flows through the reactor core and then passes to a steam generator. At least two, and sometimes three or four, independent steam generators are used (Fig. 4-2). There are various designs for steam generators, but a common design is the one described in Chapter 2. The steam drives the turbines, and the exhaust steam is condensed back into liquid water in the condensers (see Fig. 4-1). The condensate provides the feedwater for the steam generators. After passing through the tubes in the steam generator, the pressurized water is pumped back into the reactor vessel. It enters close to the top of the core and flows downward through the annular region between the core "barrel" and the reactor vessel. The core barrel, which surrounds the core, serves to separate the downward flow of the incoming water from the upward flow through the core. As the coolant water travels upward at high speed in the spaces between the fuel rods in the core, it removes the heat generated by fission in the fuel. When the heated water reaches the top of the core, it proceeds to the steam generator as described above.

It is seen from Fig. 4-1 that the PWR nuclear steam supply system, which produces steam for the turbines, consists of two separate flow circuits. One, the high-pressure circuit, made up of the reactor vessel, the inside of the tubes in the steam generators, the pumps, and associated piping, is called the *primary (coolant) circuit* or *primary (coolant) system*. Each steam generator constitutes a branch of the primary system. The other circuit, at a lower pressure, is the *secondary (coolant) system*, in which the steam is generated, condensed, and the condensate returned as feedwater. A certain amount of radioactivity may be present in the water in the primary system, but unless there is a leak in the steam generator, the secondary system (and the steam) will be nonradioactive.

High reliability and performance are demanded of the pumps, located between the steam generator outlet and the reactor vessel, in the primary coolant circuit. They must provide vital coolant to the reactor core, to prevent overheating of the fuel, and are, of course, essential to the operation of the steam generators. Because the primary coolant pumps must operate at high temperatures and pressures, stringent design and manufacturing standards are required, as indeed they are for all other components of the nuclear installation.

For stable operation of the plant, the pressure in the primary system must be maintained within certain prescribed limits. This is achieved by means of a pressurizer, connected into one of the outlet pipes from the reactor vessel to one of the steam generators (see Fig. 4-2). The pressurizer is a cylindrical steel tank containing roughly 60 percent by volume of liquid water and 40 percent steam during normal operation. The water in the lower section of the tank can be heated with electric immersion heaters, whereas cold water can be sprayed into the upper (steam) section to provide cooling when required. Both the heaters and the cooling sprays are operated by pressure signals. If the system pressure should drop, a low-pressure signal would actuate the heaters, steam would be generated, and the system pressure would be increased. On the other hand, if the pressure should rise above normal, a high-pressure signal would operate the cooling water spray. Some steam in the pressurizer would be condensed, and the system pressure would decrease. If the system pressure should increase beyond the capability of the spray to reduce it, one or more relief valves would vent steam to a water (quench) tank in the containment.

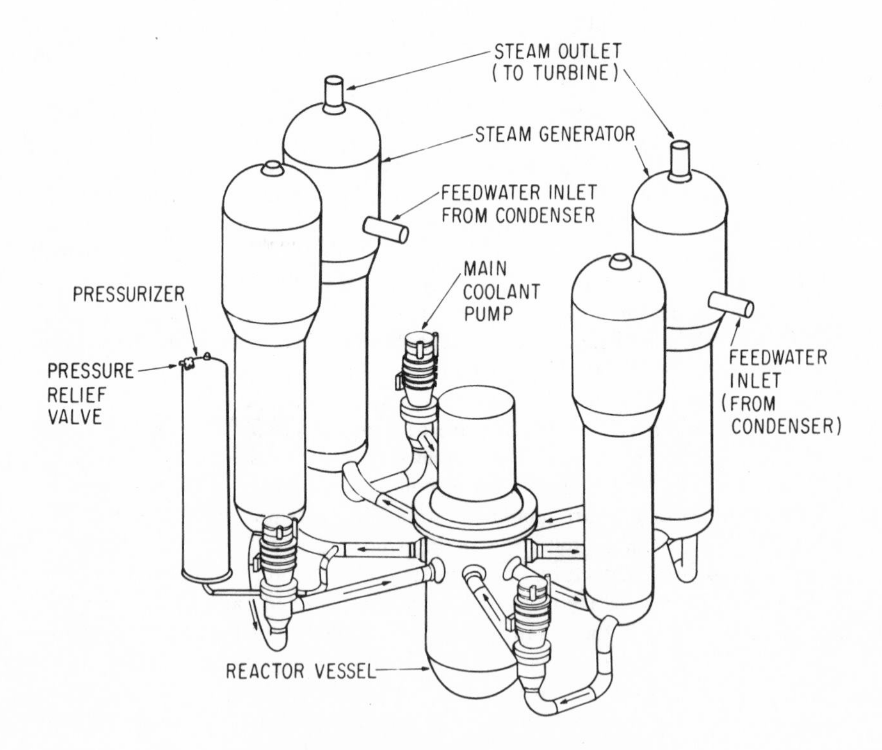

FIG. 4-2. PWR with four steam generators.

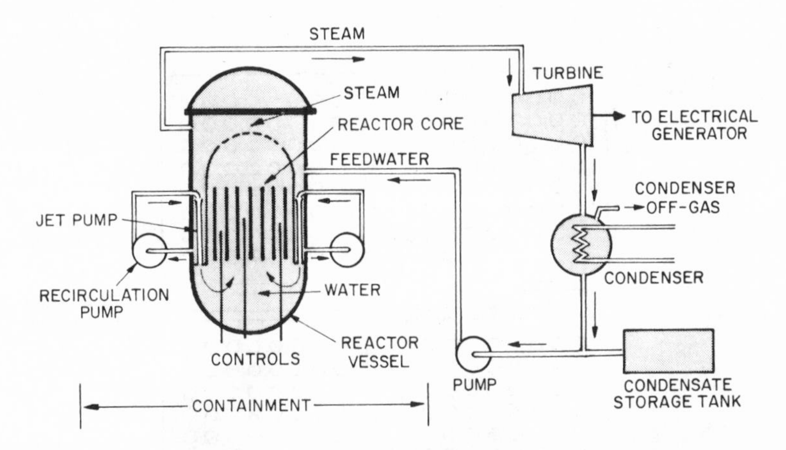

FIG. 4-3. Outline of a BWR power plant.

An important feature of a nuclear power plant is the containment structure. It is not shown in Fig. 4-1, but is described later (p. 81). The containment structure of a PWR contains within it all the components of the nuclear steam supply system. The turbines, condensers, feedwater heaters, and most of the standby cooling system, for use in the event that the main system is unable to function, are outside the containment.

Boiling Water Reactors. The essential feature of the BWR (Fig. 4-3) is that the water boils and steam is produced within the reactor core. Thus, there is no requirement for a separate steam generator. To permit the water to boil, the pressure in the vessel is less than in a PWR vessel—namely, about 7 MPa (1000 psi). At this pressure, water boils at about 285 °C (545 °F). The temperature and pressure of the steam leaving the reactor vessel are much the same as from a PWR steam generator.

A typical reactor vessel for a large BWR would be some 22 m (72 ft) high and 6.4 m (21 ft) in internal diameter; the wall thickness is 15 to 18 cm (6 to 7 in.). The core, consisting of vertical fuel rods similar to those in a PWR, is cylindrical in shape, about 3.7 m (12 ft) high and 4.6 m (15 ft) across. The fuel rods are in assemblies of square cross section, each containing 49 (in the older BWRs) or 62[a] (in the most recent designs) rods. The total number of rods in 700 to 800 assemblies is roughly 40 000, and the total weight of uranium dioxide fuel in the core is around 160 000 kg (350 000 lb). Each rod assembly is enclosed in a metal box, called a fuel channel, which is open at the top and bottom (Fig. 4-4). The purpose of the channel is to permit the coolant water to flow upward past the fuel rods but not sideways from one assembly to another.

As the water flows through the core, it is heated by the fuel rods and eventually boils. The steam-water mixture at the top of the core consists, on the average, of about 14 percent by weight of steam with the remainder being liquid water. The latter is recirculated through the core, as described below. The wet steam enters a bank of separators and then passes on to dryers in the up-

[a]The 8 × 8 array includes two hollow rods, called water rods, to provide additional moderation (see Fig. 4-4).

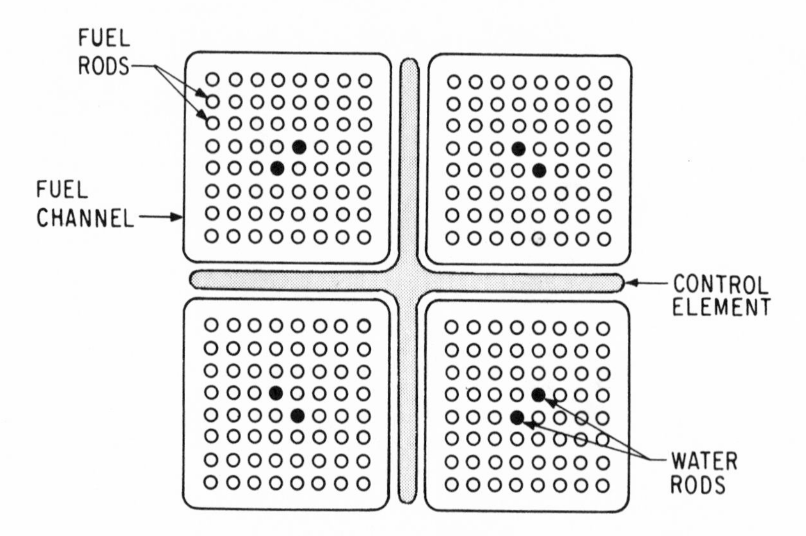

FIG. 4-4. Horizontal section through four adjacent BWR fuel bundles and associated control element.

per part of the reactor vessel. In the separators and dryers, most of the liquid water is removed from the steam and is returned for recirculation. The relatively dry steam proceeds to the turbines to generate electricity, and the exhaust steam is condensed in the usual manner. The condensate is pumped back to the reactor vessel as feedwater.

The water that has not been vaporized in its passage through the core, together with the feedwater, is recirculated in the following manner. Two (or more) recirculation pumps withdraw water from the annular region around the core (Fig. 4-5) and force it through jet pumps located in this region. Each recirculation pump services several jet pumps. The latter take in additional water at the top and deliver it at a high rate to the volume at the bottom of the reactor vessel. The water is thus forced to recirculate continuously through the core. The shroud surrounding the core serves to separate the upward flow from the downward annular flow.

Because the steam separators and dryers occupy the upper volume of the reactor vessel, the control elements are inserted and operated from below the core. This has some advantages since, by decreasing the reactivity in the lower part of the core, the control elements can balance, to some extent, the effect of steam in decreasing the reactivity in the upper part.

The main control elements in a BWR have a cruciform cross section (see Fig. 4-4); each element consists of four blades containing a neutron absorber (boron as boron carbide). The blades can move up and down in the spaces between fuel assemblies, with one control element between four assemblies in most cases. Roughly 170 control elements are distributed uniformly throughout the reactor core. They are actuated by hydrostatic pressure and can be moved up (or down) by applying water pressure either below (or above) a piston associated with each control element. If the reactor has to be shut down rapidly, the full hydrostatic

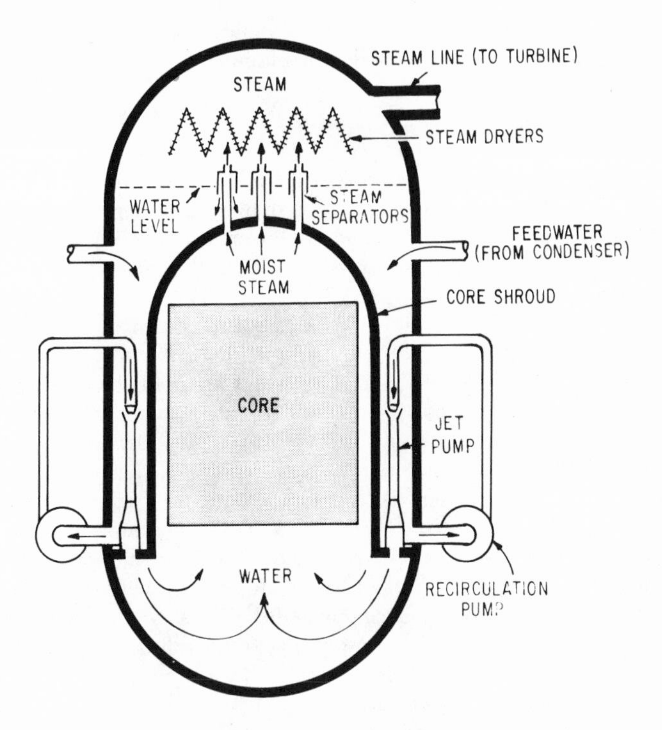

FIG. 4-5. Operation of recirculation pumps (and steam separator) in a BWR.

pressure is applied automatically to force all the control elements upward into the core. When close to its full operating power level, a BWR can be regulated by varying the rate of flow from the recirculation pumps. An increase in the flow rate of water through the core is accompanied by an increase in reactivity and power output.

The water in BWRs is not borated as it is in PWRs, but reactivity compensation is obtained by adding the neutron absorber gadolinium (as its oxide) to a small number of fuel elements. The gadolinium acts as a burnable poison, which is gradually consumed as the reactor operates. The reactivity compensation thus decreases steadily, as required to balance (roughly) the normal decrease in reactivity that occurs during operation.

Apart from the condenser cooling water, there is only one fluid circuit in a BWR. Thus, in addition to the reactor vessel, the turbines and condensers are part of the primary (coolant) system. Because of pinhole leaks in the cladding of a small fraction of the fuel rods, the steam that drives the turbines will then contain some gaseous and volatile fission products. But steps are taken to remove them or reduce their activity before discharge to the environment (Chapter 9).

The containment system of a BWR differs in many respects from that of a PWR, as noted in due course, although the ultimate purpose is the same. As a general rule, BWRs have a primary containment and a secondary containment. The major plant components in the primary containment are the reactor vessel and the recirculation pumps. The backup cooling systems, turbine, and condenser are located outside the secondary containment.

Barriers to the Escape of Radioactivity

One of the most important aspects of a nuclear power plant for protecting the health and safety of the public is the provision of multiple barriers against the escape of radioactive fission products to the environment. Although the barriers are not all quite the same in PWRs and BWRs, they are sufficiently alike to permit them to be considered together.

The first barrier is the reactor fuel itself. The fuel consists of uranium dioxide formed into cylindrical pellets of precise dimensions, roughly 1.5 cm (0.6 in.) long and somewhat less in diameter. The pellets are made by compressing the powdered oxide and then heating it until it sinters to a dense, ceramic-like material with a high melting point (2860 °C [5180 °F]). Solid fission products, that are formed within the fuel cannot escape. Volatile fission products, notably radioisotopes of krypton, xenon, and iodine, are released slowly from the ceramic material, thus allowing them time to decay to some extent.

Escape of the volatile fission products to the environment is prevented by the second barrier—namely, the (intact) fuel cladding. This consists of a long tube made of a corrosion-resistant zirconium alloy (Zircaloy) into which the fuel pellets are packed (Fig. 4-6). The empty spaces, including the narrow annular gap between the pellets and the surrounding tube, as well as the space above the pellets, are filled with the chemically inert gas helium. The tube is then sealed to form a fuel rod. After inspection and testing for possible defects, the rods are assembled in the bundles that make up the reactor core.

The primary coolant system is different in certain respects for PWRs and BWRs, but essentially the boundary of the system consists of the reactor vessel and associated pumps and piping. This boundary forms the third barrier to the escape of radioactive fission products that might leak out of the fuel rods. The reactor vessel and associated components are built and tested to exceptionally high standards to assure integrity of the primary coolant boundary.

Although a break in the primary coolant boundary would not be a common event, an additional safety feature is provided in the form of a primary containment structure. The configuration and other aspects of the containment system for PWRs and BWRs are different, but in all cases the containment is made of steel surrounded by a thick layer of concrete and is pressure-tested to assure a very low leak rate. The containment structure (p. 81) represents the fourth barrier against the escape of radioactivity. In some reactor plants, especially those with BWRs, an additional barrier is provided in the form of a secondary containment.

Fuel Boundary Protection

As long as the barriers described above remain completely intact, there will be no escape of fission products. Actually, none of the barriers is 100 percent leakproof. The ceramic fuel material effectively contains solid fission products, but volatile ones do escape to some extent. The clad fuel rods, although made to very exacting specifications, occasionally develop small cracks and pinhole

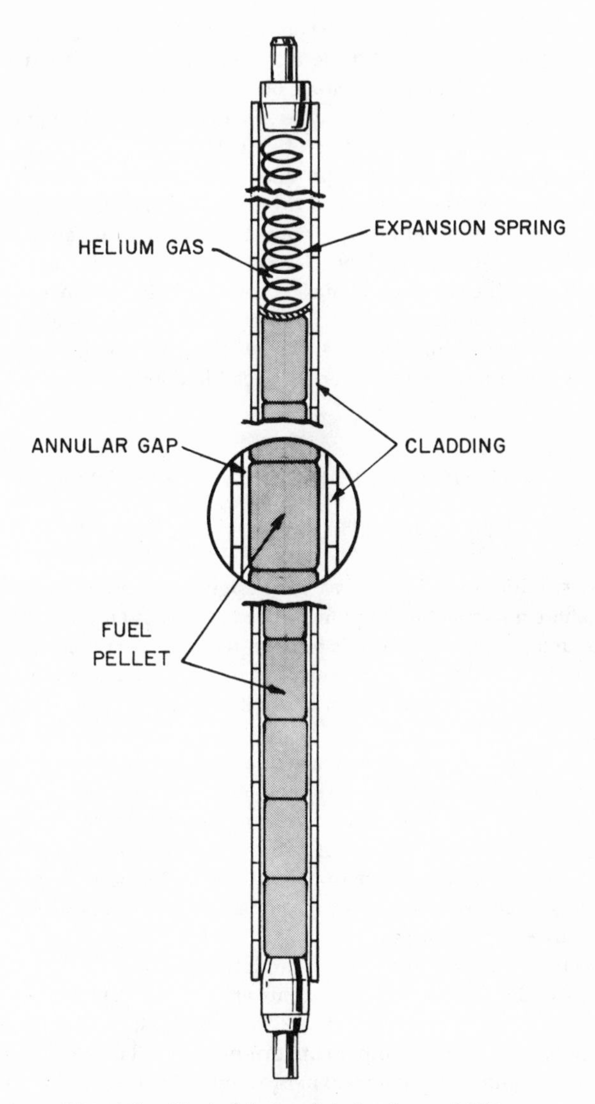

FIG. 4-6. Typical fuel rod design for an LWR.

leaks. Perhaps as many as 0.2 percent of the rods may leak by the end of their normal core lifetime of 3 or 4 years. As a result, a small quantity of radioactive fission products will enter the primary coolant circuit. Some corrosion products, which have been made radioactive by capturing neutrons in the reactor core, will also be present in the coolant water.

The coolant boundary rarely develops leaks, but there may be some leakage from valve stems and seals. Consequently, even in normal operation, small amounts of radioactive substances may escape from the plant. However, the limits set by the NRC for radioactive effluents are low enough (Chapter 5) to protect the health of the public. If, for any reason, an unusually large number of fuel rods should develop leaks near the end of their expected lifetime, the release of radioactive gases might exceed the allowed limits. This situation would be detected immediately, and the reactor would be shut down before a significant amount of radioactive material could be released to the environment.

It is apparent that the chief hazard of a nuclear power plant does not lie in a slowly increasing leakage of fission products, but rather in a sudden breach of one or more of the barriers. Provided the fuel rods and their cladding remain essentially intact, there is no possible way of releasing large amounts of radioactive material to the surroundings; hence, the primary design emphasis for safety is to protect the fuel rods against catastrophic failure. A number of protective features are therefore included in the nuclear plant design to ensure that the temperatures are maintained well below those at which the fuel element, especially the cladding, would suffer appreciable damage.

DESIGNING FOR REACTOR SAFETY

Redundancy and Diversity

A prime consideration in safe reactor design is the use of "redundancy" in situations where a particular component or system is essential to safe operation and shutdown of the reactor. In this respect, redundancy refers to the provision of two or more components or systems in parallel so that the reactor can continue to operate or can be shut down safely if one of the components or systems should fail.

Specific examples of redundancy are discussed below, but a few general instances of redundant component and system are the following: several independent instruments to determine reactor system operating parameters and to provide signals for control and protection systems; two or more pumps and separate circuits that supply cooling water to the reactor under both normal and emergency conditions; and several different sources of electrical power to operate instruments, valves, and pumps.

It is possible, of course, that two redundant components or systems may fail simultaneously. But if the two components are independent, the probability of this occurring is very much less than the probability of a single failure. However, since redundant components are often similar, they are sometimes vulnerable to a common design weakness, to operator error, or to other causes. The generic term *common-mode* (or *common-cause) failure* is used to describe largely unforeseen interactions within or between systems where a number of related failures arise from a single cause, such as an earthquake, fire, flood, or other circumstances.

A common-mode failure may occur, for example, if the cables connected to apparently redundant instruments are routed in a single tray that also carries a power cable. An overload in the power cable could cause a fire that would render all the instruments ineffective. As a result of operating experience, emphasis has been placed on separating the wiring for redundant instruments. A careful examination of electrical systems has identified several situations in which wiring for presumably separate and independent safety systems was susceptible to damage from a single event. As a result, computerized schemes have been devised for routing cables and monitoring the capacities of cable trays so as to decrease the risk of common-mode failures.

In general, the most practical way to minimize the danger of common-mode failures is by diversity, as well as redundancy, in design. This means the use of two or more completely different (and independent) methods for achieving the same objective. For example, it is desirable to have at least two different methods for shutting a reactor down and for providing cooling water in an emergency.

One goal of the reactor designer is to make all safety systems and critical components independent of one another in all foreseeable circumstances. Nevertheless, common-mode failures may occur because they are, by their very nature, largely unpredictable. The likelihood of such failures is being steadily decreased by utilizing experience to reveal situations in which common-mode failures are conceivable. However, the potential for such failures cannot be ruled out completely even in the most careful design because it is virtually impossible to predict all situations that might arise.

The Reactor Protection System

Despite the care taken to build and operate nuclear plants, there will inevitably be some failures of components. Furthermore, there will be operating errors that lead to power surges, and there may be storms or other events that cause a sudden loss of electrical power. Such deviations from normal operating conditions are referred to generally as *transients*. There are many ways of producing transients in reactor thermal power, in system pressure, or in coolant temperature. Most of such transients are accommodated by the automatic controls that adjust the position of control rods or change the flow of coolant or of steam to correct the abnormal condition. If, however, the transient is of such a nature that the power, pressure, temperature, or coolant flow rate goes beyond prescribed limits, the protection system is actuated.

The purpose of the protection system is to detect abnormal conditions in the nuclear plant and to institute automatically whatever safety action is needed. The system receives signals from instruments that monitor the important plant process variables, such as the coolant water temperature, system pressure, power level, neutron flux, coolant flow rate, and so on. If the instruments indicate a transient that cannot be corrected immediately by the adjustments referred to above, then the protection system must automatically cause the control rods to be rapidly inserted into the reactor core to terminate the fission chain. This automatic response of the protection system, which results in a very rapid shutdown of the reactor, is called a "scram" or a "trip."[b] In addition to the capability of the protection system for causing an automatic scram, a manual scram is available to the reactor operator in the control room. If there are indications of a potentially unsafe condition, the operator can scram the reactor independently of the protection system.

When the protection system in a PWR demands a scram, the electromagnetic clutches holding the control rods are deactivated, and all the con-

[b]The word *trip* is now used as an alternative to the historical *scram*.

trol rods drop under the influence of gravity. In a BWR, however, the control rods, which are inserted from the bottom of the reactor vessel, are forced upward into the core by the full hydrostatic pressure (10.3 MPa [1500 psi]) of water in an accumulator tank.

Some of the abnormal situations that will lead to reactor scram are the following:

1. too rapid a rate of power increase (resulting from too fast control rod withdrawal) during startup operations
2. high reading of neutron-measuring instruments, indicating an excessive power level
3. abnormal operating temperature or pressure
4. low water level in a BWR or high level in a PWR pressurizer
5. loss of coolant flow through the reactor (e.g., from a pump failure)
6. break in a steam line
7. turbine-generator trip in a BWR (e.g., from loss of load)
8. loss of power supply (for pumps, valves, instruments, etc.).

Redundancy and Diversity in Protection Systems

Transients that require a reactor scram are expected to occur only a few times a year. Since failure to scram on demand could have serious consequences, the protection (scram) system must be extremely reliable. Thus, special measures are taken to guarantee that the scram circuits will operate when needed.

Consider, for example, an instrument (thermocouple) that senses the temperature of the water as it leaves the reactor. The weak electrical output of the sensor is sent to an amplifier and then on to a relay. If the water temperature exceeds a predetermined value, the relay will release the latches that hold the control rods in place (in a PWR) and thus cause them to be inserted into the reactor core. This combination of sensor, amplifier, and actuator is called a *channel*. If any component of the channel should fail, either the relay would be actuated falsely (i.e., a fail-safe failure that results in a false scram) or the relay would not be actuated when a scram is required (unsafe failure). False scrams are costly because operating time is lost in investigating the cause of the scram and restarting the plant; failure to scram on demand would be intolerable. Consequently, redundant safety channels are always installed.

In practice, three or four redundant channels are generally used to monitor each operational process variable, and a scram results only when two or more channels call for action simultaneously. Unsafe failure must thus occur simultaneously in at least two channels to cause the scram system to fail. The chance that this will occur is very much less than failure of a single channel. This scheme also greatly reduces the rate of false scrams, since failure in a single channel will not lead to a scram. Furthermore, the availability of several redundant channels permits the operator to test each channel individually without scramming the reactor. By frequent testing, the probability of failure in any single channel is kept very low.

By careful design, construction, and regular testing, the chance that a channel will fail to operate when challenged can be reduced to less than one in a thousand challenges. The chance that two channels will fail simultaneously and independently would then be about one in a million challenges. However, a common-mode failure would change this situation.

The reactor protection system must always be ready to shut the plant down in the event of a transient that results in conditions lying outside the prescribed limits. Furthermore, the protection system must operate if a failure occurs in the control channels. Consequently, it is required that the protection channels be completely separated physically and electrically from the regular control channels, although they operate the same neutron absorber rods. It is then improbable that a single failure will put both systems out of operation.

An example of common-mode failure occurred in connection with a component of the protection system of the N Reactor at the Hanford Works in the State of Washington.[c] The reactor operator switched the release magnet of a control rod to a power supply to prevent that particular rod from being inserted when a scram was demanded. However, as the result of a multiple diode failure, all the rods were connected to the power supply so that none could be released. The fault was discovered when the reactor failed to scram after receiving the initial signal. After a short time, however, a completely independent automatic backup scram system, which dumped a quantity of boron balls into the core, operated to shut the reactor down.

The foregoing illustrates a situation in which diversity in the shutdown system was effective. Reactor protection systems always employ a type of diversity, called *functional diversity*, that is achieved by scramming the reactor on a number of process variables, such as neutron flux, water temperature, and pressure. If somehow the redundant neutron flux channels should fail to operate following an abnormal increase in power, then the water temperature channels should scram the reactor. However, if all signals are combined in a common logic circuit, the system is still subject to a common-mode failure. True diversity is achieved only by completely independent signal channels, logic circuits, and control rod drives, all of different design.

Many instances of common-mode failure have been reported, but the consequences have usually not been significant. However, common-mode failures were involved in the two most serious accidents at commercial nuclear power plants in the United States. The fire at Browns Ferry in March 1975 destroyed redundant electrical cables to engineered safety systems. And at Three Mile Island in March 1979 the valves in the lines from all three emergency feedwater pumps were left closed after a maintenance check. Redundant high-pressure coolant injection systems were subsequently shut down by operator action, and this led to uncovering of the core and severe damage to the fuel elements (p. 107).

Redundancy in Other Systems

Special emphasis has been placed on the need for extreme reliability of the automatic shutdown (scram) system. It is expected that this system will be called upon to operate many times during the 40-year life of a nuclear power plant. On the other hand, there are other protection systems that may be required to func-

[c]The N Reactor is not a typical, commercial LWR, but is a graphite-moderated, water-cooled reactor originally designed for plutonium production.

tion only in an extreme emergency, possibly not even once in the plant's operating lifetime. Although these systems will rarely be challenged, they must be highly reliable; they do not, however, require the extreme reliability of the shutdown system.

Such additional protection systems use redundant channels, but in many instances there are only two channels rather than the three or four in the scram system. Consequently, there will be occasions when only one channel is available because the other channel is being serviced or tested. This situation is tolerable for limited periods in a system that, although important to safety, may never be challenged.

The need for redundancy also applies to the actual emergency systems. For example, there must be at least two ways of providing cooling water to the core in the event of a break in the primary coolant boundary. Two methods of removing heat from the containment vessel are required in case one should fail. Double isolation valves are always provided for steam pipes to prevent radioactive steam from spreading from one portion of the plant to another. Furthermore, electric power for operating these safety systems must be available from diverse sources. In each instance, the design must, as far as possible, guard against common-mode failure.

Electric Power Supplies

Most of the instruments and other components required for control and for safe shutdown in the event of a severe transient are operated by batteries, which are provided in duplicate. Pumps for supplying water to the core, even after shutdown, and blowers for air purification that may be needed in an emergency require electricity for operation. Electric power to the redundant safety system is supplied from two bus bars; these are connected so that if one were to fail, the other would keep one of the redundant safety systems operating. Normally, the bus bars would be energized by the plant's turbine-generator.

To provide a reliable source of power when the plant is shut down and the turbine-generator is not operating, at least two transmission lines are available from two different offsite power supplies. As a backup, in case these should both fail, electricity can be generated on the reactor site by means of diesel generators. There must be at least two such generators, each of which is alone capable of providing the required electric power. The generators should have separate fuel supplies and bus bar connections, and ideally should be at different locations so that fire or flood would not cause simultaneous failures. The diesel engines are tested periodically to ensure that they are in operating condition.

If all off site (and on site) electric power were to be lost, the plant would be without a power source for about half a minute (or less) while the diesel generators were coming on line. Even if the diesel engines failed to start as projected and there were a delay in the restoration of electrical power, the plant would be in no immediate danger. Only in the event of a major break in the primary coolant boundary, leading to what is known as a loss-of-coolant accident (LOCA) (p. 77), would it be imperative that power be restored quickly to operate the pumps for the emergency core-cooling system (ECCS) and the blowers for cleaning the air in the containment vessel. This situation would constitute an emergency that might never arise, but if it did and if there were a

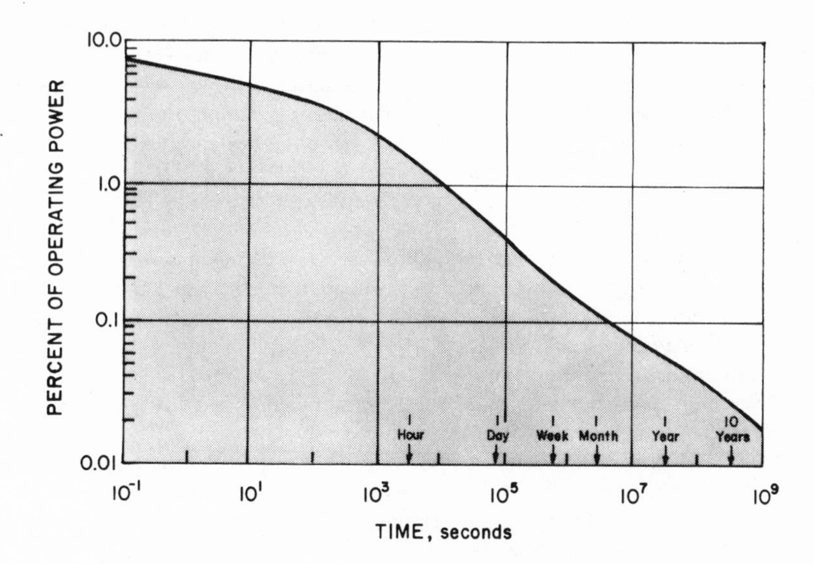

FIG. 4-7. Decay heat rate after shutdown of a reactor that has been operating for some time.

simultaneous failure in the two offsite power supplies, one of the diesel generators would have to come on line within less than 30 seconds.

Shutdown Heat Removal

When a reactor is shut down, no matter for what reason, the self-sustaining fission chain is terminated, but a considerable amount of heat energy, called the *stored heat,* is still present in the fuel. Furthermore, heat continues to be generated by the decay of the radioactive fission products that are present in the fuel rods; this is referred to as the *decay heat.*[d] Immediately after shutdown, the decay heat rate in a reactor that has been operating for some time is more than 7 percent of the total heat rate (or thermal power level) during operation; the proportion is down to just over 2 percent after 15 minutes, to roughly 0.4 percent after a day, and 0.2 after a week (Fig. 4-7). For example, consider an LWR (i.e., PWR or BWR) generating 1000 megawatts (MW) of electricity; the thermal power level during operation would then be around 3100 MW. Consequently, the decay heat rate immediately after shutdown is some 220 MW, decreasing to about 12 MW at the end of a day and 6 MW at the end of a week.

Because heat continues to be generated in the fuel rods at a substantial rate, there must be adequate provision for heat removal for several months after a reactor is shut down. Failure of the shutdown heat removal system could result in overheating of the fuel with possible disruption of the core. Since a reactor is shut down several times a year, either deliberately or following a transient, the shutdown cooling system must be highly reliable.

[d]Some fission heat is also produced by delayed neutrons emitted by certain fission products, but this decreases much more rapidly than the decay heat.

Two phases of heat removal after shutdown may be distinguished: namely, hot standby and cold shutdown (or cooldown). In the first phase, the reactor is maintained in the hot-standby condition in which the temperature and pressure are not greatly different from those during normal operation, although the thermal power level is much lower. The reactor can then be readily restarted, by gradual removal of the control rods, when it is safe to do so. However, if the system has to be cooled sufficiently to permit inspection or repairs, cold shutdown would be required.

For hot-standby cooling, the reactor continues to produce steam at the decay heat power, but the steam is *dumped* to the condenser, bypassing the turbines. The condensate is then returned by the feedwater pumps to the steam generators (in a PWR) or to the reactor vessel (in a BWR). The hot-standby condition can thus be maintained for many days.

If the main condenser should fail, steam would be dumped directly to the atmosphere in a PWR or to the pressure-suppression pool in a BWR (p. 84). Makeup feedwater would then be provided from a large condensate tank, with a capacity for at least 36 hours. Water from the tank is pumped, by means of redundant auxiliary pumps, to the steam generator in a PWR or to the reactor vessel in a BWR. Should it be required, additional water of lesser purity is available for a BWR from the pressure-suppression chamber (see Fig. 4-9). The auxiliary pumps may be electric, using offsite or onsite power, or they may be driven directly by individual diesel engines or steam turbines.

If cold shutdown is necessary, the system pressure is lowered by dumping steam to the condenser (or to the atmosphere) at an increased rate. When the pressure drops to a few hundred pounds per square inch (i.e., a few megapascals), the *residual heat removal* (or RHR) system comes into operation. The primary coolant is now pumped through the RHR heat exchangers, where it is cooled, and then back to the reactor. The secondary cooling water for the heat exchangers is obtained from the ultimate heat sink (see below). The RHR system must continue to function as long as is necessary to remove heat from the fuel rods.

Some of the equipment of the RHR system is used in normal nuclear power plant operation. This may add to the probability of failure when the system is required for cold shutdown. It has been proposed, therefore, that future plants be equipped with a dedicated (or "bunkered") RHR system to be used exclusively for cold shutdown. Research directed at the development of such a system has been initiated. A bunkered RHR system should operate automatically and should ideally be a passive one, not requiring pumps or electric power. If this is not feasible, each pump should have its own independent power supply.

The capability for both hot standby and cold shutdown must be assured in the event of a severe natural phenomenon (e.g., an earthquake, tornado, flood, etc.). Furthermore, operation must still be possible if there is a failure in a single system, such as offsite power, one diesel generator, one direct-current supply, one feedwater pump, or any component that would prevent heat removal from a single steam generator (in a PWR).

In accordance with Title 10 of the *Code of Federal Regulations* Part 50 (10CFR50), Appendix A, Criterion 44, a nuclear power plant must be provided with an *ultimate heat sink*, to which heat can be transferred under all normal operating and accident conditions. The main function of the sink is to provide cooling water for the condensers and for the RHR system heat exchangers. The NRC recommends that, unless a guaranteed and reliable source of cooling water (e.g., an ocean, large river, or large lake) is available, the ultimate heat sink should consist of two separate sources. Acceptable sinks might be two independent cooling towers or spray ponds, or a tower and a spray pond, or one of these together with a reservoir, a river, or other source of cooling water (see Chapter 13).

Each source should by itself be capable of performing all the safety functions of the ultimate heat sink for a period of at least 30 days. The design must provide assurance that these functions will not be impaired, directly or indirectly, by the most severe natural phenomenon that could be expected at the plant site.

ENGINEERED SAFETY FEATURES

General Description

The automatic reactor protection system is designed to shut the plant down safely under any anticipated transient condition, such as loss of turbine load, loss of offsite power, failure of the feedwater pumps, and accidental withdrawal of a control element. In each instance, the protection system will cause the reactor to be scrammed, and core-cooling will be provided by the shutdown heat removal system. There will then be no damage to the fuel rods and, hence, no release of radioactivity. The plant can be returned to operation after the cause of the scram has been identified and the situation corrected.

A well-designed and well-operated nuclear power plant will also be able to cope with more serious events, such as complete loss of power (other than batteries), small leaks in pipes or valves, inadvertent operation of pressure relief valves, and cessation of main coolant flow. These events, which are expected to occur much less often than once a year, may result in some damage to the fuel rods and a small release of radioactivity to the containment vessel.

In addition, current LWR plants are designed to deal with rare but potentially more serious events, such as sudden expulsion of control rods, steam line breaks, and loss of condenser vacuum, and to withstand severe natural phenomena (e.g., tornadoes, hurricanes, and earthquakes). Moreover, the plant must include provisions for protecting the public against the consequences of the rupture of a major coolant pipe and the resulting complete loss of coolant from the core. Protection against such unlikely events is provided by the engineered safety features (p. 61). These include the ECCS, to supply water to the core in the event of a complete loss of coolant; the containment vessel, to prevent (or limit) the escape of radioactivity; various auxiliary components to clean up and cool the containment atmosphere; and auxiliary feedwater pumps to maintain a water supply to the steam generator of a PWR.

Perhaps the most severe design-basis accident, which the reactor must be able to control, is the LOCA resulting from a large break in the primary coolant circuit. If such a break occurred, the pressure would decrease suddenly, and most of the coolant water would be flashed into steam. The steam would expand and fill the containment vessel, and a substantial amount of the heat energy originally present in the coolant would be transferred to the containment atmosphere, thereby raising the temperature and pressure.

In a sense, the containment vessel is the single most important engineered safety feature of an LWR because it is the last barrier to the escape of radioactivity into the surrounding environment. However, the containment vessel (p. 81) alone is not sufficient to control the consequences of a LOCA. Various supplementary features are provided as well as means for protection against mechanical damage.

In the event of a LOCA, the escaping steam must be condensed as quickly as possible so as to decrease the temperature; this will reduce the pressure in the containment and thereby limit leakage. A cleanup system is provided to decrease the radioactivity of the containment atmosphere and, hence, the amount of radioactivity that could leak from the containment vessel into the environment. Furthermore, hydrogen gas may be released into the containment atmosphere, as explained later. This hydrogen must be controlled to prevent its concentration reaching the point at which an explosion might result.

Many pipes, which might be ruptured if an accident were to occur, penetrate the walls of the containment vessel. These pipes are provided with duplicate valves that shut off automatically in an emergency. Such valves, one on each side of the containment wall, are particularly important on steam lines to prevent the possible spread of radioactivity into other parts of the plant in a LOCA. Finally, barriers are placed at suitable locations to protect the containment structure from damage by missiles and pipe whip.

The Emergency Core-Cooling System

The engineered safety feature that has received the most attention (and publicity) is the ECCS. If there should be a complete loss of coolant, it is vital that the reactor core be prevented from overheating; the ECCS is designed to do this. If the core were to suffer substantial melting, it probably could not be contained. In view of the importance of the ECCS to overall reactor safety, it is described here in some detail. The objectives of the ECCS are the same in all LWRs, but there are some differences in the manner whereby these objectives are achieved in PWRs and BWRs.

Pressurized Water Reactors. The ECCS for a PWR is shown schematically in Fig. 4-8. There are usually three independent subsystems that become operative at different system pressures. Each subsystem is characterized by redundancy of equipment and flow path. This redundancy is intended to assure reliability of operation and continued core-cooling in the event that any single component fails to perform its design function.

A small break in the primary coolant circuit of a PWR would result in some loss of coolant leading to a moderate decrease in system pressure. A pressure drop from the normal operating value of 15.5 MPa (2250 psi) to about 11 MPa (1600 psi) would actuate the pumps of the ECCS subsystem called the *high-pressure injection system*. Borated water would then be forced into the reactor vessel to make up for the loss through the break. The high-pressure injection system would operate whenever the pressure falls to 11 MPa from any cause, such as failure of a pressurizer relief valve to close after the normal system pressure has been attained (p. 106).

Small break LOCAs present less of a challenge to the ECCS than do larger breaks, but they are likely to occur more frequently. Consequently, the reliabili-

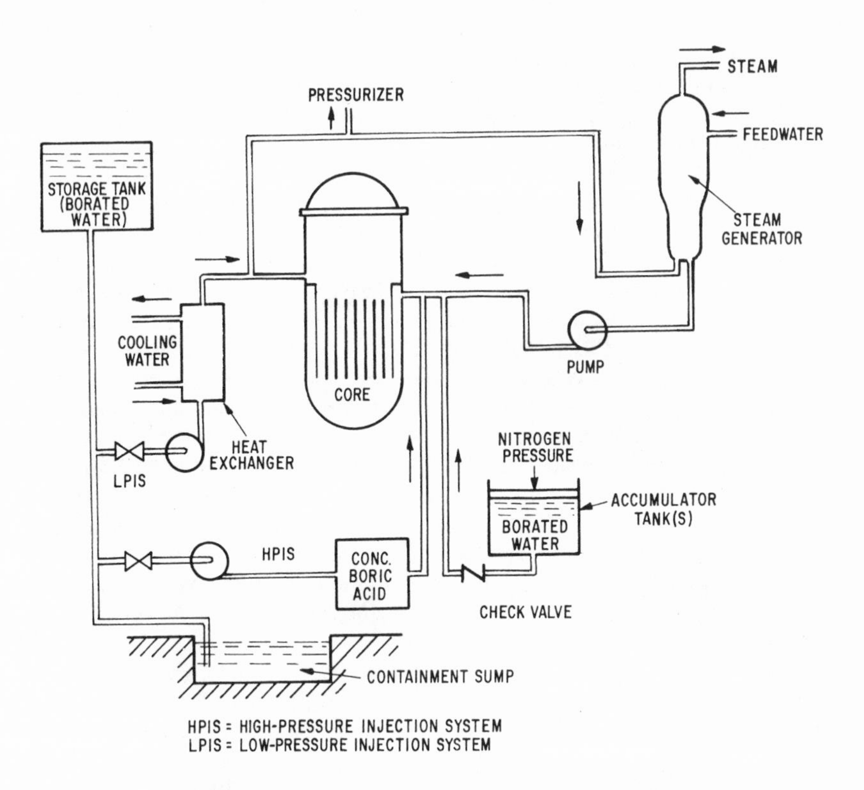

FIG. 4-8. PWR emergency core-cooling system.

ty of the high-pressure injection system to respond to frequent challenges is an important design consideration.

If a large break were to occur in the reactor coolant circuit, the expulsion of steam and high-temperature water would then be accompanied by a rapid drop in the system pressure. Another subsystem of the ECCS, the *accumulator injection system*, would then be actuated. This subsystem consists of two or more independent tanks containing cool, borated water stored under nitrogen gas at a pressure in the range of 1.4 to 4.5 MPa (200 to 650 psi), according to the system design. These tanks are connected through check valves and piping to the main coolant lines between the primary coolant pumps and the reactor vessel or directly into the vessel itself. If the system pressure were to fall below the gas pressure in the accumulator tanks, the check valves would open, and large volumes of borated water would flood the reactor core. The neutron absorber boron assures that the core remains subcritical and does not continue to generate heat by fission.

If the system pressure were to decrease further, as the result of a large break in the primary coolant circuit, the third ECCS subsystem, the *low-pressure injection system*, would come into operation. This low-pressure subsystem, which utilizes the components of the auxiliary (shutdown) cooling system

(p. 75), should continue to pump borated water from a storage tank into the reactor vessel for a long time after the accumulators are empty.

The accumulator injection system is referred to as a *passive* system because it functions automatically without pumps, motor-driven valves, or other equipment requiring a power supply. The high- and low-pressure injection subsystems, on the other hand, are *active* systems, with pumps and valves that must operate if the systems are to function. Power for the active subsystems of the ECCS is available from different sources, both on and off the site. The subsystems are activated by signals from reactor system pressure and level-sensing instruments; reliability and redundancy of instrumentation are thus essential.

The arrangement and design of the ECCS for PWRs may vary from plant to plant, depending on the manufacturer of the reactor and its components. All PWR nuclear power facilities, however, employ both accumulator-injection and pump-injection systems with redundancy of equipment and power supplies to assure operation when needed.

Boiling Water Reactors. Like PWRs, BWRs have multiple provisions for cooling the fuel in the reactor core in the event of a loss of coolant. The details may differ from one BWR plant to another, but all have several independent systems to achieve flooding and/or spraying of the core with coolant upon receipt of a signal of either low water level in the reactor vessel or high pressure in the containment structure that surrounds the reactor vessel. A typical ECCS for a BWR is represented in Fig. 4-9.

All BWRs are equipped with a high-pressure coolant injection system (in older BWRs) or a high-pressure core spray system (in newer BWRs) to provide adequate core cooling in the event of a small leak or break. If such a leak should

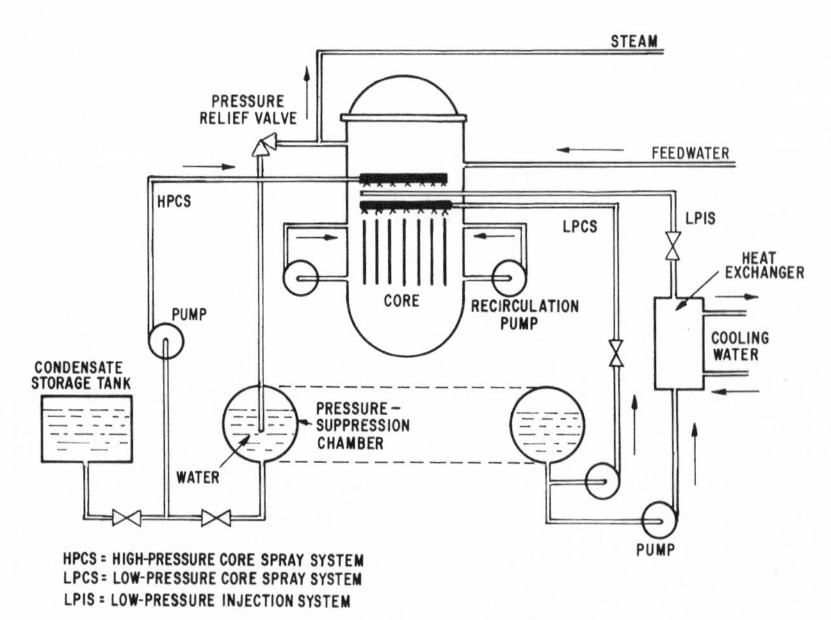

FIG. 4-9. BWR emergency core-cooling system.

occur, the reactor would be scrammed upon receipt of a signal of low water level in the reactor vessel. If the feedwater pumps are unable to maintain the water at a preselected level or if a high pressure were detected in the primary containment (drywell), the high-pressure core-spray (or injection) system will be automatically activated. Water for this system is obtained partly from the condensed water storage tank and partly from the pool in the *pressure-suppression chamber* (see Fig. 4-9 and page 84). The high-pressure coolant pump is driven either by a steam turbine, with steam generated by the heat remaining in the reactor core and massive structure, or by electricity (in the latest BWRs). In the latter case, both normal and emergency power supplies are available.

If a high-pressure system should fail to operate or if it (and the feedwater pumps) should be unable to maintain the preselected water level in the reactor vessel, a pressure-relief valve would open automatically to discharge steam from the vessel. As a result, the pressure in the reactor vessel would be decreased sufficiently to initiate automatic operation of the low-pressure emergency cooling systems. In the event of a large break in the coolant circuit, the pressure would drop even without help from the automatic depressurization system.

The low-pressure system includes both core-spray and coolant-injection systems. By spraying water onto the fuel rods from above, the core spray alone should be able to cool the core. Nevertheless, the low-pressure injection system is provided to supplement the core spray and flood the core with water. This system has sufficient capacity to protect the core following a large break in the primary circuit piping. Both low-pressure cooling systems draw water from the pressure-suppression pool.

In the older BWRs, the low-pressure core-spray system consists of two independent loops. Each loop has an electrically driven pump, each with its own normal and auxiliary power sources, connected through independent piping to a separate spray header above the core. There are also two independent injection systems that use the pumps and piping of the residual heat removal system. The more recent BWRs have a single low-pressure core spray, but three coolant-injection systems. One emergency power source is available for the low-pressure core spray and for one injection system, and a second power source serves the other two injection systems.

Containment Systems

Containment systems are designed to contain essentially all the steam and radioactivity that might be released in the event of a LOCA, assuming that the ECCS functions adequately. The nature of the containment structure and the means for conditioning the containment atmosphere are different for PWRs and BWRs; they are described here separately.

Pressurized Water Reactors. Most present-day containment structures for PWRs are cylindrical in form with a domed top. They are made of reinforced concrete, typically 1.07 m (3.5 ft) thick, with an internal steel liner. In a PWR plant, the entire primary coolant system, including the reactor vessel, steam generators, pressurizer, and pumps, are enclosed in the containment structure (Fig. 4-10). The size and design of the structure are such that it is capable of withstanding the maximum temperature and pressure that would be expected

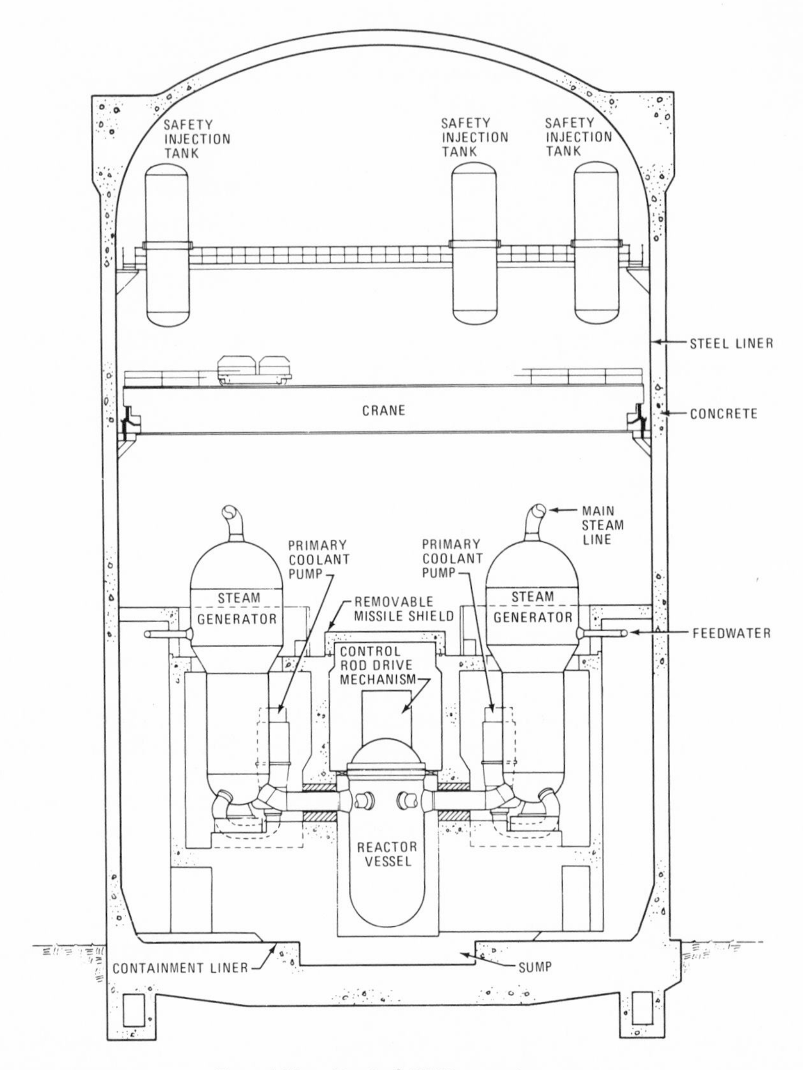

FIG. 4-10. Typical PWR containment.

from the steam produced if all the water in the primary system of an operating reactor were expelled into the containment. In some PWR plants, the containment structure is surrounded by a second structure, called the *shield building*, which provides an additional barrier against the escape of radioactive material.

If a LOCA should occur, much of the heat contained in the coolant and in the fuel elements prior to the accident would be released to the containment atmosphere. In addition, radioactive fission products would probably be present from damaged fuel rods. As indicated earlier, various engineered safety features are included in the containment system to cool and clean up the atmosphere after a LOCA.

In most PWR plants, cold water from a tank would be forced through nozzles in the upper part of the containment structure to produce a spray for condensing part (at least) of the steam. The water drains into a sump at the bottom of the containment and is recirculated to the nozzles after passing through the heat exchangers connected to the service water supply. In some cases, the containment air is cooled by blowing it across cooling coils. These procedures lower the pressure and thus protect the containment from being overpressurized (p. 95).

In a later design, the PWR containment volume is divided into separate upper and lower sections. The lower section contains the reactor vessel, steam generators, pumps, and pressurizer. The interior walls of the upper section are lined with a lattice framework filled with pieces of ice. Access from one chamber to the other is possible only through "trap doors," which are normally closed. If the pressure in the lower chamber should increase as a result of the accumulation of steam, the doors would open; the steam would then flow through the ice bed where it would be condensed. Water sprays are provided to maintain cooling after the ice melts. With the inclusion of an ice bed, the design pressure of the containment structure can be lower and the volume less than would otherwise be required.

A highly important engineered safety feature is the cleanup system for reducing the amount of radioactive fission products present in the containment atmosphere. The radioiodines are generally regarded as the most hazardous of the volatile fission products that might be released as the result of a reactor accident because they are taken up rapidly by the thyroid gland (Chapter 6). Thus, the radioiodines are usually the controlling factor in determining the hazard from radioactive releases to the atmosphere. Sodium hydroxide or sodium thiosulfate, either of which can remove iodine, is added to the water that provides the cooling spray for the PWR containment structure. If blowers are used to circulate the air, charcoal and high-efficiency particulate air (HEPA) filters are included to remove iodine and particulate matter, respectively. Iodine is also effectively removed from steam condensed in an ice bed.

The sprays and filters will greatly reduce the amounts of radioiodines and particulates in the containment atmosphere, and the rate at which they can escape to the environment is decreased correspondingly. The noble gases krypton and xenon cannot be filtered out, but the containment vessel will hold them for a sufficient time to permit substantial decay of the radioisotopes of short (and moderately short) half-life to form harmless, stable species before being released. The remaining radioactive gases that leak from the containment do so very slowly and are diluted by mixing with large volumes of surrounding air. Thus, there would be several factors that serve to minimize the radiation doses to people in the surrounding area following a LOCA.

When a shield building surrounds the PWR primary containment structure, there is a substantial air space between the inner and outer containments. Within 10 minutes (often less) after a LOCA, exhaust fans would reduce the pressure in this space. The exhaust air from the fans is recirculated through filters to retain iodine and particulate matter; these filters would remove some 95 percent or more of the radioiodines that may have escaped from the inner containment. The dual-containment system will reduce the whole-body radiation exposure from radioisotopes of krypton and xenon to some extent, but the radioiodines, usually the controlling factor, as seen above, would be greatly decreased.

Boiling Water Reactors. Containment systems of the more recent BWRs generally provide both primary and secondary containment. The primary containment structure consists of the drywell and pressure-suppression chamber, referred to earlier in the description of the ECCS. In the majority of BWRs operating in the 1970s, the drywell is a steel pressure vessel, shaped like an electric light bulb, which is enclosed in a thick structure of reinforced concrete, as seen in the cutaway drawing in Fig. 4-11. The drywell contains the reactor vessel, the recirculation pumps, and the associated piping.

Below the drywell and encircling it is the ring-shaped pressure-supression chamber (or wetwell), which is kept about half full of water; this is the pressure-suppression pool. Steam escaping from the reactor vessel as the result of an accident would be forced into the water in the pool by the increased pressure in the drywell and would be condensed. The heat capacity of the water in the suppression pool must be adequate to absorb the heat that would be discharged to the drywell atmosphere if there were a LOCA. The temperature and pressure inside the primary containment after such an accident would thus be determined by the pool water temperature.

The residual heat removal system, which would normally be used, if required, for cooling the core after shutdown (p. 75), would also serve to cool the

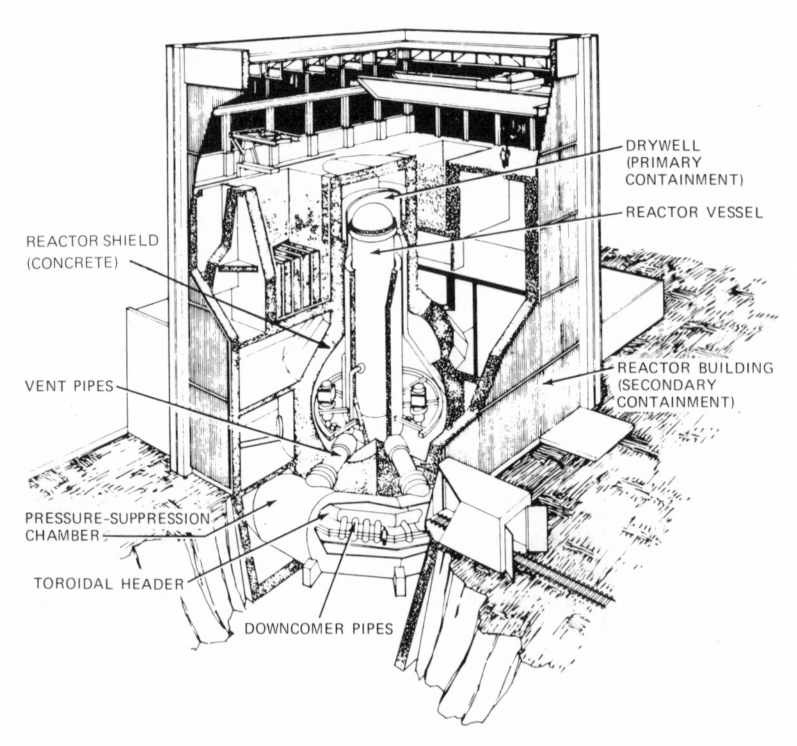

FIG. 4-11. Cutaway representation of a BWR and containment.

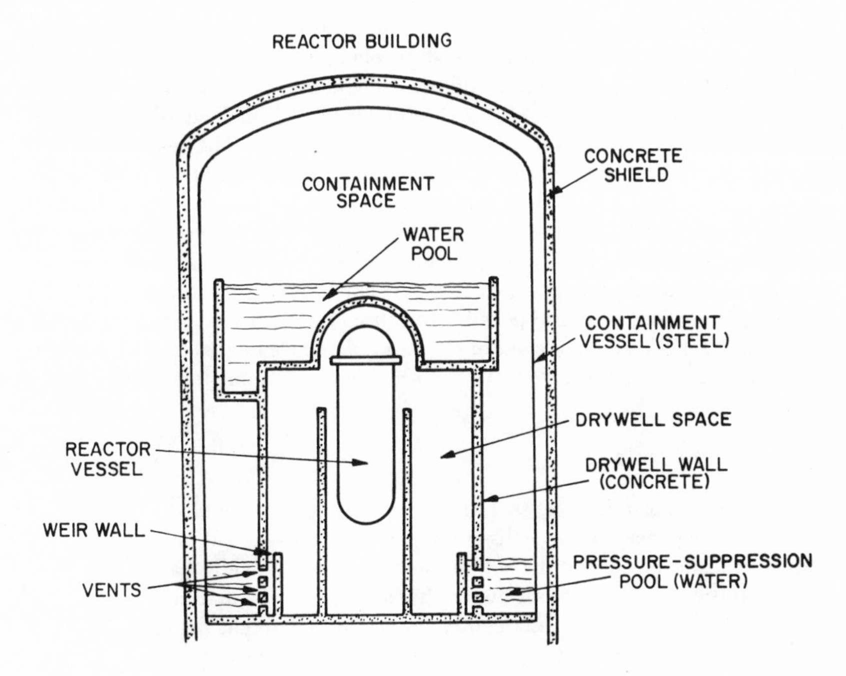

FIG. 4-12. Outline of recent (Mark III) BWR building.

suppression pool water. The water is pumped through heat exchangers in which cooling is provided by the plant service water system. Redundant systems of pumps and heat exchangers are included.

The pressure-suppression pool not only would condense steam that escapes from the reactor vessel, but also would remove part of the iodine that might be entrained in the steam. The amount of radioiodine that could leak out into the secondary containment would thus be somewhat decreased. The secondary containment structure, also called the reactor building, is made of reinforced concrete and is generally rectangular in form in the older BWRs (Fig. 4-11).

In the most recent BWRs, the drywell is a concrete cylinder with a domed top. The ring-shaped, pressure-suppression chamber (wetwell) containing water surrounds the drywell. An increase in pressure accompanying the release of steam into the drywell would depress the level of the water contained by the weir wall forming the inner wall of the pressure-suppression chamber (Fig. 4-12). Horizontal vents in the lower wall of the drywell would be uncovered, thereby forcing the steam to condense in the water. This design is said to have several advantages, including a greater reduction in steam pressure than in earlier designs.

The drywell and suppression chamber, which constitute the primary containment, are enclosed either in a rectangular concrete structure (Fig. 4-11) or in a concrete shield building (Fig. 4-12). The whole structure is called the reactor building, which serves as the secondary containment. An attached auxiliary building houses components of the residual heat removal system and the ECCS.

Regardless of the details of the pressure-suppression and containment systems, during normal operation the atmosphere in the secondary containment (or reactor building) is circulated and filtered and eventually exhausted through roof vents or a stack. In case of an accident, the normal ventilation system would be shut down and the vents closed automatically. The reactor building would be exhaust-ventilated to maintain the pressure slightly below that of the surrounding atmosphere. The building air would be circulated and filtered. Before being discharged, the air would be passed through charcoal beds and particulate filters, available in duplicate, which would remove a large fraction of the radioactive iodine and particles that may have escaped into the reactor building. The delay will also permit substantial decay of some radioisotopes of the noble gases. The effectiveness of the atmospheric cleanup system is similar to that described above for a PWR with a secondary containment (shield building) structure.

Containment Isolation

In the event of a LOCA, there is a possibility that some of the lines (pipes) that normally penetrate the walls of the primary containment might be sheared off either inside or outside the containment, thereby providing a leakage path. This would be particularly serious in the case of a break in a large steam pipe leading to the turbine. Consequently, any line that might provide a leakage path is provided with isolation valves that close automatically with an increase in the primary containment pressure. For most lines there are two valves; one valve is installed inside the vessel just before the line penetrates the containment wall, whereas the second valve is just outside the wall. The two valves, with their separate signal detectors and actuators, provide the necessary redundancy. Both valves should close in case of an accident to the reactor, but even if one should fail, the containment vessel would be sealed.

In nuclear power plants now in operation, automatic isolation of the containment is triggered by any one of a diverse set of abnormal conditions, such as initiation of the ECCS, containment overpressure, or high radiation level in the containment atmosphere. In the Three Mile Island plant, a single parameter—namely, an increase in the containment pressure of about 0.028 MPa (4 psi)—caused closure of the isolation valves. When a loss of coolant occurred through a stuck pressurizer relief valve, the pressure increase was not sufficient to cause immediate isolation. As a result, radioactive water was pumped from the containment sump to an auxiliary building, from which radioactive gases escaped to the environment (p. 107). Following the Three Mile Island accident, the NRC called for a review of containment isolation by all nuclear plant licensees.

Hydrogen Control

There are several ways in which hydrogen gas could be generated following a LOCA. The most important are:

1. chemical interaction of the containment spray water with exposed metals
2. interaction of the water (steam) with the zirconium in the hot Zircaloy cladding

3. decomposition of water by radiations, especially gamma rays from the fission products.

The amount of hydrogen from the first of these sources is kept very small, either by avoiding certain chemicals (e.g., sodium hydroxide) in the spray or by painting the exposed surface of the reactive metals, such as aluminum. Hydrogen generation from the second source tends to be greater the higher the temperature of the cladding. But even if the ECCS operated with minimum efficiency (e.g., if one diesel generator failed to start), the amount of hydrogen present in the containment atmosphere would be less than 1 percent by volume; this is well below the minimum of approximately 4 percent necessary to ignite a mixture with air.[e] Finally, the decomposition of water by radiation generates hydrogen slowly and it would be a matter of several days before enough was produced to make an explosion possible. (In the Three Mile Island accident, the ECCS failed completely because it was mistakenly shut off; this resulted in a high cladding temperature and the production of a large amount of hydrogen.)

By the time such a situation developed, the radioactivity in the containment atmosphere would have decayed to a considerable extent. Hence, in some PWRs part of the air would be filtered, discharged, and replaced by air from the surroundings, thereby reducing the hydrogen content. In some BWRs, the containment atmosphere is rendered inert by replacing the air with nitrogen during normal plant operation. In the more recent BWR containment designs, as shown in Fig. 4-12, the drywell and containment air spaces can be joined and the atmosphere can be circulated by mixing fans. The hydrogen content would then be too low for ignition to occur. The preferred method for controlling hydrogen in the containment atmosphere is to recombine the hydrogen with oxygen gas long before the ignition level is reached. Both catalytic and flame recombiners are being used for this purpose.

Habitability Systems

An important safety feature of nuclear power plants is the attention to the design and layout of all the essential safety systems so that they can be monitored and operated from the control room. Following a serious accident, such as a LOCA, many areas of the plant would have to be evacuated. The control room,however, must be shielded, and there must be provisions for cleaning the air so that the room can remain occupied. Thus, the reactor operator in the control room will have the information needed to determine if all safety systems are activated automatically as they should be; in the event of a failure, appropriate remedial action can be taken by using manual controls.

EVALUATION OF POTENTIAL ACCIDENTS

Safety Analyses

As noted in Chapter 3, the applications for a permit to construct and a license to operate a nuclear plant must include Safety Analysis Reports. One sec-

[e]The minimum hydrogen content for ignition may be greater than 4 percent in the presence of water vapor.

tion of these reports is devoted to the discussion of a number of possible accidents; each such accident must be analyzed, particularly with respect to potential damage to any of the barriers to the escape of radioactivity. Furthermore, an estimate is made of the radioactivity that might escape from the plant and the maximum radiation dose that might be received by an individual at the plant boundary under various weather conditions, including those that rarely arise.

In making the safety analyses, the license applicant is faced with many uncertainties. He must extrapolate from small-scale experiments and make assumptions concerning the event that initiates the accident, the effectiveness of the engineered safety features, and the weather conditions. In practice, the applicant is required to make two separate estimates of the consequences of postulated accidents.

One is a "realistic" estimate of radiation doses required for the Environmental Report (Chapter 3). In this estimate, the various quantities and conditions that provide the bases for evaluating the consequence of postulated accidents are assigned their most probable values. For the Safety Analysis Report, however, conservative (or "worst-case") estimates are required based on certain assumptions. These include a higher than normal rate of heat release in the fuel, failure of one of the redundant engineered safety systems,[f] failure of offsite power, weather conditions unfavorable to mixing of escaping radioactive gases with large volumes of air, and wind blowing directly toward the nearest boundary. The criteria to be used in making the conservative analysis of radiation exposure are stated by the NRC in its regulations; acceptable procedures for complying with the criteria are given in Regulatory Guides (p. 33), and the NRC staff reviews the analysis used to verify that it conforms with the criteria.

The purpose of the conservative (worst-case) analysis is to provide assurance that the public will be adequately protected should a design-basis accident occur. This analysis forms the basis for assessing the adequacy of the design of the engineered safety features, as noted below. For example, the leak rate of the containment vessel, as built and at each test during the plant lifetime, must be less than the value used in the conservative analysis.

Reasonably Possible Accidents

Potential accidents to a reactor plant fall into a number of categories according to the degree of probability that such accidents will occur. Both analyses and experience with nuclear power plants have demonstrated that certain events, called *anticipated transients*, may be expected to occur frequently during plant operation. Examples of such anticipated transients are loss of turbine load, misoperation of pressure-relief valves, and loss of offsite power.

If an anticipated transient should occur, the reactor protection system should operate automatically to prevent damage to the fuel rods. All barriers to the escape of fission products should remain intact, and the release of radioactivity, if any, would be trivial. Experience has demonstrated, in fact, that the

[f]The assumption of complete failure of the engineered safety system is not required; this should lead to what is known as a Class 9 accident, which is discussed later.

great majority of failures in nuclear power plants have resulted in only minor releases of radioactivity.

In addition to the expected transients, consideration must be given in the Safety Analysis Reports to events that, although deemed to be improbable, are nevertheless reasonably possible. Among such infrequent events are complete loss of electric power (station blackout), complete loss of forced coolant water flow through the reactor, and small leaks or breaks in pipes. These incidents have the potential for producing damage to the fuel elements and the release of radioactivity into the coolant.

The amount of this radioactivity that will escape into the containment vessel of a PWR depends on the integrity of the primary coolant loop. Leaks in valve-stem packings or pump seals may result in small radioactive releases to the containment structure.

A small leak in the tubing of a PWR steam generator (Fig. 4-1) would cause some radioactivity to be present in the steam. This would result in an increased radioactive emission from the condenser off-gas system. Before venting to the atmosphere, the off-gas is normally passed through an iodine absorber and a particulate filter (Chapter 9). However, license specifications place limits on the permitted leakage rates in steam generators and consequently on the radioactivity that could reach the off-gas system. If the radiation doses in the environment should exceed the requirements for normal operation, as specified in the NRC regulations, the plant would be shut down for necessary repairs to the steam generator.

If the fuel elements in a BWR were damaged as the result of an accident, excess radioactivity would be immediately detected in the steam. The reactor would then be scrammed, and valves in the steam lines would close, thereby preventing further radioactivity from reaching the turbine. Radioactive gases in the turbine and condenser would thus be limited to the quantity present prior to closing the valves. These gases would be held up and treated before release so that the radioactive material entering the environment by this route would be limited. During the cooldown and depressurization of the reactor vessel after the scram, minor escape of radioactive steam may occur into the containment vessel (drywell). However, leakage to the surroundings would be negligible.

Design-Basis Accidents

Finally, the license applicant must consider the possible consequences of serious but highly unlikely accidents, such as the following: a large break in a reactor coolant line; the dropping of a spent fuel rod cluster, containing large amounts of fission products, during refueling when the reactor vessel is open; and an earthquake, flood, or tornado. These are some of the design-basis accidents referred to earlier; the engineered safety features of the nuclear plant are designed to minimize the consequences of such accidents. In the Environmental Reports described in Chapter 3, they are referred to as Class 8 accidents.

In a serious design-basis accident, the fuel elements may be damaged and the primary coolant boundary breached. The engineered safety features are then called upon to limit the escape of radioactivity. The greatest release is to be expected from a LOCA resulting from a complete break in a large pipe in the primary coolant circuit.[g] This is generally regarded as the most important design-basis accident. The consequences of such a major LOCA and the operation of the engineered safety features are therefore treated in some detail in the next section.

Major Loss-of-Coolant Accident

Pressurized Water Reactors. The design and operation of the engineered safety features of PWR and BWR plants are somewhat different, as noted earlier in this chapter. The results of a severe LOCA, in terms of possible fission product release, would be much the same, however, in both reactor types. The following discussion refers in particular to a PWR; the differences that arise in a BWR system are indicated later.

If a major break should occur in a coolant pipe, water and steam would be expelled rapidly from the reactor vessel; this phase of the LOCA is called *blowdown*. The large drop of pressure in the reactor vessel and the accompanying rise in pressure in the containment vessel would be detected by redundant sensors. The scram circuits would then be triggered, and all the control rods would quickly be inserted into the reactor core. The ECCS would be started automatically, and water would be injected into the reactor vessel. At the same time, the containment spray system (p. 83) would be activated, and the cold water would serve to condense the steam pouring out of the break and thereby reduce containment pressure. The main steam isolation valves would close, thus preventing contaminated steam from reaching the turbine.

As the steam continued to escape from the break, pressure in the primary circuit would decrease rapidly, and borated water from the accumulator tanks would start pouring into the reactor vessel. However, the continued flashing of the coolant into the steam by the hot fuel rods could leave the fuel uncovered for a fraction of a minute. During this time, the fuel could become so hot that the cladding of a considerable number of rods would be damaged. The persistent injection of water from the accumulator tanks supplemented by the low-pressure injection system would, however, soon reflood the reactor core and quickly reduce the temperature of the fuel. This water would be kept cool by recirculation through heat exchangers, as described on page 76.

Boiling Water Reactors. The sequence of events in the case of a LOCA in a BWR would be much the same as in a PWR. For example, a break in a line connecting the reactor vessel and one of the coolant recirculation pumps would result in a blowdown (i.e., the rapid escape of steam and hot water into the drywell). In view of the relatively small volume of the latter, the steam and normal drywell atmosphere would be forced through the pipes (or channels) leading below the water level in the pressure-suppression pool (see Fig. 4-12). The steam would be condensed, and the remaining gas would be cooled before bubbling to the surface of the water. Additional cooling is provided by containment sprays. In this way, the pressure would be limited in both the drywell and the suppression chamber.

[g]Such a break is sometimes referred to as a *double-ended guillotine break;* it is as if the pipe (possibly 1 m [3 ft] in diameter) were cut through by a guillotine.

Following blowdown from the reactor vessel, the core must be reflooded with water. However, because of the larger mass of fuel in a BWR than in a PWR operating at the same power level, the temperature of the fuel elements would not rise as rapidly following the loss of coolant. Consequently, more time is available for injecting water into the reactor core by means of a pump; accumulator tanks, such as are used in PWRs, are thus not required.

If the break in the BWR coolant circuit were small, the level of water in the core would be maintained by the high-pressure (spray or injection) system. As the pressure fell, either by way of the pipe break or by depressurization of the reactor vessel (i.e., by venting steam to the suppression pool), the low-pressure (spray and injection) system would provide an ample supply of water to flood and cool the core (p. 81). Since the pumps are operated mostly by electric motors, there could be as much as a half-minute delay in the event that offsite power were not available and it was necessary to start onsite (diesel) power generators. This situation could arise regardless of the size of the coolant line break.

Escape of Radioactive Material

The chief function of the ECCS in both PWRs and BWRs is to limit the damage to the fuel elements and thereby minimize the amount of fission product activity that would escape from the reactor fuel in the event of a LOCA. The radioactive material that would escape from the core is chiefly volatile iodine and the chemically inactive (or noble) gases krypton and xenon. Most of the iodine would be absorbed by water droplets in the containment spray.

Although the noble gas activity cannot be dissolved out or held by chemical combination, the containment vessel would reduce the amount that escapes. Much of the activity is associated with isotopes of short half-life, such as krypton-87 (half-life 76 minutes) and xenon-137 (half-life 4 minutes), which decay in the containment vessel, leaving solid products. Futhermore, since the containment is designed to have a low leak rate, the rate of release of radioactive material to the surroundings (i.e., the amount escaping in unit time) is minimized. This is important since the concentration of radioactivity (and the radiation dose) downwind is proportional to the release rate.

It is apparent that even in the unlikely event of a very serious accident, such as a large break in the reactor coolant circuit, the amount and rate of escape of radioactive material should be held to low values by proper operation of the engineered safety features.

If radioactive fission products leak from the containment vessel following a postulated accident, they would be mixed with the large volume of air around the plant. The radioactivity would then be carried downwind in the form of a radioactive cloud, or plume, similar to the plume from a smokestack, but closer to ground level. As the plume moves toward the plant perimeter, it would mix with ever larger volumes of air and so become more and more dilute. A person standing in the path of the plume would receive a radiation exposure to the entire body, chiefly from gamma rays emitted by the fission products in the plume. This is called the *whole-body radiation dose*, and the most immediate injury would be to the blood-forming organs.

A person immersed in the radioactive cloud would inhale radioactive material into the lungs from which it can be taken up by the blood and carried to other parts of the body. As far as the gases krypton and xenon are concerned, the internal radiation dose to the whole body that would result from inhalation would be much less than the external dose from immersion in the cloud. However, if radioactive isotopes of iodine should be present, they would enter the blood stream by way of the lungs and would tend to be concentrated in the thyroid gland. This gland is relatively insensitive to radiation and so plant site criteria allow for thyroid doses that are larger than the whole-body doses.

Radiation Exposure from a Loss-of-Coolant Accident

As mentioned earlier, in the procedures for obtaining a license to operate a nuclear power plant, the applicant must make both a realistic and a conservative (worst-case) estimate of the consequences of various accidents. In the realistic estimate for a LOCA, it is postulated that a break has occurred large enough to permit escape of most of the high-pressure water. However, it is assumed, as is reasonable, that offsite electric power is available to operate the pumps for the ECCS and containment spray system. If both of the redundant channels of the ECCS function, as can be expected, damage to the fuel elements would be limited, and only a small fraction of the fission products would be released. The containment spray would then absorb most of the iodine. Moreover, the containment vessel is assumed to leak at its low design rate, and the atmospheric conditions are such as to provide a reasonable degree of mixing.

Under these realistic conditions, the whole body radiation dose received by a person in the downwind direction outside the plant exclusion area would be a small fraction of a rem. The thyroid dose would be correspondingly small. Thus, even a severe nuclear power plant accident would not result in appreciable radiation exposure to the surrounding population, provided the engineered safety features work as expected and the most adverse weather conditions do not occur.

In making the worst-case estimate of the consequence of a LOCA, however, the following highly conservative assumptions are generally made:

1. There is a complete rupture of the largest coolant pipe.
2. Offsite power is not available.
3. Only one of the redundant ECCS and containment spray systems is operable.
4. Unfavorable atmospheric conditions, such as occur less than 5 percent of the time, exist (e.g., an inversion that prevents the radioactive cloud from diffusing upward and a light breeze that produces very little atmospheric turbulence and mixing).
5. All the gaseous fission products krypton and xenon and half of the volatile iodine present in the fuel elements escape into the containment.[h]

The last of these assumptions appears to be inconsistent with the NRC requirement that, if a LOCA should occur, ECCS must prevent the fuel cladding

[h]Half of the escaping iodine, including organic iodides, which are difficult to remove, are assumed to be immediately available for leakage from the primary containment. The remainder is supposed to *plate out* (i.e., deposit) on solid surfaces or be absorbed.

from exceeding 1204 °C (2200 °F) at the hottest point. If this requirement is met, fuel element damage would not be sufficient to permit the stipulated release of fission products.

Other Design-Basis Accidents

The large LOCA has been considered in some detail because it has the potential for releasing a significant amount of radioactivity to the environment. There are, however, other failure modes that are also regarded as design-basis accidents since they can result in some fission product release. Among such accidents are the following: ejection of a control rod, an accident in the handling of spent fuel, and a steam-line break. These are examined in turn.

If a control rod were somehow to be accidentally ejected from the reactor, there would be a surge in the power level before the remaining control rods could be driven into the core. Such an incident would undoubtedly result in damage to a considerable number of fuel rods and release of fission products to the coolant. However, it is not likely that the primary coolant boundary would be breached. The radioactivity release would be relatively small, and even a worst-case analysis indicates a maximum offsite dose of less than 1 rem, both to the whole body and to the thyroid gland.

When a reactor is being refueled, on the average about once a year, the assemblies of highly radioactive spent fuel rods are removed from the reactor and are transferred to a water-filled storage pit outside the containment. The transfer system is so designed that the fuel assemblies are under water at all times. There is a chance that during the unloading operation a fuel assembly may be dropped. If the spent fuel elements are damaged, fission products would be released outside the containment vessel.

In practice, fuel elements are never handled immediately after the reactor is shut down. Usually, several days have elapsed prior to removal, and during this period the activity of the short-lived fission products will have decreased to a small fraction of the value at shutdown. However, one of the more potentially hazardous fission products, iodine-131, will still be present, and, as a result of an accident, some could escape from the fuel rods. Since the spent fuel is under water at all times, much of the iodine will be trapped and held in the water. The krypton and xenon isotopes that are still present will also be released from the fuel. These are not retained by the water, and so they will escape into the surrounding atmosphere.

A typical worst-case analysis of a fuel assembly drop accident indicates a maximum whole-body dose of 2 rems and perhaps 5 rems to the thyroid of residents of the low-population zone. A more realistic analysis predicts substantially smaller doses. The radiation dose from a spent-fuel handling accident would generally be less than that from a large LOCA. On the other hand, the former may have a greater probability of occurring; moreover, the engineered safety features would not be available if the accident occurs outside the containment. The chief protection lies in careful design of the equipment and attention to the procedures used in handling the spent fuel assemblies.

Another type of accident that would result in the release of radioactivity outside the containment would be a break in one of the steam lines connecting the steam generator (in a PWR) or the reactor (in a BWR) to the turbine. Such a break would cause the steam isolation valves, one on each side of the containment vessel wall, to close quickly. However, in the intervening few seconds, some radioactive steam would be released in the turbine building, which is by no means leaktight. If the worst-case assumptions are made concerning the amount of radioactivity in the steam and the atmospheric conditions, the estimated whole-body dose is 1 rem or so, and the thyroid dose might be as high as about 30 rems for some power plants (BWRs), although it would generally be smaller. These are substantially less than the maximum doses specified in the NRC regulations.

PROBABILITIES AND CONSEQUENCES OF CLASS 9 ACCIDENTS

Introduction

A Class 9 accident is one that is more severe than a design-basis accident.[i] It could result from a design-basis accident followed by a failure or malfunction of one or more of the engineered safety features or in other ways, as described later. In a Class 9 accident, all four barriers to the escape of radioactivity (i.e., the ceramic fuel material, the fuel cladding, the primary coolant system, and the containment structure) would be breached (or bypassed), at least to some extent. The amounts of radioactivity released to the environment could then result in lethal radiation exposures to some of the people living in the general vicinity of the nuclear power plant.

Any accident leading to substantial core damage accompanied by failure of the containment isolation system could result in an unacceptably large release of radioactivity. The Class 9 accidents of greatest concern, however, are those in which core melting is followed by containment penetration in one way or another. The subsequent discussion is concerned in particular with accidents of this type.

Core Meltdown Phenomena

A core meltdown could result from insufficient coolant being available to remove the decay heat after a reactor is shut down or from failure of the protection system to shut the reactor down when a mismatch occurs between the cooling rate and the rate of heat generation. The quantity of radioactive material escaping to the environment would depend on the chain of events following the core melting and the subsequent modes and times of failure of the containment.

In the event of a major core meltdown, the molten mixture of uranium dioxide fuel and Zircaloy (zirconium alloy) cladding would fall into the water remaining at the bottom of the reactor (pressure) vessel. If the melt should hit the water as a rain of small drops, steam would form extremely rapidly. In some cases, the sudden large increase of pressure could result in what is called a *steam explosion.* In a steam explosion, much of the molten core material would be dispersed into the reactor vessel and the containment.

[i]The classification (Class 9) is that used in the Environmental Reports (Chapter 3).

If a steam explosion did not occur in the reactor vessel, the hot core material would probably melt through this vessel and drop into the water in the floor sump of the containment structure. Here also a steam explosion might occur if the melt were in a finely divided form. Eventually the core material, in which decay heat is being generated continuously, would melt through the steel liner, penetrate the concrete bed at the bottom of the containment structure, and sink into the ground possibly to a depth of 18 to 27 m (60 to 90 ft).

Modes of Containment Failure

Prior to a comprehensive study of reactor safety, which is described shortly, it was generally assumed that core melting would inevitably result in an accident having serious consequences to the public. The study showed, however, that insufficient credit had been given to the effectiveness of the containment vessel. It has been found that the potential radioactive release following a Class 9 accident would depend strongly on the mode and time of containment failure and also on the operability of the containment cleanup system.

The modes of containment failure and escape of radioactivity following a core meltdown fall into three general classes: (a) melt-through of the structure, (b) overpressure, and (c) failure of the isolation valves to close. In the last case, radioactivity could be liberated to the surroundings even though the containment structure remained intact. The other failure modes are reviewed below.

For large containment structures, melt-through is considered to provide the most likely means for the escape of radioactivity following a Class 9 accident. Containment melt-through is predicted to occur about one-half to one day after the initiating event leading to a core meltdown. During this period, there is ample time for radioactive decay, washout by the containment spray, plateout (i.e., deposition on surfaces), etc., to reduce the radioactivity available for escape. The releases and consequences following containment melt-through are thought to be much less than from other modes of containment failure providing the containment cleanup system functions properly.

An internal overpressure may arise in three different ways:

1. a steam explosion (in the reactor vessel or the containment)
2. gas generation
3. an increase in temperature of the containment atmosphere.

If a steam explosion occurs in the reactor vessel, the flying debris could penetrate the containment structure, regardless of its volume. On the other hand, a steam explosion outside the reactor vessel is unlikely to cause rupture of a containment with a relatively large volume, although it might if the structure had a small volume. As the result of a destructive steam explosion, substantial amounts of radioactivity would be released to the environment. In addition to fission product gases and vapors, such as the noble gases (krypton and xenon) and iodine, normally solid fission products would be released as fine particles or smoke. However, the probability of containment failure caused by a steam explosion is moderately low.

At high temperatures, the zirconium in the fuel-rod cladding and molten iron from the reactor-core support structure would react readily with water (or steam) to generate large volumes of hydrogen gas. Furthermore, in penetrating the bottom of the containment structure, the molten core material would decompose the concrete and thereby produce considerable amounts of carbon dioxide gas. For small containment structures, as in BWRs, the high pressure arising from these two gases is the most probable cause for failure. This would not be likely to occur for several hours after meltdown, thus allowing time for a decrease in the amount of radioactivity that might be released. For large containment structures, the overall probability of failure due to high gas pressure is expected to be low.

If the containment cooling system should fail after a core meltdown accident, the temperature of the containment gases, including large amounts of water vapor, would rise. An increase in temperature of the gases could also occur as a result of hydrogen burning. In each case, the increase in temperature would cause an increase in the interior pressure; if this pressure became high enough, the containment would be disrupted.

The Reactor Safety Study

In 1972, the U.S. Atomic Energy Commission (AEC) (as it then was) sponsored a comprehensive study to assess the public risks from potential accidents in commercial (PWR and BWR) nuclear power plants of the type operating or under construction in the early 1970s. A report, entitled *Reactor Safety Study: An Assessment of Accident Risks in U.S. Commercial Nuclear Power Plants,*[j] was submitted to the NRC in October 1975. It has five major elements:

1. identification of accident sequences that lead to core meltdown and the probability of occurrence of each such sequence
2. the consequences of core meltdown in terms of the radioactive releases in several categories for various containment failure modes
3. dispersion of released radioactivity in the atmosphere and the paths leading to radiation exposures to the public
4. assessment of the health effects (and property damage) resulting from radiation exposures
5. fatalities and injuries as a function of the probability of reactor accidents.

Determination of Accident Probabilities

The procedure used in the *Reactor Safety Study (RSS)* to estimate the probability of a rare accident is to consider the chain of events that could lead to the accident. Suppose the probability of each event in the chain is known (or can be inferred) from previous (possibly nonnuclear) experience as P_1, P_2, P_3, etc. Then, if the individual events are independent, the probability of the rare accident is $P_1 \times P_2 \times P_3 \times \ldots$. This is the basic principle of an *event tree.*

An event tree starts with an initiating event, such as a break in the primary coolant system; this may (or may not) be followed by a failure in one or more of the engineered safety features; and finally the containment may (or may not) be ruptured in one or more of the modes described earlier. By considering various initiating events and various possible subsequent events, the event tree leads to a

[j]Published as NRC Report WASH-1400 (NUREG-75/014), it is commonly referred to as WASH-1400, or as the Rasmussen Report, after N. C. Rasmussen, the director of the study. In this book, it is referred to as the *RSS* (for *Reactor Safety Study*).

final outcome that may (or may not) have serious consequences. From the probabilities of the various stages in the chain of events, the probability of each final outcome can be estimated.

At some stages in the event chain (or sequence), the probability of failure is known from past experience. Otherwise, the failure (or success) probability at each stage of the event sequence is derived from a *fault tree*. Here the procedure is the reverse of that used in developing an event tree. The fault tree starts with a particular failure of a system or component, called the *top event*. This event is then traced back, step by step, to identify all the failures that could lead to the top event. From the known (or estimated) probabilities of these failures, the probability of the top event could be determined for use in the event tree. A fundamental difficulty has been in estimating the probabilities of failures of which there has been little or no experience.

The risk associated with any event may be defined as the product of the probability of that event and the magnitude of the consequences. Several hundred event trees were examined by the *RSS*, but many were eliminated as having insignificant risks, either because of their very low probability or their minor consequences. The remaining events were then subjected to a detailed examination, and some of the more important results of this examination are given shortly.

Review of the Reactor Safety Study

In July 1977, the NRC organized a Risk Assessment Review Group, consisting of a number of independent experts, to review the *RSS*. Among the objectives of the review group were to clarify the achievements and limitations of the *RSS* and to evaluate comments on the report (WASH-1400). The review group presented its report in September 1978,[k] and its findings were accepted by the NRC in January 1979.

The review group found the *RSS* to represent a substantial advance over previous attempts to determine the risks of nuclear power plant operation. The study was also successful in giving the assessment of reactor safety a rational basis, in establishing the chain of events in many accident sequences, and in indicating how the probabilities can be derived for those sequences for which there is an adequate data base.

The fault-tree/event-tree methodology was deemed to be sound, but its implementation in the *RSS* was regarded by the review group as being deficient in several respects. These include the use of inadequate data bases, the inability to quantify common-mode failures and human adaptability during the course of an accident, and questionable statistical procedures. Furthermore, insufficient emphasis was placed on the possible initiation of accidents by earthquakes, fires, and human errors. No matter how thorough the treatment, there can never be certainty that all possible initiating events have been considered.

For the foregoing reasons, the review group was not able to state whether the absolute probabilities of accidents given in the *RSS* (see below) were either

[k]"Risk Assessment Review Group Report to the U.S. Nuclear Regulatory Commission," NUREG/CR-0400, U.S. Nuclear Regulatory Commission, Washington, D.C. (1978), also called the Lewis Report, after the name of the group's chairman, H. W. Lewis.

too high or too low. It was concluded, however, that the error bounds were greater than indicated in the report.

The review group found that "despite its shortcomings, WASH-1400 provides at this time the most complete picture of accident probabilities associated with nuclear reactors. The fault-tree/event-tree approach coupled with an adequate data base is the best available tool with which to quantify these probabilities."

Reactor Safety Study Results: Accident Probabilities

Introduction

The *RSS* was limited to LWRs, since they are essentially the only commercial power reactors currently operating. A single PWR, the Surry Power Station Unit 1, with an electrical capacity of 788 MW, and a single BWR, the Peach Bottom Atomic Power Station Unit 2, electrical capacity 1065 MW, were selected for study. They were chosen because they were the largest facilities of each type that started operation in 1973 soon after the *RSS* was initiated.

It appears from what has been stated earlier that the *RSS* quantitative results are subject to considerable uncertainties. This should be borne in mind in connection with the various numbers quoted here. Nevertheless, even though the calculated accident probabilities cannot be accepted as absolute, they do provide an indication of the relative probabilities of a wide variety of meltdown accidents that might occur in LWRs.

The *RSS* has led to an increased understanding of a large spectrum of possible reactor accidents and their consequences. Furthermore, it has brought to light ways in which nuclear plant safety can be enhanced and has drawn attention to areas where further research is desirable. As a result, nuclear reactors should be safer in the future.

The main originating causes of core meltdown in PWRs and BWRs are discussed below. The estimated probability of a given accident sequence is expressed as a certain (small) number per reactor-year of operation.[l] For example, if the probability of an accident is 5×10^{-5}/reactor-year (i.e., 0.00005 per reactor-year), then there is a better than even (63 percent) chance that such an accident will occur once in $1/(5 \times 10^{-5})$ reactor-years of operation, or 2×10^4 (i.e., 20 000) reactor-years. There is a smaller chance that the accident will occur sooner.[m]

Pressurized Water Reactors

Primary System Pipe Breaks—According to the *RSS*, the most probable single cause of a PWR meltdown would be primary coolant system pipe breaks

[l]A reactor-year is the equivalent of a single reactor operating for 1 year.

[m]If the probability of a particular event is *P* per reactor-year, then the chance that this event will occur within *n* years after a reactor starts operating is

$$1 - \exp(-nP),$$

where exp is the base of natural logarithms, 2.72. Consequently, there is a 0.63 (or 63 percent chance, i.e., somewhat better than an even or 50 percent chance) that the event will occur within $1/P$ reactor-year of operation. The chances are 39 percent that it will occur in $1/2P$ and 22 percent in $1/4P$ reactor-year.

followed by failure of one or more of the engineered safety features. The total probability of such accident sequences for pipe breaks of all sizes was estimated to be about 3×10^{-5}/reactor-year.

Reactor (Pressure) Vessel Rupture—Disruption of the pressure vessel would be equivalent to a large LOCA followed by complete failure of the ECCS, since any coolant entering would be lost immediately. From the extensive experience with steam boiler drums and other pressure vessels in nonnuclear applications, and with allowance for the fact that reactor vessels are built to more exacting specifications, the *RSS* concluded that the probability of failure is 10^{-7} to 10^{-6}/reactor-year. If a failure should occur, it would be followed by a meltdown of the core.

Check Valve Failure—The low-pressure coolant injection system (LPIS), which is an essential component of the ECCS, is located outside the containment structure. It is isolated from the primary coolant system by two check valves in series in each (redundant) line. If both check valves should open while the reactor is operating, the LPIS, which is designed for a pressure of 4 MPa (600 psi) or less, would immediately be subjected to the full pressure (15.5 MPa) of the primary system. The LPIS would then be expected to fail, and the coolant water would escape through the breach outside the containment. As a result of the complete loss of coolant, the core would melt, and radioactivity would be released directly to adjacent buildings and thence to the environment, bypassing the containment.

An important fact that emerged from the *RSS* is that failure of the LPIS check valves is only 6×10^{-6}/reactor-year, but it is the most likely means of causing a large release of radioactivity from a PWR meltdown accident. If one of the check valves should fail, so that it remained open, the failure would not normally be detected. If the second valve should fail, a serious accident would occur. A regular monthly test of the valves, however, should decrease the probability of failure to an estimated 2×10^{-7}/reactor-year.

Transient Events—A transient event in the present context is any significant deviation from normal behavior that could lead to a fuel imbalance (i.e., heat is generated in the fuel faster than it can be removed). The great majority of transients are anticipated, and they have been observed to occur, on the average, about 10 times per reactor-year. Unlikely (or unanticipated) transients were estimated to have a frequency of less than 10^{-5}/reactor-year and were disregarded in the *RSS*.

Anticipated transients generally will not result in a core meltdown because of the action of backup safety features. For example, loss of feedwater pumps will cause the auxiliary pumps to start; a surge in power level will be halted by the reactor protection system; and loss of offsite power will scram the reactor and signal the diesel generators to start. Although each of the various safety systems are redundant, failure is possible. The *RSS* estimated the probability of a meltdown following the anticipated transient to be 10^{-5}/reactor-year.

If the protection system fails when a transient requires the reactor to shut down (i.e., to scram), the result is called an *anticipated transient without scram* (ATWS) event. The probability of meltdown following an ATWS event in a PWR was reported to be 4×10^{-6}/reactor-year. However, the particular PWR for which this result was found is specially protected against the effects of an ATWS event. Many others do not have the same degree of protection, and the probability of meltdown in such reactors was estimated in a later NRC staff report to be greater by a factor of possibly 10 or more. The staff therefore recommended modifications to PWRs that will reduce the probability of a core meltdown from ATWS events to 10^{-6}/reactor-year.[n]

Pressurized Water Reactor Summary—According to the *RSS*, the total probability of a meltdown in a PWR from internal causes (e.g., equipment failures, operating errors, etc.) is 6×10^{-5}/reactor-year. The most likely mode of containment failure following a meltdown was considered to be a melt-through. However, the release of (gaseous) fission products to the atmosphere following a melt-through was expected to be less than that following overpressurization of the containment structure.

Boiling Water Reactors

Primary System Pipe Breaks—The probability of a pipe break in a BWR was taken in the *RSS* to be the same as in a PWR, but the subsequent probability of a core melt was estimated to be much lower, at 10^{-6}/reactor-year, because of the greater reliability of the ECCS in a BWR.

Reactor Vessel Rupture—The probability of core meltdown as a consequence of a BWR pressure vessel rupture was thought to be the same as in a PWR—namely, 10^{-7}/reactor-year.

Transient Events—The *RSS* found that, in a BWR, transients were the most probable cause of core meltdown. Two sequences of events following a transient, which were about equally probable at 10^{-5}/reactor-year, were dominant in this respect. One possible sequence is the failure of the heat removal system after a shutdown due to any anticipated transient. The other is failure of the reactor to shut down following a loss of load and the closing of the steam valves. This would represent an ATWS event that could result in a pressure surge sufficient to rupture the reactor pressure vessel.

In the NRC staff report referred to above, it was concluded that the probability of an ATWS event in a BWR was much greater than estimated in the *RSS*. As with the PWR, design modifications were recommended that would reduce the probability of a meltdown from an ATWS in a BWR to 10^{-6}/reactor-year.

Boiling Water Reactor Summary—The total meltdown probability for a BWR from internal causes was estimated in the *RSS* to be 3×10^{-5}/reactor-year. This is only half of the meltdown probability of a PWR, but in the BWR the dominant mode of containment failure is overpressurization rather than by melt-through. As stated earlier, the release of (gaseous) fission products would be greater in the former case.

External Causes of Meltdown

Apart from internal initiating events, it is conceivable that core meltdown could occur as the result of a sequence initiated by natural (external) causes,

[n] "Anticipated Transients Without Scram for Light Water Reactors," NUREG-0460, Vols. 1 and 2 (April 1978), and Vol. 3 (December 1978), U.S. Nuclear Regulatory Commission, Washington, D.C.

such as major earthquakes, hurricanes, flood, tornadoes, etc. Nuclear power plants are constructed to withstand the most severe earthquakes and other natural phenomena that may be expected at the plant site. Hence, such events were not regarded as important causes of core meltdown.

The possibility of sabotage was also evaluated in the *RSS*. The conclusion was reached that the expected consequences of successful sabotage are small in comparison with the consequences of meltdown accidents. The safeguards that are being instituted to prevent sabotage are described at the end of this chapter. In view of these safeguards, it appears that the probability of successful sabotage is low, and it may be reduced even further in the future.

Total Meltdown Probability for Light Water Reactors

Although the *RSS* estimated the total meltdown probability to be 6×10^{-5}/reactor-year for a PWR and 3×10^{-5}/reactor-year for a BWR, the difference was not regarded as significant in the circumstances. It was concluded, therefore, that 5×10^{-5}/reactor-year would be a reasonable value for the meltdown probability for all LWRs of current design. This would mean that there is a better than even chance that a meltdown will occur once in 20 000 reactor-years of operation (p. 101). Although 5×10^{-5} was given as the most likely probability, the report stated that the upper bound for core melt probably would "almost certainly not exceed" 3×10^{-4}/reactor-year (or once in 3300 reactor-year of operation). This upper bound is close to the estimate based on the more recent review of ATWS events. (It is interesting to note that an independent study of LWR safety made in the Federal Republic of Germany concluded that the median meltdown probability is 10^{-4}/reactor-year.)

Reactor Safety Study Results: Radiological Effects

Estimation of Health Effects

A major objective of the *RSS* was to assess the effects of meltdown accidents on the health of people living in the vicinity of nuclear power plants. To do this, possible radioactive releases were divided into a number of categories of varying severity. Each potential accident sequence was then assigned to one of these categories. Assignment of an event chain to a particular category was determined by the following factors: the amount and composition of the radioactive material released from the reactor core, the effectiveness of the cleanup system (and other factors) in decreasing radioactivity in the containment, the times after the initiating event at which releases from the core and containment occur, and the height of the release.

To allow for dispersion of the released radioactive material, six atmospheric conditions were considered, ranging from a stable atmosphere (Pasquill F) with little mixing or spreading of the radioactive plume, to a highly unstable atmosphere (Pasquill A), which would cause the plume to spread widely. Wind speed was an additional variable. Allowance was also made for the possibility of washout of radioactivity from the plume by rain. The probabilities of the occurrence of various weather types were obtained from meteorological records covering at least a one-year period of six areas representative of the variability of climatic or topographical features.

By using these data, together with the compositions of and amounts of the releases in various categories, it was possible to calculate both internal and external radiation doses for each weather condition and release category in terms of distances from the nuclear power plant. The overall hazard arising from the release of radioactivity will also depend on the number of people exposed to the radiation. Hence, it was necessary to know the local population distribution. For purposes of the *RSS*, an average distribution was estimated from the population data for areas in which 68 reactors were located.

The *RSS* assumed that the people living within 20 miles of a plant in the downwind direction would be evacuated in the event of a core meltdown. A mathematical model to represent the evacuation process was developed, based on an analysis of data from a number of evacuations that have taken place in the United States. The radiation doses that would be received by people evacuated at different times were determined from calculations of the expected decrease with time of the dose rate from each release category.

The health effects of large (or moderately large) doses of radiation may be divided into two broad classes: early (or acute) effects and delayed (or latent) effects (see Chapter 7). The former are observed only for the larger doses, while the latter are considered to be possible for all who survive (or do not experience) the early effects. Early fatalities may occur within a few weeks to a few months after receipt of a large dose of radiation. Somewhat smaller doses can result in early sickness for which hospitalization may be required. Still smaller doses have no observable early effects.

The possible delayed effects, which are not observable for several years following exposure to radiation but may extend over a period of about 10 to 40 years, include various forms of cancer and also thyroid nodules. The latter are easily detected and can be readily removed surgically. In addition, genetic defects may appear in subsequent generations; these are also regarded as delayed effects. By making various assumptions, the incidence of the foregoing health effects can be related approximately to the radiation doses received.

Summary of Estimated Health Effects

The information obtained above includes the probabilities of radioactive releases in various categories, weather type, population distribution, effectiveness of evacuation, radiation doses received both externally (from the radioactive cloud emerging from the containment and radioactivity deposited from the cloud on the ground) and internally (from inhaled and ingested radioactive material), and the relationships between dose and associated health effects. In the final phase of the *RSS*, all the information was combined to derive a set of curves showing the probability per reactor-year that a meltdown accident would produce more than a certain number of early fatalities, early illness, delayed cancers, etc. Some of the results are shown in Figs. 4-13 and 4-14 (for early fatalities and illness) and Fig. 4-15 (for delayed cancer fatalities); these refer to large LWRs of the type that started operation in the mid-1970s. Similar curves were given in the *RSS* report for other health effects (and for property damage). Differences were found between PWRs and BWRs, but because of the

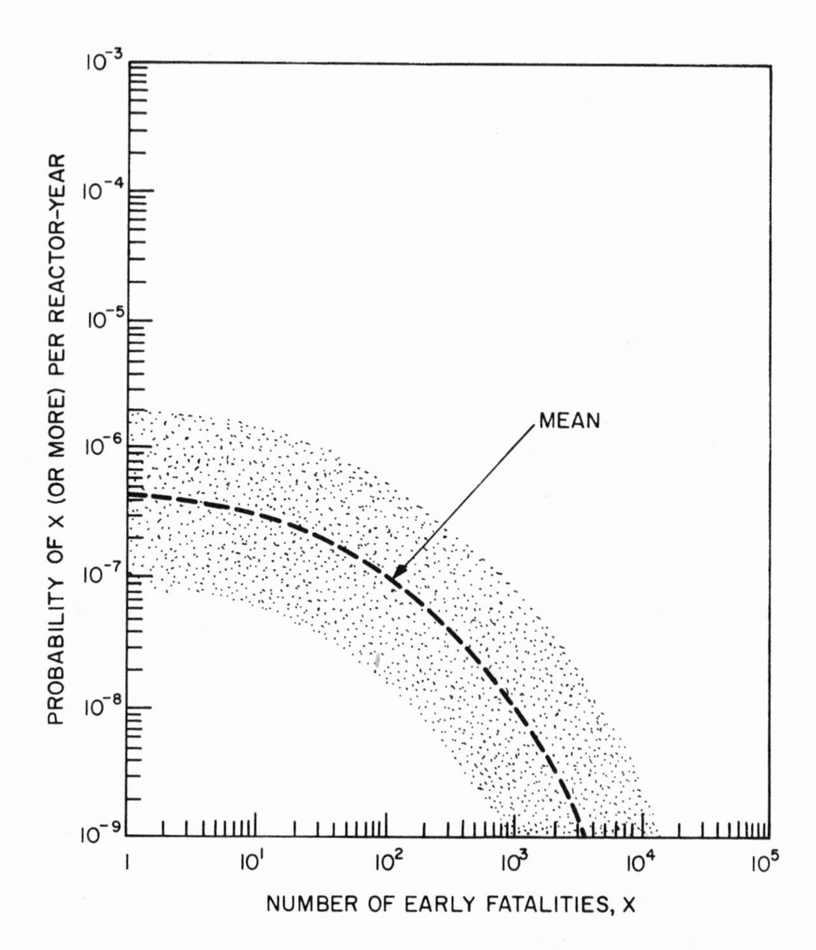

FIG. 4-13. Probability distribution for early fatalities resulting from hypothetical nuclear reactor accidents. The breadth of the curve indicates the approximate uncertainty range. (Adapted from WASH-1400, NUREG-75/014, U.S. Nuclear Regulatory Commission, Washington, D.C. [1975].)

uncertainties inherent in the calculational methods, a single average curve for each health effect, with appropriate error bands, was assumed to be applicable to all LWRs.

According to the *RSS*, the estimated uncertainties in the probabilities in Figs. 4-13 and 4-14 are represented by factors of 0.2 and 5; that is to say, the probabilities might be five times too small or five times too large. The uncertainties in the health consequences are given as 0.25 and 4 for Fig. 4-13 and 0.17 and 3 for Fig. 4-14. These ranges were meant to include uncertainties in atmospheric dispersion, effectiveness of evacuation, radiation doses corresponding to various radioactivity levels in the air and in the body, and the relation of health effects to the radiation dose.

The review group assigned to examine the *RSS* (p. 97) concluded, however, that in all these respects the uncertainties might be greater than stated. Never-

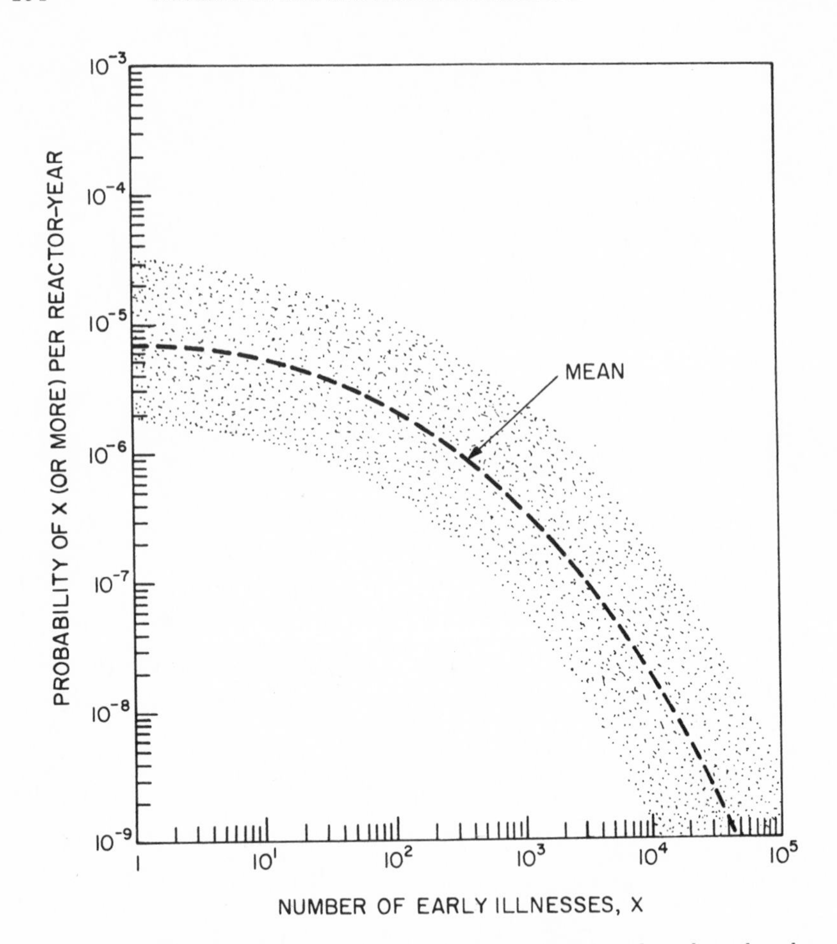

FIG. 4-14. Probability distribution for early illness resulting from hypothetical nuclear reactor accidents. The breadth of the curve indicates the approximate uncertainty range. (Adapted from WASH-1400, NUREG-75/014, U.S. Nuclear Regulatory Commission, Washington, D.C. [1975].)

theless, the data in the figures (and other similar curves) are the best available, although the numbers should be treated with caution.

The same uncertainties apply to the summary in Table 4-I, which gives the calculated health consequences of meltdown accidents having various probabilities. The delayed cancers and thyroid nodules are those expected over a period from about 10 to 40 years. The genetic effects are those that would appear per year in the first generation (i.e., in the children of exposed parents); the numbers would be smaller in subsequent generations.

The first line in Table 4-I applies to consequences of the median (or most likely) core meltdown. It is seen that if this accident were to occur, its consequences should not be very significant. The other lines in the table refer to various combinations of meltdowns followed by containment failure modes that are less likely to occur, worse weather conditions, and larger population den-

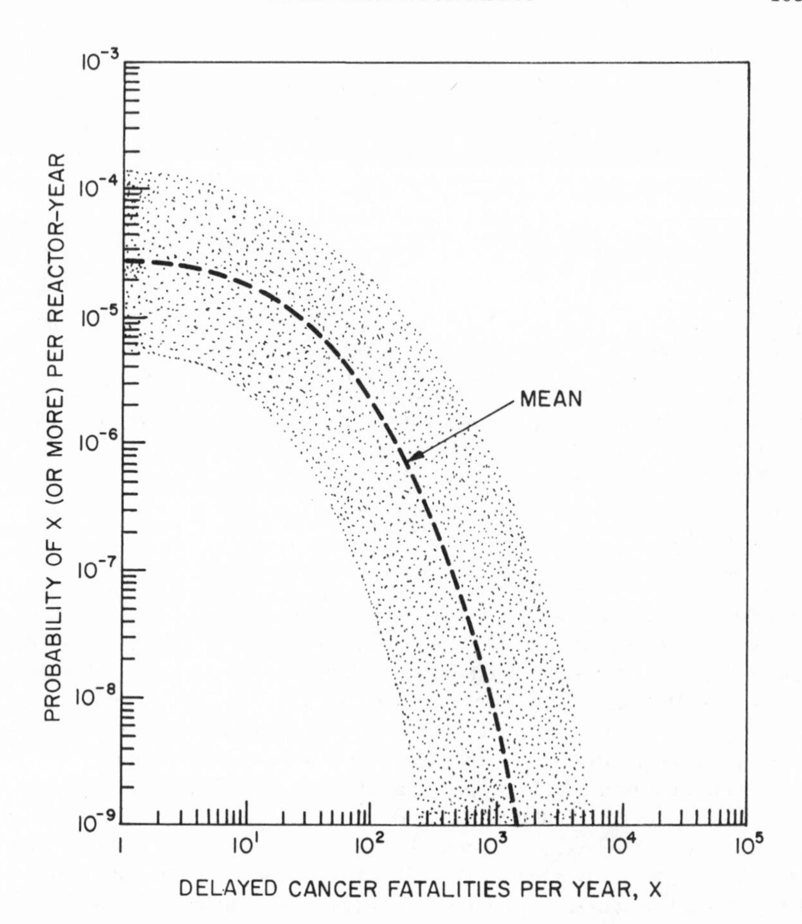

FIG. 4-15. Probability distribution for delayed cancer fatalities resulting from hypothetical nuclear reactor accidents. The breadth of the curve indicates the approximate uncertainty range. (Adapted from WASH-1400, NUREG-75/014, U.S. Nuclear Regulatory Commission, Washington, D.C. [1975].)

sities in the plant vicinity. The last line represents a core meltdown associated with the "worst case" mode of containment failure, weather, and population distribution. The total population exposed in this worst case would be approximately 10 million. The normal incidence of cancer fatalities in such a population group is about 17 000/year.

THE THREE MILE ISLAND ACCIDENT

An accident with the potential for a core meltdown occurred in the PWR at the Three Mile Island Nuclear Station Unit 2, near Harrisburg, Pennsylvania, on March 28, 1979. It was the most serious accident in a commercial nuclear

TABLE 4-I
CORRELATIONS OF HEALTH EFFECTS AND MEAN PROBABILITIES OF ACCIDENTS LEADING TO CORE MELTDOWN

Probability of Accident per Reactor-Year	Early Fatalities	Early Illness	Delayed Cancers (per year)	Thyroid Nodules (per year)	Genetic Defects (per year)
1 in 20 000 (median)[a]	1	1	1	1	1
1 in 1 000 000	1	300	170	1400	25
1 in 10 000 000	110	3 000	460	3500	60
1 in 100 000 000	900	14 000	860	6000	110
1 in 1 000 000 000	3300	45 000	1500	8000	170

[a]Corresponds to median probability of 5×10^{-5}/reactor-year; this represents the chance of the most likely core melt.
(Symbol < indicates "less than.")

power plant to that date, and the sequence of events is of great interest for nuclear reactor safety.

An anticipated transient—namely, loss of the main feedwater supply to the two steam generators—led to the automatic operation of three auxiliary pumps; two started immediately, and the third came on within 30 seconds. However, the discharge valves had been left closed, so that no feedwater reached the steam generators. With the loss of secondary coolant, the pressure (and temperature) in the primary system rose. The reactor was tripped, and a motor-operated pressurizer relief valve opened to reduce the steam pressure (p. 64).

When the pressure had dropped to a safe level, the relief valve should have closed automatically, but it failed to do so. Consequently, the pressure continued to fall to the level at which the high-pressure coolant injection subsystem (HPIS) of the ECCS was actuated (p. 78). A rapid rise in the pressurizer water-level indicator was misinterpreted by the operator as implying a high level in the reactor vessel, although this was evidently not the case. (There was no direct means of determining the actual water level in the vessel.) The HPIS pumps were therefore shut down manually. It is now known that there were substantial steam (and other) voids in the cooling system so that liquid continuity between the reactor vessel and pressurizer, which was assumed, did not actually exist.

Auxiliary feedwater flow to the steam generators was started by opening the valves, and the HPIS was operated intermittently in an attempt to compensate for the coolant loss through the still-open pressurizer relief valve. The effort was apparently based on the misleading pressurizer level indicator, and water was actually being lost. (The relief valve could have been blocked at any time by operator action, but this was not done for 2.3 hours into the accident; it was reopened periodically at later times.)

At 1.25 and 1.8 hours after initiation of the incident, the two primary coolant pumps, between the reactor and the steam generators, were shut down to prevent damage from excessive vibration. Normally, cooling would have continued by natural convective circulation, but it was prevented by the voids in the primary sytem. These voids were apparently also responsible for the vibrations in the primary coolant pumps. At this stage, with little or no cooling, the core suffered a large temperature increase.

The upper part of the core was completely uncovered for almost an hour. The extent and nature of the damage that occurred is not known, but it is possible that there was some fuel melting. Reaction of the zirconium in the very hot Zircaloy with water (and steam) caused cladding failure in some 90 percent of the fuel rods. Consequently, several feet of the upper part of the core collapsed and fell into the gaps between the fuel rods, thus causing partial blocking of steam and water flow. This blockage combined with the large hydrogen "bubble," formed by the zirconium-water reaction, inhibited cooling of the core by natural convection for an extended period.

With the disruption of the fuel rods, large amounts of radioactive fission products entered the coolant water. Contaminated liquid water and steam thus flowed through the open pressurizer relief valve to a quench tank in the containment. When this pressure in the tank reached the design level, a rupture disk allowed the water to be released to the containment sump.

Because of a containment isolation design weakness (p. 86), contaminated water was pumped from the sump to a radioactive waste tank in an auxiliary building. Some radioactive gases escaped to the environment, but the amount was relatively small. The main release occurred by way of the letdown/makeup system through which the reactor water is normally circulated to control its volume and purity. Radioactive fission gases, mainly krypton and xenon, that collected in the makeup tank header escaped to the atmosphere.

From measurements of radiation levels made at a number of locations several days after the accident, it appeared that the health effects on the general public should be minimal. At most, one or two additional delayed cancer deaths are expected over a period of years and perhaps the same number of genetic defects in the next generation in a population of about 2 million.[o] (The emotional disturbances arising from fear of the unknown may be much more serious.)

Although there was no meltdown in the Three Mile Island accident, the fact that it occurred once in approximately 800 reactor-years of LWR operation[p] suggests that the accident probability estimated in the *RSS* for existing reactors (p. 101) may be too low. It is important to note, however, that the incident has indicated where improvements in reactor operation and design should greatly decrease the probability of future accidents.

The Three Mile Island accident appears to have resulted from a combination of design deficiencies, inadequate procedures, and operator errors. The consequences will be far-reaching. In addition to design changes, the methods of operator selection and training are being reconsidered. Furthermore, many NRC criteria and regulations will undoubtedly be strengthened. Dedicated telephone lines have been installed from operating reactors to a center where ex-

[o]The following information is of interest for comparison with the requirements of 10CFR100 given in Chapter 3. No individual off the Three Mile Island plant site received an additional dose (above background) of as much as 100 mrem (0.1 rem) and the average dose to about 2 million people living within 50 miles of the plant was less than 2 mrem (0.002 rem).

[p]This is roughly the sum of the operating times of all commercial LWRs in the world at the end of March 1979; reactors on submarines and surface ships are not included.

pert advice can be obtained immediately in an emergency, at any time. In addition, there is now a requirement that nuclear plant licensees establish, in cooperation with state and local authorities, plans for evacuation procedures should they become necessary (p. 37).

The nuclear industry has responded to the Three Mile Island accident by forming an Institute of Nuclear Power Operations that is expected to have a permanent staff of some 150 persons in addition to consultants. The major objective of the institute is to assure the highest possible standards for the safe operation of nuclear reactors. The institute will advise individual utilities concerning their operations, by site visits, and will assist in improving training programs, maintenance practices, and management procedures. In addition, the Electric Power Research Institute has established a Nuclear Safety Analysis Center to review and analyze information relative to the safety of nuclear power plants.

PROTECTION AGAINST SABOTAGE

Introduction

Another aspect of nuclear reactor safety, which could have consequences similar to those of a severe accident, is the possibility of deliberate destruction by sabotage. The importance of this matter was recognized several years ago by the AEC when it established a program of *nuclear safeguards.* The objective of this program was to develop and institute plans to detect and deter both diversion or theft of fissile material and attempts to sabotage nuclear facilities. In the Energy Reorganization Act of 1974, the U.S. Congress indicated its concern by specifying the establishment of the Office of Nuclear Materials Safety and Safeguards within the NRC. The discussion here refers primarily to the prevention of sabotage; the aspect of nuclear safeguards concerned with unauthorized diversion of fissile materials is treated in Chapter 11.

The owner (licensee) of a nuclear power plant (or other nuclear facility) is required by the NRC to develop a physical security plan, details of which must be included in his application for an operating license (Chapter 3). Licensees of plants that are already operating must submit such a plan for approval by the NRC. The appropriate regulations are given in Section 73.50 of 10CFR73, "Physical Protection of Plants and Materials." Specific guidance on implementing physical security criteria is provided by NRC Regulatory Guide 1.17, "Protection of Nuclear Power Plants Against Industrial Sabotage." This guide endorses American National Standards Institute Standard N18.17, "Industrial Security for Nuclear Power Plants," but includes several additional features.

Physical Protection Regulations

To protect against sabotage of a nuclear power plant,[q] the security organization must include a specified number of guards properly trained and armed. Each guard must be requalified at least annually.

[q]The regulations for physical protection summarized here are applicable to spent-fuel reprocessing plants (Chapter 9), as well as to nuclear power plants.

In addition, there must be compliance with other requirements for physical security; the most important aspects of these requirements for protection against sabotage are summarized below:

Physical Barriers: Equipment, systems, devices, or materials, interference with which would endanger the health or safety of the public, shall be enclosed within at least two separate barriers. The space between the barriers, called the "protected area," shall be illuminated at night and monitored regularly to detect the possible presence of unauthorized persons or vehicles. An isolation zone around the protected area shall be clear of all objects that could conceal a person and shall be illuminated and continuously monitored. Means shall be provided for quickly locking all doors leading into the reactor control room.

Access Requirements: Access by people and vehicles to the protected area shall be controlled. Individuals, other than those having NRC or Department of Energy security clearance, and hand-carried packages shall be searched for firearms, explosives, and other devices that could be used to cause damage. Packages delivered into the area shall be checked for proper identification and authorization, and searches shall be made at random intervals. Barriers shall be provided to control access by vehicles to vital plant areas.

Detection Aids: Alarms shall be located at important points to indicate intrusions or unauthorized use of emergency exits. At least two continuously manned alarm stations are required so that a single act cannot remove the capability of responding to an alarm. The type of alarm (e.g., intrusion, emergency exit, etc.) shall be indicated at a manned central alarm station within the protected area. Alarms shall be designed to indicate tampering, and they should be tested at least once every seven days.

Communications Requirements: The capability of continuous communication shall be maintained between guards and watchmen on duty and the central alarm station. In addition to conventional telephone service, the central station shall have two-way voice communication with local law enforcement authorities by radio or microwave transmission. Equipment used for communication within the plant site shall be tested at least once at the beginning of each security personnel shift. For communications offsite, testing is required at least once per day.

Response Requirement: The plant licensee shall establish liaison with local law enforcement authorities so that they can be called upon to provide assistance, should it be deemed necessary. In developing the physical security plan for the facility, the licensee shall take into account the probable size and response time of the local law enforcement authority assistance.

In a nuclear power plant there are two areas where sabotage by means of an explosive would be of particular concern; these are the reactor itself and the deep pool of water in which the highly radioactive spent fuel elements are held prior to final storage (Chapter 10). As far as the reactor is concerned, the danger is extremely small. In the first place, the reactor is surrounded by a concrete radiation shield 1.8 to 2.4 m (6 to 8 ft) thick, and the whole is enclosed in the concrete containment structure, to which access is difficult. A considerable quantity of explosive would be required to damage the reactor or its vital equipment. The spent-fuel storage pool is not readily accessible because of the hazard to operating personnel. For a potential saboteur to reach the pool, he would have to penetrate at least two barriers without detection. In any event, the large volume of water in the pool would probably minimize the consequences of an explosion.

Nuclear reactor safeguards are under continuous study and review by the NRC. As new and better techniques are developed, plant licensees are required to adopt them. If the NRC regulations are properly implemented, the hazard to the general public from the sabotage of nuclear installations is extremely small.

Bibliography—Chapter 4

American National Standards Institute. *Industrial Security for Nuclear Power Plants* (Standard N18.17). New York: American National Standards Institute, 1973.

Chester, C. V. "Estimates of Threats to the Public from Terrorist Acts Against Nuclear Facilities." *Nuclear Safety* 17 (1976): 659.

Cooper, W. E., and Langer, B. F. "The Safety of Reactor Pressure Vessels." *Nuclear Safety* 17 (1976): 55.

Farmer, F. R., ed. *Nuclear Reactor Safety*. New York: Academic Press, 1977.

Hagen, E. W. "Anticipated Transients Without Scram: Status Quo." *Nuclear Safety* 17 (1976): 43.

Hetrick, D. L., ed. *Proceedings of the Topical Meeting on Nuclear Safety*. La Grange Park, Ill.: American Nuclear Society, 1977.

Leeper, C. K. "How Safe Are Reactor Emergency Cooling Systems?" *Physics Today* 26, no. 8 (1973): 30.

Lewis, E. E. *Nuclear Power Reactor Safety*. New York: John Wiley and Sons, Inc., Publishers, 1977.

Lewis, H. W. "The Safety of Fission Reactors." *Scientific American* 242, no. 3 (1980): 53.

Lipuralo, M. J.; Tinklar, C. G.; and George, J. A. "The Ice-Condenser System for Containment Pressure Suppression." *Nuclear Safety* 17 (1977): 710.

Mehta, D. S., *et al.* "Trends in the Design of Pressurized-Water-Reactor Containment Structures and Systems." *Nuclear Safety* 18 (1977): 189.

Nuclear Safety Research Institute. *Analysis of the Three Mile Island-Unit 2 Accident* (NSAC-1 and Supplement). Palo Alto, Calif.: Electric Power Research Institute, 1979.

Report of the President's Commission on the Accident at Three Mile Island (Kemeny Report). Washington, D.C.: U.S. Government Printing Office, 1979.

"Report to the American Physical Society by the Study Group on Water Reactor Safety." *Review of Modern Physics* 14, Suppl. 1 (1975).

Row, T. H. "Reactor Containment-Building Spray Systems for Fission Product Removal." *Nuclear Safety* 12 (1971): 516.

Rust, J. H., and Weaver, L. E., eds. *Nuclear Power Safety*. Elmsford, N.Y.: Pergamon Press, 1977.

Sandia Laboratories. *Safety and Security at Nuclear Power Reactors to Acts of Sabotage* (SAND-75-0504). Albuquerque, N. M.: Sandia Laboratories, 1976.

U.S., Atomic Energy Commission. *Analysis of Pressure Vessel Statistics from Fossil-Fueled Power Plant Service and Assessment of Reactor Vessel Reliability in Nuclear Power Plant Service* (WASH-1318). Washington, D. C., 1974.

______. *Integrity of Reactor Vessels for Light Water Power Reactors* (WASH-1285). Washington, D. C., 1974.

______. *The Safety of Nuclear Power Reactors and Related Facilities* (WASH-1250). Washington, D. C., 1973.

U.S., Atomic Energy Commission, Advisory Committee on Reactor Safeguards. *Report on the Integrity of Reactor Vessels for Light-Water Reactors* (WASH-1285). Washington, D. C., 1974.

U.S., *Code of Federal Regulations*, Title 10 (Energy), Washington, D. C.
Part 50. Licensing of Production and Utilization Facilities, Appendix A (General Design Criteria for Nuclear Power Plants);
Part 73. Physical Protection of Plants and Materials.

U.S., Nuclear Regulatory Commission. *Anticipated Transients Without Scram for Light Water Reactors* (NUREG-0460), Vols. 1, 2, and 3. Washington, D. C., 1978. See also, *Nuclear Safety* 20 (1979): 422.

______. *Environmental Survey of the Reprocessing and Waste Management Portions of the Fuel Cycle* (NUREG-0116), Sect. 4.10 (Sabotage). Washington, D. C., 1976.

______. "Materials and Plant Protection" (Regulatory Guides Div. 5). Washington, D.C.

______. "Protection of Nuclear Plants Against Industrial Sabotage" (Regulatory Guide 1.17). Washington, D.C.

______. *Reactor Safety Study: An Assessment of Accident Risks in U.S. Commercial Nuclear Power Plants* (WASH-1400) (NUREG-75/014), Main Report and Technical Appendices. Washington, D. C., 1975.

______. *Risk Assessment Review Group Report to the U.S. Nuclear Regulatory Commission* (NUREG/CR-0400). Washington, D. C., 1978.

______. *Three Mile Island: A Report to the Commissioners and to the Public* (Rogovin Report). Washington, D.C., 1980.

______. *TMI-2 Lessons Learned, Task Force Status Report* (NUREG-0578). Washington, D.C., 1979. See also, *Nuclear Safety* 20 (1979): 735.

Wade, G. E. "Evolution and Current Status of the BWR Containment System." *Nuclear Safety* 15 (1974): 163.

5

Radiation Protection Standards

INTRODUCTION

Ionizing Radiation Exposure in Perspective

As stated in Chapter 4, stringent precautions are taken in the design and operation of nuclear power plants to restrict the escape of radioactive materials to the environment. Nevertheless, during normal operation very small quantities of such materials are released, partly to the atmosphere and partly to an adjacent body of water. Because the ionizing radiation emitted by radioactive materials (p. 11) has the potential for causing harm to living organisms, these effluents have attracted interest as an environmental aspect of nuclear power generation. It is important, however, to view these radiations in their proper perspective.

In the first place, the release of radioactive material is not limited to nuclear power plants. Coal contains small amounts of such material, part of which is discharged to the atmosphere in the fly ash when the coal is burned. The radioactive species in the fly ash are not the same as those in nuclear power plant effluents. Nevertheless, it is possible to compare the radiations from coal-burning and nuclear plants on the basis of their expected biological effectiveness. From a number of independent studies, it appears that in this respect, radioactive emissions from modern coal-fired plants (with 97.5 percent removal of particulate matter) and nuclear plants are, on the average, not greatly different. This is not meant to imply that the emissions of radioactive material from coal-burning (or nuclear) plants are a significant hazard, because they probably are not. Nevertheless, it should be understood that the discharge of radioactivity to the environment is not an exclusive characteristic of nuclear power plants.

Furthermore, ionizing radiations and radioactive materials are actually much more widespread than the effluents from coal-burning and nuclear installations. The human race has evolved and developed in the presence of what is called *natural background radiation*. This radiation, to which people are (and always have been) continuously subjected, arises from interactions in the atmosphere of electrically charged particles from outer space (cosmic rays) and from radioactive materials in the ground and within the body itself (terrestrial radiation). More is said in Chapter 6 about natural radiation, but for the present it is sufficent to state that in amount it greatly exceeds ionizing radiations from nuclear power plant effluents.

Another major source of radiation to which a large proportion of the population is exposed is the x rays used in medical diagnosis and treatment. More than half the people in the United States have at least one x-ray examination each year. Although the portion of the body receiving the radiation and the amount received vary from one person to another, the annual average per exposed individual approaches that of natural background radiation.

Less important, but nevertheless significant, common sources of exposure to ionizing radiations are television receivers and jet aircraft flights. Even luminous watch dials emit such radiations, although the amounts are now less than they were in the past.

Studies of the Biological Effects of Radiation

Soon after the discovery by W. C. Roentgen, toward the end of 1895, of x-rays (also called roentgen rays), users of these ionizing radiations realized that they could produce an initial reddening of the skin (erythema), followed by more serious skin lesions if the exposures were large enough. The ionizing radiations from radioactive substances were found, shortly thereafter, to cause similar effects. The early observations, particularly on x rays, are of interest because they led to the use of these radiations and later of radium for the treatment of malignant growths. The radiations destroyed the cancerous cells but could also damage healthy cells in the body. Thus, during the early years of the present century, it became apparent to radiologists, who used x rays for diagnostic purposes and both x rays and radium for therapy, that sufficient exposure to ionizing radiations could cause significant injury.

Observations on the effects of these radiations on small experimental animals were reported as long ago as 1901. Since that time, the biological effects of ionizing radiations have been studied in laboratories and hospitals in many countries. These effects are, however, not unique. Large amounts of radiation can result in an increased incidence of cancer, but so also can many chemicals that are present in industrial effluents. Even an increase in the normal frequency of genetic changes, which has been known since 1927 to be a consequence of large radiation exposures, can also be caused by many substances, including caffeine, alcohol, and heat. As a result of numerous studies, there is now a better understanding of the biological effects of radiation than of most (if not all) other potential environmental hazards, such as sulfur and nitrogen oxides (produced in the combustion of coal and oil), tobacco and other smoke, pesticides, and chemical additives in food products.

ORGANIZATIONS FOR RADIATION PROTECTION

Introduction

The need for protecting x-ray workers and patients receiving x-ray treatments from excessive exposure to radiation became apparent even before the end of the 19th century. An attempt to define acceptable exposure conditions for the occupational use of x rays was made in 1902, and by 1906 most of the basic principles of radiation protection had been described. Unfortunately, many radiologists were either not aware of the need for protection or they disregarded it.

In 1916, the British Roentgen Society adopted a resolution recognizing the hazards of x rays and recommended protective measures; a formal statement of these recommendations, entitled "X-Ray and Radiation Protection," was published in the Society's journal in 1921. In the following year, the American Roentgen Ray Society independently adopted a similiar set of recommendations. By the mid-1920s, significant contributions to radiation protection had also been made in France, Germany, Holland, the USSR, and, especially, Sweden.

International Commission on Radiological Protection

The need for sharing information and advancing knowledge on radiation protection and related matters among radiologists and scientists in several countries led to the First International Congress of Radiology, held in London in 1925. One of the difficulties in formulating reasonably precise conditions for radiation protection was that, at the time, radiation exposures were expressed in several different ways. What was needed was a clearly defined basic unit of radiation exposure that could be used by everyone interested in radiation protection. Consequently, the Radiology Congress appointed a group, which eventually (after several name changes) became the International Commission on Radiation Units and Measurements (ICRU), to study the problem of radiation units.

The Second International Congress of Radiology, held in Stockholm, Sweden, in 1928, was significant in at least two respects. First, the committee on radiation units presented a proposal for the definition of a unit, now called the *roentgen*. This unit, based on the *ionization unit* suggested in 1908 by P. Villard, who first identified gamma rays, was adopted officially in 1931 and is still in use. The roentgen is discussed in more detail later, but for the present it is sufficient to state that it provides a means for expressing quantitatively the exposure of an individual to x rays (or to gamma rays from a radioactive substance) in terms of the ionization produced in air.

The second important action taken at the Stockholm meeting was to approve the formation of an international committee to study the problems of radiation protection. This group was called the International Committee on X-Ray and Radium Protection, later renamed the International Commission on Radiological Protection (ICRP). Initially, the committee was concerned with making general recommendations relating to working conditions, shielding of x-ray machines, and general precautions to be taken by radiologists and other workers who might be exposed to ionizing radiations. In 1934, the ICRP made a quantitative recommendation in terms of the roentgen unit of what was then called the "tolerance dose" of radiation for those who were occupationally exposed to x rays. This represented the first of the generally accepted standards (p. 123) that have played an important role in radiation protection.

Membership of the ICRP was increased in 1950 to permit more nations to participate in its activities. At present, there are some 40 individuals from 14 countries in the ICRP and its four committees. The recommendations of the

ICRP now include proposed standards for protection of the general public, as well as those occupationally exposed, from the possible effects of radiation.

National Council on Radiation Protection and Measurements

To present a unified position by the United States on various aspects of radiological safety, the American Roentgen Ray Society, the Radiological Society of North America, and the Radium Society agreed in the late 1920s to consolidate their activities in this area. The suggestion was made that the National Bureau of Standards (NBS) (U. S. Department of Commerce) assume responsibility for organizing a committee that would develop and formulate the United States point of view to the ICRP. The NBS was selected because it was expected to retain an independent and objective position; moreover, it was the only organization in the United States that had established a long-range program in the general field of radiation protection.

In 1929, the Advisory Committee on X-Ray and Radium Protection was formed under the auspices of the NBS, but it was not then, nor has it ever been, an agency of the United States government. Membership of the Advisory Committee consisted mainly of radiologists, physicists, and representatives from industry. Some members were government employees, but they served on the committee as private individuals with the necessary expertise. One of the earliest actions of the Advisory Committee was to suggest in 1934 a tolerance dose for occupational exposure to x rays, similar to the standard recommended in the same year by the ICRP.

As a consequence of developments in the radiation field, the Advisory Committee on X-Ray and Radium Protection decided in 1946 that the scope of its activities should be extended. In addition to an executive committee, to serve as the main committee, several subcommittees were appointed to study special aspects of radiation protection. At the same time, the Advisory Committee was renamed the National Committee on Radiation Protection (NCRP). In 1956, in order to indicate more fully the range of its interests, the name of the organization was changed to the National Committee on Radiation Protection and Measurements.

Until the early 1960s, NCRP reports had been published as part of the NBS Handbook Series. Despite repeated disclaimers, this gave the impression that the reports represented official United States government recommendations, whereas they were actually those of an independent body having no legal authority. For this and other reasons, the NCRP decided to sever its association with the NBS and to seek a federal charter through the United States Congress.

In 1964, the charter was granted and the word *Committee* was changed to *Council;* the body is now designated the National Council on Radiation Protection and Measurements, with the abbreviation still remaining NCRP. The NCRP now operates as an independent organization financed by voluntary contributions from the United States government, scientific and professional societies, and industry. Its reports are published under its own imprint with no reference to any governmental agency.

According to its charter, some of the objectives of the NCRP are to collect, develop, and disseminate, in the public interest, information and recommendations relating to radiation protection and to radiation measurements involved in radiation protection. The NCRP is also required to cooperate with the ICRP, the ICRU, and other organizations, both national and international, concerned with radiation protection and measurements. There are now almost 30 such organizations that collaborate with the NCRP.

Federal Radiation Council

In the course of hearings on radioactive fallout from nuclear weapons tests in the atmosphere, held in 1957 under the auspices of the Joint (Congressional) Committee on Atomic Energy, it was realized that despite the federal government's basic responsibility for radiological safety through the Atomic Energy Act (and its amendments), the setting of radiation protection standards (athough not their implementation) was in the hands of private organizations—specifically, the ICRP and the NCRP. As a result of this realization, an executive order was issued by President Eisenhower in 1959 creating a Federal Radiation Council (FRC). The responsibility of the Council was to "advise the President with respect to radiation matters, directly or indirectly affecting health, including guidance for all Federal agencies in the formulation of radiation standards . . . ." By Congressional legislation passed later in 1959, the FRC was given a statutory basis.

In 1960 and 1961, the FRC issued recommendations on radiation protection standards for both occupationally exposed individuals and the general population. Subsequently, reports dealing with fallout from nuclear weapons testing and other radiation matters were published. The FRC was also concerned with the development of safety standards for uranium miners who were susceptible to serious lung injury caused by radioactive gas emitted by uranium ores. More is said about this specialized topic in Chapter 8.

Environmental Protection Agency

The FRC was abolished by President Nixon in December 1970 as part of a general reorganization of federal environmental programs. Its functions were then transferred to the U.S. Environmental Protection Agency (EPA), specifically to the Office of Radiation Programs of the EPA. As originally envisaged, the activities of this office were to include four main areas:

1. collection and analysis of radiation exposure data, including those from natural background radiation, medical practice, occupational exposures, and radioactive fallout
2. reexamination of the scientific bases used to estimate the risks associated with exposures to different quantities of radiation
3. reexamination of the various benefits derivable from activities associated with exposure to radiation and how these can be judged
4. derivation of appropriate balances between benefits and risks.

A number of studies relating to radiation and the environment have been made by the EPA, and several reports have been published.

Atomic Energy Commission/Nuclear Regulatory Commission

The function of the U.S. Nuclear Regulatory Commission, formerly that of the U.S. Atomic Energy Commission, in relation to radiation protection is somewhat different from that of the organizations referred to above. The ICRP, NCRP, and FRC/EPA review the large body of information concerning the effects of ionizing radiation on people (and animals) and recommend standards for limiting exposures to such radiations. The AEC/NRC, on the other hand, is responsible for actually implementing (as well as formulating) radiation safety standards in connection with activities under licenses issued for the operation of plants concerned with the production and utilization of nuclear energy (Chapter 3). Furthermore, since its inception, the AEC has sponsored numerous research activities at its national laboratories and at universities and hospitals on the biological effects of ionizing radiations. This activity was transferred to the U.S. Energy Research and Development Administration (ERDA) in January 1975, when the AEC ceased to exist, and to the U.S. Department of Energy (DOE) in October 1977.

In the Atomic Energy Act of 1946, the U.S. Congress assigned to the AEC the responsibility for assuring safety with regard to radiation exposures resulting from nuclear energy operations. This safety responsibility extends to the general public as well as to those working in licensed plants. Rather than have the individual states responsible for safety, Congress made the AEC (a federal agency) responsible for several reasons; among these were the following:

1. At the time, specialists in radiation biology and protection were mostly associated with government laboratories.
2. Congress realized the basic interstate nature of the risks associated with radiation exposure.
3. Much of the information concerning materials used in the nuclear energy program was then classified.

However, the language of the Atomic Energy Act (and its subsequent modifications) was interpreted as requiring that the responsibility for the health of uranium miners remain with the individual states, as it was prior to 1946 (see Chapter 8).

In 1957, before the first demonstration nuclear power plant at Shippingport, Pennsylvania, started operating, the AEC issued Title 10 of the *Code of Federal Regulations*, Part 20 (10CFR20), entitled "Standards for Protection Against Radiation." As stated in Chapter 3, these standards (as well as those of the EPA) have the force and effect of law, and, in this respect, they differ from the standards of the ICRP and NCRP, which are only recommendations or guides for radiation protection. However, the requirements of 10CFR20 are based on these recommendations and are either equivalent or, in certain respects, more restrictive. Details of the 10CFR20 (and other) standards, especially as they affect the general public, are given later.

National Academy of Sciences

The U.S. National Academy of Sciences–National Research Council (NAS-NRC) is not directly involved in setting radiation protection standards. Nevertheless, it has played an important role in the development of these standards; this role has included the preparation of comprehensive, critical reviews of the effects of ionizing radiation on living organisms.

In the early 1950s, as a result of the testing of nuclear weapons in the atmosphere, there was public concern in the United States over the potential effects of ionizing radiation on the human population. In 1955, in response to this concern, the NAS-NRC appointed a Committee on the Biological Effects of Atomic Radiation, commonly referred to as the BEAR Committee.[a] The committee's first report, entitled "The Biological Effects of Atomic Radiation," was issued in 1956 (and in revised form in 1960). Several other reports on various aspects of ionizing radiation effects were prepared by subcommittees of the BEAR Committee and published between 1950 and 1963.

At the request of the FRC, in 1964 the NAS-NRC established an Advisory Committee to the Federal Radiation Council. This committee has continued to issue reports bearing on a variety of problems of radiation exposure and protection. With the development of peacetime applications of nuclear energy, and allegations that the existing radiation protection guides were inadequate, the FRC asked the NAS-NRC Advisory Committee to undertake a complete review and reevaluation of the scientific knowledge concerning the effects on people of exposure to ionizing radiations. In March 1970, the NAS-NRC accepted the task proposed by the FRC. When the activities of the FRC were transferred to the EPA at the end of 1970, the NAS-NRC Advisory Committee to the FRC was renamed the Advisory Committee on the Biological Effects of Ionizing Radiation (BEIR). Its functions and activities, however, remained unchanged.

The report of the BEIR Committee, entitled "The Effects on Populations of Exposure to Low Levels of Ionizing Radiation," was published in November 1972. More is said later about some of the findings in the report. It is sufficient to state here, however, that it contained no surprises that would have required substantial changes in existing radiation standards. While realizing the formidable problems involved in making a useful cost-benefit analysis, the report recommended that numerical standards for each major type of radiation exposure be based on the results of such an analysis.

Joint Committee on Atomic Energy of the U. S. Congress

A highly important factor in focusing attention in the United States on the problems of radiation protection from many divergent viewpoints has been the extensive hearings conducted by the Congressional Joint Committee on Atomic Energy and its subcommittees. The hearings in 1957 led to establishment of the FRC; those in 1969 and 1970 on the environmental effects of electric power generation were largely responsible for the FRC's request to the NAS-NRC for the complete review of radiation effects referred to above.

The following hearings, of which complete records were issued by the U.S. Government Printing Office, may be cited because of their significance for radiation protection:

The Nature of Radioactive Fallout and Its Effects on Man (1957).
Industrial Radioactive Waste Disposal (1959).
Fallout from Nuclear Weapons Tests (1959).
Radiation Protection Criteria and Standards: Their Basis and Use (1960).
Radiation Standards, Including Fallout (1962).

[a]The term *atomic radiation* was commonly used at one time for what is more properly called *ionizing radiation*.

Fallout, Radiation Standards, and Countermeasures (1963).
Radiation Exposure of Uranium Mining (1967).
Radiation Standards for Uranium Miners (1969).
Environmental Effects of Producing Electric Power (1969/70).

United Nations

Following a debate on the implication of the radioactive fallout from the atmospheric testing of nuclear weapons, the General Assembly of the United Nations (U.N.) adopted a resolution in December 1955 establishing a U.N. Scientific Committee on the Effects of Atomic Radiation (UNSCEAR). This committee has made no quantitative recommendations concerning radiation protection standards. Nevertheless, the scientific and medical information collected and reviewed by UNSCEAR has influenced the discussions of the bodies responsible for proposing the standards.

Seven UNSCEAR reports have been published so far in 1958, 1962, 1964, 1966, 1969, 1972, and 1977. These are very thorough critical reviews of all available information on such topics as the biological effects of ionizing radiation, effects of external sources of radiation and of radioactive substances within the body, natural (background) radiation levels, radiation exposures from medical and occupational activities, and environmental (i.e., fallout) contamination. The 1972 report, entitled *Ionizing Radiation: Levels (Vol. I) and Effects (Vol. II)*,[b] is a complete review of essentially all biological aspects of ionizing radiation.

International Atomic Energy Agency

The International Atomic Energy Agency (IAEA) was established in 1957 as an autonomous intergovernmental agency under the aegis of the U.N. At present, there are more than 100 member nations of the IAEA. One of the prime functions of the agency is to "accelerate and enlarge the contribution of atomic energy to peace, health and prosperity throughout the world." Among its many activities, the IAEA sponsors symposiums, seminars, and panels at which reports on particular aspects of nuclear energy are presented and discussed. These reports and discussions are published, generally in the IAEA proceedings series. Among the symposiums concerned with understanding radiation effects on living organisms, the following may be mentioned:

Biological Effects of Ionizing Radiation at the Molecular Level (1962).
Genetical Effects of Radiosensitivity: Mechanism of Repair (1966).
Effects of Radiation on Cellular Proliferation and Differentiation (1968).
Biophysical Aspects of Radiation Quality (1971).
Radiological Safety Evaluation of Population Doses and Application of Radiological Safety Standards to Man and the Environment (1974).
Biological Effects of Low-Level Radiation Pertinent to Protection of Man and His Environment (1975).

[b]Official records of the U.N. General Assembly, 27th Session, Suppl. 25, A/8725 (1972).

Health Physics Societies

Health physics has been defined by K. Z. Morgan, one of its pioneers, as "the science that deals with problems of protection from the hazards of ionizing radiation or prevention of damage from exposure to this radiation . . . . It deals with mechanisms of radiation damage, methods of measuring and assessing radiation dose, . . . , and the establishment of maximum permissible exposure levels." The science and profession of health physics started with a small group in 1942 and has now grown to include several thousand individuals in all parts of the world.

In June 1955, the Health Physics Society was organized in the United States, with chapters in several other countries. In addition, a number of independent professional societies and associations concerned with radiation protection were formed in Europe and elsewhere. A truly international organization of health physicists, called the International Radiation Protection Association, was established in 1964 and its First International Congress was held in 1966.[c]

Summary

The major events relating to the formation and activities of the more important international and United States organizations concerned with radiation protection are listed in chronological order in Table 5-I. In addition, there are many similar national organizations in other countries that cooperate with those included in the table.

RADIATION QUANTITIES AND UNITS

The Roentgen

The first unit for expressing the amount (or dose) of exposure to radiation that received international acceptance was the roentgen, based on the ionization produced in air (p. 11). Since 1928, when the original proposal was made to the International Congress of Radiology, the definition of the roentgen has been altered to some extent, in the interest of precision, but its significance has remained unchanged. The most recent form is based on the internationally accepted unit of electric charge called the *coulomb*. The *roentgen* (R) is then defined as the exposure (or quantity) of x rays or gamma rays that will produce (by primary and secondary ionization[d]) in 1 kilogram (kg) of dry air charged particles of one sign carrying a total charge of 2.58×10^{-4} coulomb when all the electrons liberated are stopped in air.

The roentgen was initially described as the dose unit for x rays and gamma rays. On the basis of recommendations made by the ICRU, the term *dose* is now restricted to two other units described below. The roentgen is now said to be the

[c]The International Radiation Protection Association, which is an association of health physics societies, should not be confused with the International Commission on Radiological Protection (ICRP) described earlier.

[d]Primary ionization is that caused by the radiation; secondary ionization refers to that produced by the charged particles liberated by direct (primary) ionization.

TABLE 5-I

MAJOR LANDMARKS FOR RADIATION PROTECTION ORGANIZATIONS

1925	First International Congress on Radiology meets. Appoints group, which eventually becomes the International Commission on Radiation Units and Measurements (ICRU), to study radiation units.
1928	Second International Congress on Radiology meets. Proposal made for the roentgen unit (adopted officially in 1931). Committee, later called the International Commission on Radiological Protection (ICRP), formed to study problems of radiation protection.
1929	Advisory Committee on X-Ray and Radium Protection formed in the United States.
1934	First numerical recommendations of radiation protection standards made in terms of roentgens. (Developments and changes in these standards are discussed in the text.)
1946	The U.S. Advisory Committee extends its activities and is renamed the National Committee on Radiation Protection, later changed to the National Committee on Radiation Protection and Measurements (NCRP).
1955	The U.S. National Academy of Sciences–National Research Council (NAS-NRC) forms a Committee on the Biological Effects of Atomic Radiation (BEAR).
	The United Nations forms its Scientific Committee on the Effects of Atomic Radiation (UNSCEAR).
	The Health Physics Society is formed.
1957	The U.S. Atomic Energy Commission first issues 10CFR20, "Standards for Protection Against Radiation."
1959	Federal Radiation Council (FRC) established in the United States.
1964	International Radiation Protection Association (of Health Physics Societies) organized.
	The NCRP receives a federal charter and is renamed the National Council on Radiation Protection and Measurements.
1970	The U.S. Environmental Protection Agency is formed, and the functions of the FRC are transferred to it.
	The NAS-NRC establishes an Advisory Committee on the Biological Effects of Ionizing Radiations (BEIR).

unit of *exposure* to x rays and gamma rays. As stated on page 9, x rays and gamma rays are basically the same, although they originate in different ways.

The roentgen is a satisfactory unit for gamma rays and x rays, which are electromagnetic wave radiations similar to visible light, but of much shorter wavelength. It is not suitable, however, for particle-type radiations, such as alpha and beta particles and neutrons. Consequently, there was a need for a unit (or units) that would be applicable to ionizing radiations of all kinds.

The Rad and Rem

The roentgen measures the amount of ionization produced in air, but an important quantity, as far as potential effects of radiation are concerned, is the amount of energy absorbed from the radiation. This led to the concept of an *absorbed dose,* which is expressed in terms of a unit called the *rad*, an acronym for *R*adiation *A*bsorbed *D*ose. It is defined as a radiation dose that deposits 100 erg (or 10^{-5} joule) of energy per gram of absorbing material, such as flesh, bone, etc. The rad can be applied to ionizing radiation of all kinds.

For x rays, an exposure of 1 R results in the deposition of approximately 1 rad in soft body tissue (e.g., muscle). Hence, for the latter, the exposure in roentgens is roughly the same as the absorbed dose in rads. In bone, however, 1 R corresponds to an energy desposition of more than 1 rad, especially for x rays (or gamma rays) of low energy. This fact indicates the necessity for using the rad for x rays as well as for other ionizing radiations to which the roentgen is not applicable.

The absorbed dose (in rads) is an important (and measurable) quantity, but the biological response associated with a given absorbed dose depends on the nature and energy of the radiation and on the irradiation conditions. To allow for the characteristics of different radiations, the absorbed dose is multiplied by certain modifying factors, as explained below. The resulting quantity is called the *dose equivalent*, and the unit is the *rem*, an acronym for its original (but not present) use as *R*oentgen *E*quivalent (in) *M*an. Thus, the dose equivalent in rems is defined by

$$\text{Dose equivalent in rems} = \text{Absorbed dose (in rads)} \times \text{Modifying factors} .$$

The introduction of the modifying factors, which depend on the nature and energy of the radiation and other considerations, makes it possible to state dosages received from different radiations and under different conditions on a common basis. Thus, 1 rem of any ionizing radiation is essentially equivalent biologically to 1 rem of another kind of radiation. Dose equivalents in rems received from different radiations can then be added to provide an indication of the total biological response.[e]

One of the modifying factors in the definition of the dose equivalent is the *quality factor* of the particular radiation. Its purpose is to allow for the biological effectiveness of that radiation for a given amount of absorbed energy (i.e., dose in rad). In assigning a numerical value to the quality factor, the ICRU recommended that it be related to what is called the *linear energy transfer* (LET). This can be regarded as the rate of energy deposition (per unit length) measured along the track of the ionizing radiation; it is commonly expressed in terms of the loss of energy per unit path length in water. The higher the LET, the larger the quality factor, and, hence, the more biologically effective is a given absorbed dose of the specified radiation.

Additional modifying factors may sometimes be required to allow for nonuniform distribution of the absorbed dose in space and time throughout the body (or part of the body). These factors may be important for situations in which some radioactive species, when present within the body, tend to accumulate preferentially in certain tissues. For external sources of radiation, however, such modifying factors are not important.

For x rays, gamma rays, and beta particles, the quality factors may be taken as unity. In other words, for these (low-LET) ionizing radiations from external sources, the dose equivalent in rems is equal numerically to the absorbed dose in rads. When x rays and gamma rays are absorbed in soft tissue, the exposure in roentgens is approximately the same as the absorbed dose in rads, as seen above, and consequently it is roughly the same as the dose equivalent in rems.

[e]In the international (SI) system, the unit of absorbed dose is the *gray* (Gy), representing the deposition of 1 joule of energy in 1 kg of material; it is equal to 100 rads. The corresponding unit dose equivalent is the *sievert* (Sv), which is equal to 100 rems. Since the rad and the rem are used in the existing literature, they are retained here.

Alpha particles (high-LET) from radioactive sources outside the body are not a significant hazard, unless they are in direct contact with the skin. If the sources should enter the body, however, the situation could be quite different, especially if the alpha-particle emitter should tend to concentrate in certain tissue (e.g., plutonium and radium in bone). For alpha particles within the body, the quality factor might be in the range from 10 to 20, and there may be, in addition, other modifying factors—e.g., a *distribution factor.* In these circumstances, the dose equivalent would be very much larger than the absorbed dose.

The quantities actually measured by instruments are exposures in roentgens (for x rays and gamma rays) or absorbed doses in rads (for any radiation). However, for radiation protection purposes, the important quantity is the dose equivalent in rems. Hence, as far as possible, standards described below are given in rems. When the early standards, which apply only to x rays and gamma rays, are considered, the exposures are stated in roentgens. The same unit is also used in connection with certain measurements in which x rays or gamma rays are used for experimental purposes.

The dose equivalents of radiation involved in protection standards are often fractions of a rem. Consequently, it has been found convenient for this and other purposes to introduce a smaller unit, called the *millirem* (mrem), which is a one-thousandth part of a rem; thus,

$$1 \text{ rem} = 1000 \text{ mrems.}$$

A dose equivalent of 0.17 rem, for example, would be the same as 170 mrems.

For convenient reference, the main characteristics of the important radiation units are outlined in Table 5-II. The word *dose* is often used for brevity without any qualification, but the nature of the unit indicates whether absorbed dose (in rads) or dose equivalent (in rems) is implied. In the remainder of this chapter, *dose* refers to dose equivalent.

PROTECTION STANDARDS FOR OCCUPATIONAL EXPOSURE

Basis of Radiation Protection Standards

Standards for protection against radiation are generally expressed as the dose of ionizing radiation, in excess of that arising from background sources and from medical diagnosis and treatment, that an individual may be allowed to receive in a specified time (e.g., a day, a week, or a year).[f] The standards are under constant review and are subject to change from time to time, based on increased knowledge of the biological effects of radiation and on experience of what can reasonably be achieved in reducing exposure to radiation.

The setting of numerical standards for radiation protection is difficult, because it is largely a matter of judgment based on many considerations. One of the problems is that the limiting (or maximum allowable) radiation doses proposed in the standards are so low that no adverse effects have been identified

[f]In a few instances where the radiation arises mainly from certain radioactive substances deposited within the body, the protection standards are given in terms of the maximum amount (or activity) of the radioactive species (p. 132).

TABLE 5-II
CHARACTERISTICS OF RADIATION UNITS

Name	Unit of	Applies to	Indicates
Roentgen (R)	Radiation exposure	X rays and gamma rays	Ionization in air
Rad	Absorbed dose	All ionizing radiations	Energy absorbed per unit mass
Rem (1000 mrems)	Dose equivalent	All ionizing radiations	Biological effectiveness of the absorbed dose

that can be specifically attributed to them. Nevertheless, it is not known with absolute certainty that there are no such effects. Consequently, as a precautionary measure, the organizations responsible for proposing standards have always operated on the principle that there should be no exposure to radiation without some accompanying benefits. For this reason (and others that are given later) there are different radiation protection standards (a) for those occupationally exposed, who derive direct economic benefits, and (b) for the general population. This section is concerned with the standards for people who may be exposed to radiation in the course of their work; the standards for the general population are discussed subsequently.

The earliest numerical radiation protection standard for those occupationally exposed was referred to as the *tolerance dose* (p. 114), implying that such a dose could be tolerated.[g] However, this concept of a tolerance dose erroneously came to be associated in some people's minds with the existence of a threshold value below which no adverse effect was expected to result.

To remove this misconception, the NCRP in 1946 introduced the expression *permissible dose,* defined (in 1949) as "the dose of ionizing radiation that, in the light of present knowledge, is not expected to cause appreciable bodily injury to a person . . . during his lifetime." The radiation protection standard was then expressed in terms of the *maximum permissible dose* (or MPD) that just fulfills the requirements set forth in the definition of the permissible dose. As used above, the expression *appreciable bodily injury* means "any injury or effect that the average person would regard as being objectionable and/or competent medical authorities would regard as being deleterious to the health and well-being of the individual."

In its first report, published in 1960, the FRC objected to the expression *maximum permissible dose* on the grounds that the words *maximum* and *permissible* were often misunderstood. The report noted that "there can be no single permissible acceptable level [of radiation] without regard to the reason for permitting the exposure." The FRC then proposed that the MPD be replaced by an equivalent *radiation protection guide* (RPG), defined as "the radiation dose which should not be exceeded without careful consideration of the reasons

[g]The term *tolerance dose* appears to have had two different meanings. Originally it referred to the dose that could be tolerated by a specific tissue of patients in radiation therapy, although it might cause injury if repeated frequently. Later, it came to be associated with a maximum dose that could be allowed in the daily exposures of radiation workers.

for doing so." However, any terminology may be subjected to misunderstanding, and both the NCRP and ICRP have decided to retain the term MPD for those occupationally exposed to radiation but to use *dose limit* for the general population.

Regardless of terminology, the standards are not meant to indicate an objective to be attained; rather, they are guidelines for doses that should not be exceeded for any individual. Experience has shown that, in practice, when precautions are taken to prevent a certain limit from being exceeded by any one person, the doses received by most people are much less than the permissible maximum.

Some occupational radiation exposures, such as those received by operators of x-ray machines and by workers in nuclear power plants, are at least partially controllable; they are "progressively reducible with progressively increasing difficulty, technologically, economically, or both" (NCRP Report No. 39, [1971]). In this connection, the NCRP in 1954 (Report No. 17) stated explicitly what had long been understood and has since been often reiterated—namely, that "exposure to radiation be kept at the lowest practicable level in all cases." The use of the word *practicable* implies that technological and economic factors must be taken into consideration. In 1974, the ICRP introduced the expression "as low as is reasonably achievable" instead of "as low as is practicable." This change did not reflect a change in objectives of dose limitation, but rather a choice of language that would more closely describe the intent of the requirement.

For convenience and ready reference, abbreviations that are used frequently in the discussion of radiation protection standards are collected in Table 5-III.

Early Radiation Protection Standards

One of the first reported attempts to define the conditions that would provide protection for operators of x-ray machines was made by W. Rollins in the United States in 1902. He suggested that if a photographic plate, after being developed in the usual manner, showed no sign of fogging as a result of exposure for 7 minutes at the operating location, the radiation would not be harmful. In view of the variable sensitivity of photographic plates to ionizing radiation, as well as to ordinary light, this attempt at a quantitative protection criterion was somewhat uncertain.

Before the roentgen had been introduced as a unit of radiation, exposures were often expressed as multiples or fractions of the "threshold (or skin) erythema dose." This was the dose received as a single exposure (i.e., within a short time period) that would cause just a slight skin reddening, similar to a mild sunburn. In order to express radiation exposures, the threshold erythema dose was defined in the somewhat uncertain terms of the voltage and current of the x-ray machine, the distance of the exposed person from the machine, and the duration of the exposure.

In 1925, A. M. Mutscheller in the United States and R. M. Sievert in Sweden reported on studies they had made of conditions at x-ray installations where good work patterns had been established. Although their conclusions were based on different arguments, these two observers independently suggested that an exposure of one-tenth of a threshold erythema dose per year would be acceptable for those working with x rays (or radium). This tolerance dose (or exposure) was a "value judgment" made in the absence of any definite information concerning

TABLE 5-III

ABBREVIATIONS USED IN CONNECTION WITH RADIATION PROTECTION STANDARDS

Abbreviation	Meaning
AEC	U.S. Atomic Energy Commission
NRC	U.S. Nuclear Regulatory Commission
AEC/NRC	AEC regulatory activities taken over by the NRC in 1975
EPA	U.S. Environmental Protection Agency
ICRP	International Commission on Radiological Protection
NCRP	National Council on Radiation Protection and Measurements
FRC	Federal Radiation Council (activities transferred to the EPA in 1970)
NAS-NRC	National Academy of Sciences–National Research Council
BEAR	NAS-NRC Committee on the Biological Effects of Atomic Radiation
BEIR	NAS-NRC Advisory Committee on the Biological Effects of Ionizing Radiations
10CFR20	Title 10 (Energy) *Code of Federal Regulations*, Part 20
MPD	Maximum permissible dose
MPC	Maximum permissible concentration
RPG	Radiation protection guide (FRC)

the effects of such an exposure. Thus, it was largely a matter of coincidence that Mutscheller and Sievert made the same recommendation.

NCRP and ICRP Standards for Occupational Exposures

The radiation exposures in terms of the threshold erythema dose could not be measured directly, but had to be estimated from the working conditions. When the introduction of the roentgen unit was being considered in 1927, H. Kustner in Germany obtained the opinions of several radiologists concerning the relationship between the threshold erythema dose and the roentgen. Based on the information collected in this manner, he concluded that the threshold erythema dose was equivalent to about 550 R, and this value came to be generally accepted.

The first standard for radiation workers, expressed in directly measurable form, was proposed in the United States in 1934 by the Advisory Committee on X-Ray and Radium Protection, now the NCRP. To derive this standard, Kustner's value of 550 R/threshold erythema dose was rounded off to 600 R; thus, the Mutscheller-Sievert tolerance dose of one-tenth of a threshold erythema dose was roughly 60 R/year. On the assumption that there are 250 working days in a year, this becomes 0.24 R/day. Because of uncertainties in the threshold erythema dose and its conversion into roentgens (and in the interpretation of radiation measurements[h]), the Advisory Committee decided to reduce the value to 0.1 R/day, and this was the accepted standard for occupational exposure in the United States for several years.

[h]An ionization meter indicates a smaller number of roentgens in the free air away from the body than if it is held near the body, because in the latter case, some of the radiation incident on the body is reflected back to the instrument and thus increases the reading.

At about the same time, in 1934, the committee (which later became the ICRP) formed at the Second International Congress of Radiology (p. 114) to study radiation protection, using much the same arguments as outlined above, recommended a tolerance dose of 0.2 R/day for people occupationally exposed to x rays. This standard was generally adopted in European practice. Because of the uncertain ways in which radiation measurements were made, the difference between the NCRP and ICRP standards may not have been as great as appears at first sight.

The foregoing values of the tolerance dose for radiation workers remained in effect until 1949. In that year the NCRP, and in the following year the ICRP, recommended a decrease in the maximum permissible dose (or MPD), as it was then called. A primary consideration was that, with the expected developments in nuclear energy, many more people would be occupationally exposed. Another change made was allowance for the possibility of exposure to ionizing radiations other than x rays (and gamma rays). Furthermore, the MPD was expressed in terms of a weekly, rather than daily, dose limit.

Prior to 1949, the NCRP maximum permissible dose of 0.1 R/day was equivalent to roughly 0.5 R per working week. This was reduced in the new NCRP and ICRP standards to 0.3 R measured in air, for occupational exposure to x rays during any seven consecutive days. The MPD was extended to other types of ionizing radiation by stating the value as 0.3 rem (or 300 mrems) per week.[i] This particular MPD referred to absorption of radiation by the whole body or by the blood-forming organs, such as bone marrow, the spleen, and lymph nodes. When only certain portions of the body (for example, the hands and feet) were exposed, the permissible dose to these organs could be higher (see Table 5-V).

The decrease in the MPD mentioned above was not based on any new biomedical information that showed ionizing radiation to be more harmful than previously supposed. It was recognized, however, that there were still considerable uncertainties about the effects of radiation on the body at low doses (and dose rates). Furthermore, since it appeared that no unreasonable economic cost would be involved in decreasing the MPD by a factor of 2, the change was made as a step in the direction of added safety for occupationally exposed individuals. Consideration was also given to the probability of a large future increase in radiation uses.

The NCRP report in which the reduction in the MPD was proposed was completed in 1949, but was not published until 1954. In the intervening period it had been submitted to a number of scientific and medical authorities for review and was the subject of several international discussions. Only those matters on which complete agreement had been reached were included in the final report.

The Gonad (and Whole-Body) Dose

Injuries to living organisms caused by substantial doses of ionizing radiations fall into two broad categories: somatic and genetic. More is said in Chapter 7 about these two types of injuries, but, for the present, it is sufficient to state that

[i]The *week* in connection with radiation protection standards refers to any seven consecutive days, and not necessarily a calendar week.

somatic effects are those that are apparent, either in a short time or after some delay, in the organism that has been exposed to radiation. Genetic effects, on the other hand, cause no observable injury to the exposed individual, but they appear as new characteristics (or mutations) in subsequent generations. Mutations induced by radiation are indistinguishable from "spontaneous" mutations that occur fairly frequently in all life forms in nature, without any definitely assignable cause or human intervention. Many chemicals and even heat can also cause mutations.[j]

Mutations, no matter how they arise, can be deleterious (i.e., harmful) or beneficial. Since the great majority are deleterious, the effects of radiation exposure of a population could be serious over a period of generations. It is for this reason that the genetic effects of ionizing radiation must be taken into consideration in the development of protection standards.

The frequency of mutations induced by radiation in a population group is related to the total accumulated dose to the reproductive organs (gonads) of the parents at the time of conception. It is important, therefore, that particular attention be paid to the gonad dose; for external sources of ionizing radiation, the gonad dose may be taken to be almost the same as the whole-body dose.

In 1935, the ICRP referred to the necessity for protecting internal organs, "particularly the generative organs," from exposure to radiation but made no specific numerical recommendations. Such recommendations were, however, included in the 1954 NCRP report, where an MPD of 0.3 rem (300 mrems)/week (i.e., any seven consecutive days) was suggested for the gonad dose to occupationally exposed individuals. Larger doses would be permitted, however, to certain other parts of the body.

An important development in radiation protection standards as relating to the gonad dose resulted from the publication in 1956 of the report of the BEAR Committee of the NAS-NRC, entitled *The Biological Effects of Atomic Radiation*. This report was partly concerned with the genetic effects of ionizing radiation, and the conclusions were in substantial agreement with those reached at about the same time by the Medical Research Council in the United Kingdom. Bearing in mind that radiation workers constitute only a small proportion of the population, the BEAR Committee recommended that the total dose to the gonads received during the first 30 years of life should not exceed 50 rems,[k] and that between 30 and 40 years, the additional dose from occupational exposure should not exceed another 50 rems. The reason for choosing the ages of 30 and 40 years is that these represent the ages of the parents when, on the average, about half and nearly all, respectively, of their children are conceived.

Before making these recommendations for occupationally exposed individuals, the BEAR Committee had examined the records of the AEC and its contractors for the doses actually received by radiation workers. It was found that, as a result of the precautions taken to avoid all unnecessary exposures, very few individuals received more than one-tenth of the whole-body MPD, which was 15 rems/year at the time. In other words, for the great majority of people

[j]It has been estimated (by L. Ehrenberg *et al.* [1957]) that as much as 50 percent of the mutations that occur normally in "civilized" man may be due to an increase in the temperature of the male gonads that results from wearing trousers!

[k]The original recommendations were stated in terms of roentgens, but the value in rems is more significant.

who were occupationally involved with radiation sources, the annual whole-body dose was less than 1.5 rems.[l] Hence, it appeared that there would be no difficulty in implementing the recommendations based on the expected genetic effects of radiation.

To reduce the frequency of deleterious mutations and of potential delayed somatic injuries, such as leukemia, and other forms of cancer (Chapter 7), the NCRP suggested in 1956 that the whole-body MPD for those occupationally exposed be decreased by a factor of 3 and that it be expressed on an annual basis. The MPD for radiation would thus be 5 rems/year to the whole body (including the gonads).

In order to allow for the possibility of occasional small annual overruns, while complying with the recommendations of the NAS-NRC Committee, the NCRP expressed its recommendations in 1957 (published in 1958) in a somewhat different manner. The change was also adopted by the ICRP, so that it became the international standard for the protection of occupationally exposed individuals. The proposed MPD was such that the whole-body dose accumulated by N years of age should not exceed $5(N - 18)$ rems.[m] This formula implies a maximum average dose of 5 rems/year to the gonads while occupationally exposed. Thus, the accumulated gonad dose would not be more than 60 rems by 30 years of age and 110 rems by 40 years of age. These values are close to the 50 and 100 rems, respectively, recommended by the 1956 BEAR Committee.

In 1958, new information, which has been confirmed by subsequent studies, became available concerning the genetic effects of radiation in mice (Chapter 7). This prompted the NAS-NRC to reevaluate its 1956 recommendations based on earlier genetic data. In a report issued in 1960, the BEAR Committee recognized that the genetic effects of radiation were probably significantly less than had been thought earlier. That is to say, in setting radiation protection standards, the genetic hazard had been overestimated. Nevertheless, it was decided that, since the existing MPD standards for radiation workers could be met, they should be retained.

The main changes made since 1934 in the NCRP recommendations for maximum permissible whole-body (and gonad) radiation doses for occupationally exposed individuals are given in Table 5-IV. The ICRP values are essentially the same and need not be repeated. The various dates are those when the recommendations were made but, when different, the dates of official publication of the reports are noted in parentheses.

The last column, giving the approximate average weekly dose equivalent, shows that the whole-body MPD has been decreased over the years. But this decrease is not the result of any evidence of adverse effects. The changes were made partly in the general interest of reducing any somatic risks that might exist and partly to decrease the probability of deleterious genetic mutations. It may be noted, too, that x-ray tubes in the 1930s generally operated at relatively low voltages so that the radiations could not penetrate far into the body. Hence, the permissible exposure of 0.5 R/week was probably substantially less than a dose equivalent of 0.5 rem to internal tissues.

[l]During the years 1973 through 1975, the average was close to 0.8 rem.

[m]Individuals less than 18 years of age are not usually subject to occupational radiation exposure. If they are, the radiation protection standards are the same as for the general population (p. 137 *et seq.*).

TABLE 5-IV
MAXIMUM PERMISSIBLE WHOLE-BODY DOSES FOR OCCUPATIONALLY EXPOSED PERSONS

Date	Recommendation	Approximate Average Weekly Equivalent
1934	0.1 R/day (measured in air)	0.5 R
1949 (1954)	0.3 rem in any seven consecutive days	0.3 rem
1957 (1958)	$5(N - 18)$ rems accumulated at N years of age	0.1 rem

Summary of Maximum Occupational Doses

Radiologists and others concerned with the biological uses and effects of ionizing radiations have long known that the consequences of doses restricted to certain parts of the body are much less than when the whole body receives an equal dose. Furthermore, some tissues are known to be more sensitive to radiation than others. Hence, in 1949 both the NCRP and the ICRP suggested different MPDs for the whole body (and gonads) and for certain specific parts of the body.

The radiation protection guides proposed by the FRC in 1960 for occupationally exposed individuals were essentially the same as the NCRP and ICRP MPDs. For the protection of people employed in licensed nuclear plants, the AEC/NRC has promulgated standards, published in 10CFR20, that are similar to, but not identical with, those proposed by the other organizations. The latest recommendations of the NCRP, FRC (now EPA), and 10CFR20 are summarized in Table 5-V. All values are, however, subject to revision from time to time. Furthermore, the paramount consideration, that radiation doses be kept as low as is reasonably achievable (p. 159), should always be borne in mind.

During 1979, the NRC proposed to replace the accumulated whole-body dose limit in 10CFR20 by a more restrictive annual limit of 5 rems. In other words, the limit of 5 rems per annum averaged over the years would be replaced by an absolute limit of 5 rems/year. By March 1980, no final action had been taken on the proposed change.

Attention may be called to the MPD to the thyroid gland. If radioactive isotopes of iodine should enter the body, the dose to this gland could be significantly larger than the whole-body dose. Such a larger dose would be permissible, for a single organ, provided the whole-body dose does not exceed the limit in Table 5-V.

Standards for Alpha-Particle Emitters

Standard for Radium. Alpha-particle emitters outside the body are not a significant (external) hazard because these particles are so readily absorbed that they cannot penetrate the unbroken skin. However, because their energy is dissipated within a short distance, alpha particles have a large quality factor (p. 123). In other words, a certain absorbed dose (in rads) of alpha particles is much more biologically effective than the same absorbed dose of beta particles

TABLE 5-V
SUMMARY OF MAXIMUM OCCUPATIONAL DOSES

	NCRP (1971)	FRC (1960/1961)	AEC/NRC (10CFR20)
Whole body			
Annual dose	5 rems		
Accumulated dose	$5(N - 18)$ rems at N years of age	$5(N - 18)$ rems at N years of age	$5(N - 18)$ rems at N years of age[a]
Dose per quarter	3 rems	3 rems	1¼ rems[b]
Gonads, active blood-forming organs, or lens of eye	Same as above	Same as above	1¼ rems/quarter
Head and trunk		Same as above	Same as above
Thyroid gland	15 rems/year (5 rems/quarter)	30 rems/year (10 rems/quarter)	
Skin of whole body	15 rems/year	Same as above	7½ rems/quarter
Hands	75 rems/year (25 rems/quarter)	75 rems/year (25 rems/quarter)	18¾ rems/quarter
Forearms	30 rems/year (10 rems/quarter)	Same as above	Same as above
Feet and ankles		Same as above	Same as above
Other tissue or organ	15 rems/year (5 rems/quarter)	15 rems/year (5 rems/quarter)	

[a]The accumulated dose may be replaced by an annual dose limit of 5 rems.

[b]A whole-body dose of 3 rems/quarter is permitted provided the accumulated dose does not exceed $5(N - 18)$ rems.

or gamma rays. For this reason, alpha-particle emitters may constitute an important potential hazard if they should enter the body.

The need for setting protection standards for alpha-particle emitters became apparent in the 1930s when a number of persons who had ingested radium in the course of their work as luminous dial painters developed bone cancers. The element radium is similar chemically to calcium, an important constituent of bone, and when radium enters the bloodstream it tends, like calcium, to deposit in the skeleton. The continuous emission of alpha particles, with their high quality factor, then causes cell damage, which may lead to cancer in some individuals.

To deal with this problem, an Advisory Committee on the Safe Handling of Radioactive Luminous Compounds was appointed in the United States. After studying the available information (p. 188) relating the incidence of bone cancer in radium workers to the amounts of radium accumulated in the body, the committee recommended in 1941 that a maximum permissible body burden be set at 0.1 microgram (μg)[n] of radium-226 (and its decay products). This standard, which was accepted by the NCRP and the ICRP, has been reviewed on several occasions, but has remained unchanged.[o]

[n]A microgram is a one-millionth part of a gram.

[o]The maximum permissible body burden must not be exceeded during the individual lifetime, even if there is repeated exposure to the radioactive material (p. 136).

The equivalence of the amount of radium in the body to the dose in rems is not well established. If it is assumed that 0.1 μg of radium-226 is distributed uniformly throughout the skeleton of a person weighing 70 kg (154 lb), the absorbed dose would be 3 rads (i.e., 300 ergs/g of bone) per year. The quality factor of the alpha particles is probably about 10. Hence, the protection standard for occupational exposure to radium would represent a dose equivalent of 30 rems/year. This is the maximum permissible bone dose recommended by the ICRP.

At the present time, the maximum permissible body burden for radium is expressed in terms of the activity—that is, the rate of alpha-particle emission, expressed in curies (Ci) (see below) rather than of the mass in grams. One reason for this change is that the rate of energy absorption in the body (i.e., the dose in rads) is related to the rate of particle emission and the energy of the particles. The energies of alpha particles from radioactive sources of interest are much the same; hence, the rate of energy absorption is essentially dependent on the activity. The protection standard for a given element in terms of activity is consequently applicable to all alpha-emitting isotopes of that element.

The unit of activity that hitherto has been in common use is the *curie*.[p] It is defined as the activity of a quantity of radionuclide that decays at a rate of 3.7×10^{10} nuclear disintegrations per second (dis/s). This number was chosen because it is (approximately) the disintegration rate of 1 gram of radium-226. Hence, a mass of 0.1 μg of radium-226 is equivalent to an activity of 0.1 microcurie (μCi) of any alpha-emitting isotope of radium, where a microcurie is a one-millionth part of a curie. The maximum permissible body (or skeleton) burden of radium is thus 0.1 μCi.

Standards for Transuranium Elements. The protection standards for radium provide the basis for standards for other bone-seeking elements. These include plutonium and other transuranium elements found in spent nuclear fuel. The term *transuranium* refers to the man-made elements that are heavier—that is, they have higher atomic numbers (p. 8)—than uranium. The group includes neptunium (atomic number 93), plutonium (94), americium (95), and curium (96). Various isotopes of these elements are produced in nuclear reactor fuel, starting with uranium, as a result of a series of neutron captures and radioactive decays (p. 266). With the exception of plutonium-241, which is a beta-particle emitter, all the isotopes of the aforementioned elements with appreciably long half-lives decay by the emission of alpha particles.

The transuranium elements have similar chemical properties and appear to behave in the same way in the bodies of experimental animals. Since plutonium is more common and has been studied more extensively than the other transuranium elements, the following discussion is mainly concerned with plutonium.

In 1944, soon after plutonium-239 became available in quantity, tests showed that when this substance was ingested by animals, part of the plutonium entering the bloodstream was deposited in the skeleton. Some was also retained in the liver, but the bone deposits appeared to be more significant. Comparison

[p]The SI unit of activity is called the *becquerel*, which represents a disintegration rate of 1/s. One curie is equivalent to 3.7×10^{10} becquerels.

with radium led to a preliminary proposal of a maximum permissible body (skeleton) burden of 0.3 μCi of plutonium-239.

Later, however, it was found that plutonium is deposited in a different, more sensitive portion of bone than radium. Consequently, the maximum permissible burden was decreased to less than that for radium. Since 1949, the internationally accepted value for the maximum permissible activity of alpha-emitting isotopes of plutonium in the skeleton of occupationally involved persons has been 0.04 μCi.[q] This is also the maximum permissible skeleton burden for other alpha-emitting transuranium isotopes. An activity of 0.04 μCi of any of these isotopes is estimated to correspond to a dose equivalent of roughly 30 rems/year.

Plutonium and other transuranium elements can also enter the body by the inhalation of air containing particles of a chemical compound—for example, plutonium dioxide. Some of the particulate matter entering the lung is removed rapidly, but the remainder, roughly one-third of the total inhaled, which is deposited in the deeper parts of the lung, may be retained for several years. There is a gradual transfer of the transuranium material to the adjacent lymph nodes, but it appears from animal studies that the deposit in the lung is the more important biologically. It is desirable, therefore, to establish a maximum permissible burden for transuranium activity in the human lung.

The present occupational standard for radiation exposure of the lungs, as given in the category of "other tissue or organ" in Table 5-V, is 15 rems/year. Based on the assumption that the alpha-particle energy from transuranium isotopes is absorbed evenly throughout the lung and that the alpha particles have a quality factor of 10, the activity corresponding to a lung dose of 15 rems/year is 0.016 μCi. This is currently the recommended maximum lung burden for occupational exposure to transuranium alpha-particle emitters. For the few beta emitters, such as plutonium-241, the permissible maximum would be at least 10 times as large.

Nonuniform Distribution of Alpha Emitters. On several occasions, dating back to before the discovery of plutonium, the standards-setting agencies have considered the possibility that the biological effects of an alpha-particle emitter, when distributed nonuniformly in a given tissue or organ, might be quantitatively different than when uniformly distributed. Repeatedly, the conclusion has been reached that evaluation of the absorbed dose by averaging the alpha-particle energy deposition over the whole organ, as is done, for example, in calculating the maximum permissible lung burden, is a reasonable procedure. That is to say, in estimating the biological effect, there is little difference between uniform and nonuniform distribution of an alpha-particle emitter.

There has been, nevertheless, a revival of interest in the so-called "hot-particle" hypothesis in connection with insoluble plutonium (as dioxide) particles in the lung. When particulate matter containing plutonium or other transuranium element is inhaled, the particles are deposited nonuniformly and may then aggregate as a result of cellular action, thereby increasing the nonuniform distribution.

Alpha particles from transuranium elements have a range of 40 to 45 micrometers (μm) (1 μm = 1 millionth part of a meter) in soft body tissue. Hence.

[q]For the beta-emitting plutonium-241, the maximum permissible activity is 0.9 μCi.

all the energy is deposited (and absorbed) within a short distance of the plutonium (or other) aggregate in the lung. The radiation dose (in rads) in the immediate vicinity of an aggregate would then be many times greater than the dose based on the assumption that the alpha-particle energy is absorbed uniformly over the whole lung. It is then assumed that such "hot particles" would constitute a greater risk of incidence of lung cancer than a uniform distribution of the same amount of plutonium. If this were indeed the case, the maximum permissible lung burden of 0.016 μCi of transuranium isotopes would be too high.

A substantial body of experimental animal data, however, is now available to refute this speculation. The results indicate that transuranium elements in particulate form in the lung are not a greater hazard than the same amount of the alpha emitter distributed uniformly. In fact, in many instances it appears that nonuniform distribution leads to a lower incidence of lung cancer in test animals. There are several possible reasons for this difference.

Calculations show that, except for extremely small particulates, fewer lung cells would be exposed to alpha particles when the plutonium is distributed nonuniformly than for a uniform distribution. Furthermore, in nonuniform distribution, much of the alpha-particle energy—that is, the large dose in the vicinity of a "hot particle"—is wasted in killing the cells. Once killed, the cells cannot become cancerous. It is possible that lung cancer incidence is related to the number of cells that interact with radiation but survive. This number is expected to be smaller when the distribution of the alpha-particle emitter is nonuniform.

Human Experience. Many people have been working with plutonium in the nuclear weapons program for more than 30 years, but there is no clear evidence of cancer arising from plutonium deposited in the body of a human being. In one case, a worker had small particles of metallic plutonium-239 (with an activity of 0.005 μCi) embedded in his hand for over four years. He developed a small lesion, smaller than a pinhead, that exhibited changes having a "similarity to known precancerous epidermal cytolic changes."

A group of 26 men who had worked at the Los Alamos Scientific Laboratory (LASL) during 1944 and 1945 were selected for study because they had the highest body burdens of plutonium. Except for one who died from a heart attack in 1959 and another who died as the result of an accident in 1975, the medical history of these men has been followed regularly since 1952-53. The plutonium body burdens in 1977, as estimated from urinalysis, ranged from 0.007 to 0.23 μCi; the initial values were undoubtedly larger.

Consideration of the working conditions in 1944-45 and of measurements of plutonium levels in the nasal passages of workers made at the time suggests that inhalation was the principal mode of exposure. This is confirmed by postmortem tissue analysis of the individual who died in 1975; more than 70 percent of the plutonium body burden of 0.019 μCi was found in the lung and adjacent lymph nodes.

The latest study, completed in 1978, showed that there had been no deaths from cancer. Moreover, no cases had been diagnosed, except for two skin cancers that do not appear to be related to plutonium exposure. Both skin cancers were excised in the early 1970s and have not reappeared. The observed diseases and physical changes in the group were characteristic of a male population with ages in their fifties and sixties. There was no evidence suggesting that any adverse health effects had resulted from the more than 30-year exposure to internally deposited plutonium.

Another study involved 224 males who had worked with plutonium at LASL since the middle and late 1940s and who had body burdens of 0.010 μCi or more at the end of 1973. Mortality data through June 1976 showed a total of 33 deaths from all causes, compared with 61 expected in the same age group. There were seven cancer deaths of all types, compared with 11 expected. Only one death was ascribed to lung cancer, although three were expected, and there were no reported cases of bone cancer.

To obtain more information on the possible health effects of plutonium, a comprehensive epidemiological study has been initiated at LASL. The study is to include more than 20 000 former and present workers at the following United States government facilities where plutonium is (or was) handled: LASL, Mound Laboratory, Rocky Flats Plant, Oak Ridge Plants, Savannah River Plant, and Hanford Plant.

The total includes about 4800 who have records of plutonium exposure and about 9000 who worked with plutonium but have no recorded exposure; the remainder worked at the installations but were not exposed to plutonium. Genetic effects are to be included in the study, and plans are being made to extend it to the wives of male plutonium workers. Since 85 percent of the subjects are still alive, several years must elapse before significant conclusions can be drawn.

Maximum Permissible Concentrations in Air and Water

The amounts of radioactive materials that can enter the body by way of air and water are limited by the recommended *maximum permissible concentrations* (MPCs) for various radioactive species (or radionuclides). The subject of MPCs of radionuclides in air and water was discussed at a tripartite conference among representatives of the United States, the United Kingdom, and Canada held in 1949 and 1950. At the meeting in 1950, such concentrations for occupationally exposed persons were agreed upon for a few important radionuclides, and the values were published in an appendix to an ICRP report. In 1953, the NCRP issued a more complete list, as did the ICRP in 1955. Revisions and extensions of these lists, now covering about 240 radionuclides, were published by both the NCRP and the ICRP in 1959. They were based on the new MPD (maximum permissible dose) standards that had been accepted in the intervening years.

A brief outline is given here of the essential basis of the principles used in calculating MPCs. In this respect, it should be noted that many elements entering the bloodstream through inhalation and ingestion soon become distributed more or less uniformly throughout the body. Other elements, however, tend to become concentrated in particular organs or tissues of the body; examples are iodine in the thyroid gland and strontium, radium, and transuranium elements in the skeleton. All of the isotopes of a given element behave in the same way in this respect.

In deriving the MPC for air or water for a particular radionuclide, the first step is to determine the *critical organ* or tissue in the body. For internal exposures, it is defined by the NCRP as "the organ [or tissue] in which damage from a given internally deposited radionuclide results in the greatest bodily injury. This is usually, but not always, assumed to be the body organ [or tissue] in which there is the greatest concentration of radionuclide." The thyroid gland, for example, is the critical organ for internal exposure from radioisotopes of iodine. For elements that are distributed fairly uniformly throughout the body, the critical organ is the whole body, including the gonads. In the following discussion, the term *critical organ* will be used, regardless of whether a specific organ or the whole body is critical.

The next step is to calculate the maximum permissible body burden of a given radionuclide in the critical organ. This quantity has been referred to earlier in connection with alpha-particle emitters, but it may be defined in a general sense as the quantity (or activity) of the radionuclide that would expose a body (or critical organ) to its accepted MPD of radiation as given in Table 5-V. It may take some time, usually ranging from days to years, before the maximum permissible burden (and, hence, the MPD) is attained. The dose arising from a specified internal source depends on several factors, including the rate of radioactive decay of the source material, the rate of biological elimination from the body, the nature and energy of the particles (or gamma rays) emitted, and the extent of their absorption in the organ. These factors are taken into consideration in calculating the maximum permissible organ burden. For radionuclides that tend to concentrate in the skeleton (e.g., strontium isotopes), the maximum permissible burdens are equivalent to the accepted value of 0.1 μCi for the bone-seeker radium.

By prescribing a standard man and devising models based on actual observations for the various stages in the transportation of different elements through the body, it has been found possible to relate the maximum critical organ burden, expressed in microcuries, to the rate of intake of many different radionuclides. The intake rate corresponding to the maximum permissible organ burden can then be determined. By assuming that a person consumes water and breathes air at specified rates [r] during a working week (e.g., 40 hours in seven consecutive days), it is possible to calculate the appropriate MPCs in microcuries per millilitre (μCi/ml) (p. 132) for the various radionuclides. These MPCs are such that continuous use of the water or air over 50 years of occupational exposure would result in an annual radiation dose to the critical organ that would not be expected to exceed the accepted MPD for that organ.

When a person regularly drinks water and/or breathes air containing radioactive material, the amount of the latter in the organ increases gradually toward a limiting value. The reason is that, although there is a regular intake into the organ from water and air, there is also a continuous loss through radioactive decay and normal biological elimination. The amount of a specified radionuclide remaining in a particular organ, and hence the radiation dose received from it, will ultimately reach a steady-state (or equilibrium) value when the rate of loss from the organ just balances the rate of intake. If the radioactive species decays rapidly and/or is readily eliminated from the organ, the steady state is closely approached within a relatively short time (e.g., about seven weeks for iodine-131 in the thyroid gland). On the other hand, if radioactive decay and/or elimination is slow (e.g., strontium-90 and plutonium-239 in bone), the steady state may not be reached within a person's lifetime.

[r]The specified rates are 2 litres (2.2 quarts) of water and 20 000 litres of air in 24 h.

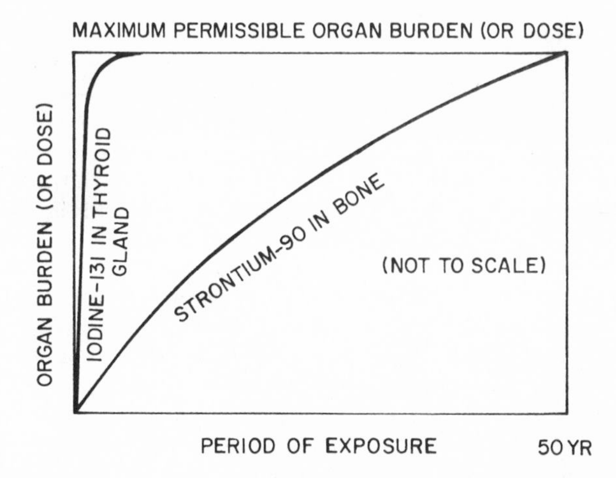

FIG. 5-1. General nature of the increase of critical organ burden (or dose) with time as a result of continued exposure to internal source.

To allow for situations of the latter type, the MPCs are based on the dose to the critical organ that would be attained in 50 years of continuous use of the air or water at the specified concentrations. The dose delivered to the organ would continue to increase, assuming the same rate of intake, but this is more than offset by the annual dose having been much lower for 50 years, as seen in Fig. 5-1. The choice of 50 years as the time for the MPD to be reached is somewhat arbitrary, but it is based on the supposition that no person is likely to be occupationally exposed for a longer period of time.

The values of the MPCs in the tables published by the ICRP, NCRP, and AEC/NRC (in 10CFR20, Appendix B, Table I) refer to individual radionuclides, although air (especially) and water consumed by workers in nuclear installations usually contain several such nuclides.[s] The concentrations of the various radioactive species must then satisfy the requirement that the total annual dose to the critical organ, which may be the whole body, from all the species shall not exceed the MPD within 50 years.

RADIATION STANDARDS FOR THE GENERAL POPULATION

Genetic Considerations

The incidence of genetic effects in a population group arising from exposure to ionizing radiations is related to the total accumulated gonad doses of all individuals in the population group weighted according to the expected future number of children of each individual. Because the general population greatly exceeds the number occupationally exposed, it is desirable, from genetic considerations alone, that the radiation protection standards for the former be set considerably lower than for the latter. Hence, in addition to proposing max-

[s]Drinking water in a nuclear power plant is obtained from outside sources and is the same as that supplied to the general population in the plant vicinity.

imum accumulated gonad doses for occupational exposure (p. 129), the BEAR Committee in 1956 introduced the concept of regulating the average radiation dose to the general population.

In the same year, both the NCRP and the ICRP considered this matter. The NCRP recommended that "radiation or radioactive material outside a controlled area [e.g., the exclusion area of a nuclear power plant (p. 34)] attributable to normal operations within the . . . area shall be such that it is improbable that any individual will receive a [whole-body or gonad] dose of more than 0.5 rem [i.e., 500 mrems] in any year from external radiation." This is one-tenth of the average of 5 rems/year as the MPD for those working within the controlled (or exclusion) area. The NCRP also proposed that "the maximum permissible average body burden of radionuclides in persons outside of the controlled area . . . shall not exceed one-tenth of that for radiation workers."

The suggestion made by the ICRP was that "the genetic dose to the whole population from all sources, additional to natural background, shall not exceed 5 rems plus the lowest practicable contribution from medical exposures." The genetic dose is effectively that which would be accumulated by the gonads (or whole body) during the first 30 years of life, since 30 years is the average age of parents when children are conceived. An accumulated dose of 5 rems (i.e., 5000 mrems) in 30 years corresponds to an average of about 170 mrems/year per individual in the general population. The ICRP suggestion appears, at first sight, to differ from the proposal of the NCRP as given above. The apparent difference was reconciled, however, in 1960 by the Federal Radiation Council, as noted below.

Somatic Considerations for the General Population

The original motivation for specifying lower radiation standards for members of the general population than for those occupationally exposed came from genetic considerations. To provide a clearer understanding of the overall problem of dose limits for the general population, the NCRP in 1958 appointed an ad hoc committee to examine the somatic aspects. Although no specific reduction factor was suggested in its report published in 1960, the committee gave a number of reasons, based on somatic considerations, as to why the dose to individuals in the general population should be less than that for occupationally exposed persons. Some of the reasons are given below:

1. The general population is much larger in number, and if they all receive the same dose as radiation workers, many more individuals would be subject to possible injury.
2. Employment involving a certain degree of occupational hazard is voluntary, and any risk can, in principle, be foreseen.
3. Workers who may be exposed to radiation are (or should be) screened medically before employment, the radiation doses they receive are measured and recorded, and they have periodic medical examinations.
4. Children and embryos, who are particularly sensitive to radiation, are included in the general population, but not among radiation workers.

5. The time of exposure to radiation for occupational reasons (i.e., about 40 hours per week during 45 years) would be less than for the general public (i.e., 168 hours per week for a lifetime).
6. Potential industrial hazards, of which radiation is one, should not be spread beyond the individuals involved in that industry. Otherwise, the risk to the population would be unacceptably high because of contributions from several kinds of industrial hazards.

In the absence of any definite information, one way or the other, the ad hoc committee assumed that even the smallest dose of ionizing radiation might be associated with some risk. In these circumstances, the exposure of the population to any man-made radiation should not be permitted unless there is reason to expect some compensating benefit. It was not possible, however, to make an accurate estimate of the risks, on the one hand, and of the benefits, on the other hand, of specific low levels of radiation. Consequently, the committee recommended that, as a guide, the population dose limit for man-made radiation, excluding medical and dental uses, should be based on (or related to) the average natural background level. As seen in Chapter 6, this ranges from about 80 to 200 (or more) mrems/year in the United States.

Current Radiation Protection Standards

In its 1960 report, the FRC expressed the view that the "population exposure resulting from background radiation is a most important starting point in the establishment of Radiation Protection Guides for the general population." After taking this and other matters into consideration, including the NCRP and ICRP proposals, the FRC recommended that "the yearly radiation exposure [i.e., dose equivalent] to the whole body of individuals in the general population (exclusive of natural background and the deliberate exposure of patients by practitioners of the healing arts) should not exceed 0.5 rem."

The FRC report pointed out that, under certain conditions, the only data available may be related to average doses received in a population group. Since the doses are expected to vary from one individual to another in such a group, it is necessary to make assumptions concerning the relationship between average and maximum doses. The arbitrary assumption made was that "the majority of individuals do not vary from the average by a factor greater than three." On this basis, with a proposed maximum of 0.5 rem (500 mrems)/year for any individual, the FRC recommended that a yearly whole-body dose of one-third of this amount--that is, 0.17 rem (170 mrems) per person—not be exceeded for the average of a population group. This average dose is identical with that suggested by the ICRP to satisfy the requirement of a 30-year accumulated genetic dose of 5 rems.

The current radiation protection standards for the general population, accepted by both the NCRP and ICRP, are in agreement with the FRC guides. The recommended maximum whole-body dose for an individual not occupationally exposed is 500 mrems/year; this is called the *dose limit*. For the population as a whole, the gonad dose should not exceed a yearly average of 170 mrems. In each case, the proposed limits are from all sources of radiation other than the natural background and from medical (including dental) exposures.

TABLE 5-VI
SUMMARY OF STANDARDS FOR THE GENERAL POPULATION
(EXCLUSIVE OF NATURAL BACKGROUND AND MEDICAL EXPOSURES)
(mrems/yr)

Organ or Tissue	FRC (1960-61)	ICRP (1966)	NCRP (1971)
Whole body (including gonads)	500	500	500
Population average	170 (5 rems/30 yr)	170 (5 rems/30 yr)	170
Thyroid, bone	1500[a]	3000[b]	
Skin		3000	
Hands, forearm, feet, ankles		7500	
Other single organs		1500	

[a]This is the FRC maximum for an individual; the population average is one-third (i.e., 500 mrems/yr).

[b]For the thyroid glands of children up to 16 years of age, the ICRP dose limit is 1500 mrems/yr.

The ICRP and the FRC have also suggested protection standards for some specific parts of the body. These are included in the summary in Table 5-VI.

Maximum Permissible Concentrations in Air and Water for the General Population

The sources of the radiation exposures that may result from the operation of a nuclear power installation are almost entirely the radioactive materials contained in the liquid and gaseous effluents. To minimize the dose levels to members of the general public, the plant licensee must make all reasonable efforts to limit the amounts of radioactive materials released to unrestricted areas in the effluents. An appendix to the regulations in 10CFR20 (Appendix B, Table II) lists maximum concentrations of many radionuclides in air and water effluents at the boundary of the restricted area. With a few exceptions, these are essentially the same as the respective MPCs recommended by the NCRP and the ICRP for individual members of the general population.

In most cases, the maximum permissible radionuclide concentrations are based on a maximum critical organ burden of one-tenth that for occupationally exposed persons. Because of the sensitivity of children's thyroid glands to radiation, the maximum thyroid gland burden for iodine isotopes is decreased still further. In addition, allowance is made for the fact that members of the general public may consume water and breathe air containing nuclear plant effluents for 168 hours per week. Consequently, except for radioiodines, the concentrations in Appendix B, Table II of 10CFR20 are generally 1/25 of 1/40 of the corresponding MPCs for occupational exposure in Appendix B, Table I (p. 137). The MPCs in air and water for the general population, like those for occupationally exposed persons, are expressed in terms of microcuries per millilitre (μCi/ml.)

A few typical MPCs for radionuclides of special interest in connection with licensed nuclear operations are given in Table 5-VII. The concentration limits for air and water apply at the boundary of the restricted area of the nuclear

TABLE 5-VII
CONCENTRATION LIMITS AT PLANT BOUNDARY
FOR SELECTED RADIONUCLIDES IN SOLUBLE FORM
(From 10CFR20)
(μCi/ml)

Radionuclide	Water	Air
Tritium (hydrogen-3)	3×10^{-3}	2×10^{-7}
Cobalt-60	5×10^{-5}	1×10^{-8}
Krypton-85 (gas)		3×10^{-7}
Strontium-90	3×10^{-7}	2×10^{-11}
Iodine-131	3×10^{-7}	1×10^{-10}
Cesium-137	2×10^{-5}	2×10^{-9}
Cerium-144	1×10^{-5}	3×10^{-10}
Plutonium-239	5×10^{-6}	6×10^{-14}

plant; they may represent values averaged over a period of time not exceeding one year. The various nuclides (other than krypton-85) are assumed to be present as compounds that are soluble in body fluids, essentially water. Appendix B of 10CFR20 also gives limiting concentrations for radionuclides in insoluble form; these may be slightly larger than, smaller than, or the same as for the soluble form, depending on the nature of the nuclide.

The concentration limits quoted are for individual radionuclides. In practice, air and water usually contain several such nuclides. The concentration of each radionuclide must then be such that the total dose received from all the radionuclides will not exceed the appropriate MPD for the critical organ, which may be the whole body. This requirement can be stated in fairly simple mathematical terms. Supposing the water (or air) contains three radionuclides, A, B, and C, with actual concentrations, in microcuries per millilitre, of C_A, C_B, and C_C, respectively. The MPCs of the three nuclides may be represented by $(\mathrm{MPC})_A$, $(\mathrm{MPC})_B$, $(\mathrm{MPC})_C$. The requirement stated above can then be expressed as

$$\frac{C_A}{(\mathrm{MPC})_A} + \frac{C_B}{(\mathrm{MPC})_C} + \frac{C_C}{(\mathrm{MPC})_C} \text{ not greater than } 1$$

for the critical organ. The latter can usually be readily identified from the nature and concentrations of the various radionuclides present in the effluent.

To apply this formula, it is neccessary to know the concentration of every radionuclide present in the air or water in a significant amount. If the radioactivity arises from fission products, many nuclides are present, and even if those making a negligible contribution to the radiation dose are omitted, analysis for the others is complicated and time-consuming. To simplify the problem, while retaining a factor of safety, the ICRP and the NCRP have proposed MPCs for a mixture of unidentified isotopes (i.e., radionuclides) such as fission products. One such MPC of immediate interest, which is included in 10CFR20, applies to water that contains insignificant quantities of iodine-129, radium-226, and radium-228, as is almost invariably the case for effluents from nuclear power installations. The appropriate MPC is then 1×10^{-7} μCi (total)/ml of water.

In addition to restricting the concentrations (i.e., the quantity per unit volume) of radioactive materials in nuclear plant effluents, the 10CFR20 regulations provide for a limitation on the total quantity of such materials that may be released in air and water over a specified time period. This could be done "if it appears that the daily intake of radioactive material from air, water, or food by a suitable sample of an exposed population group [in an unrestricted area], averaged over a period not exceeding one year, would otherwise exceed the daily intake resulting from continuous exposure to air or water containing one-third the concentration of radioactive materials specified in Appendix B, Table II."

Continuous use of air or water at the concentrations in this table would lead (in the steady state or after 50 years, whichever comes sooner) to the annual dose limit to the critical organ. If the latter is the whole body, the dose limit would be 500 mrems/year. The limitation stated in the preceding paragraph implies that, for the general population living near a licensed nuclear plant, the average individual radiation dose from air, water, or food should not exceed one-third of the appropriate dose limit. This is consistent with the FRC and other recommendations described earlier.

Note that radiation protection standards in 10CFR20 indicate limiting (or maximum permissible) conditions for the operation of nuclear plants licensed by the AEC/NRC. As a result of precautions taken by licensees to reduce the amounts of radioactive materials discharged in effluents, these quantities have in most cases been only a few percent of the permissible values. Also, 10CFR20 requires that the effluents from nuclear plants be kept "as low as is reasonably achievable." Numerical guides have been set for new plants that will assure that the dose to nearby residents will be kept to a few percent of natural background (p. 145).

EPA Standards for the Uranium Fuel Cycle

When the EPA was established in 1970, it became responsible for setting radiation standards for the protection of the public. The responsibility for enforcing these standards rests, however, with other agencies, such as the AEC/NRC for licensed nuclear plants and the U.S. Department of Energy for its contractor facilities. Before proposing possible revisions in the existing FRC standards, as given in Table 5-VI, the EPA plans to make a study of each major activity contributing to public radiation exposure.[t] As a first step in this direction, standards were promulgated in 1977 for protection against radiations associated with the uranium fuel cycle (40CFR190).

The term *uranium fuel cycle* as used in the EPA standards refers to "all facilities conducting the operations of milling uranium ore, chemical conversion of uranium [to fluoride], isotopic enrichment of uranium, fabrication of uranium [dioxide] fuel, generation of electricity by a light-water-cooled [i.e., LWR] nuclear power plant using uranium fuel, and reprocessing of spent uranium fuel, to the extent that these directly support the production of electrical power . . ." Mining, operations at waste disposal sites, transportation,

[t]The standards in Table 5-VI remain in effect pending the completion of these studies.

reuse of non-uranium recovered and by-product materials are excluded from the uranium cycle.

In developing radiation standards for the uranium fuel cycle, the EPA made an effort to strike a balance between the need to minimize the public health risk, on the one hand, and the effectiveness and cost of the technology required to mitigate these risks, on the other hand. The standards limit the annual dose equivalent to the whole body to 25 mrems, to the thyroid to 75 mrems, and to any other organ to 25 mrems as a result of exposures to planned discharges of radioactive materials (except radon and its daughters) and to direct radiation from uranium fuel cycle operation. These standards were to become effective on December 1, 1979, except that for doses arising from operations associated with the milling of uranium ores the effective date was set at December 1, 1980.

The dose equivalents referred to above would be almost entirely due to radiation from fission products of short and moderate half-lives. In addition, however, fission and other reactor product radionuclides of long half-lives are generated in the uranium fuel cycle. Because they decay slowly, these nuclides will tend to accumulate in the biosphere and ultimately may produce substantial radiation exposures, both internal and external. To prevent this accumulation, the EPA has set limits on the total quantities (or activities) of certain radionuclides that may be allowed to enter the environment from the entire uranium fuel cycle per 1000 megawatt-years of electrical energy production. (A typical nuclear power plant, with a 1000 MW electrical capacity, operating at full capacity for 365 days would produce 1000 MW-years of electrical energy.) The limits are 0.5 μCi of combined plutonium-239 and other alpha-emitting transuranic nuclides with half-lives greater than 1 year generated after December 1, 1979, and 50 000 Ci of krypton-85 and 5 μCi of iodine-129 generated after January 1, 1983.

Bibliography—Chapter 5

American Nuclear Society, Public Policy Statement. "Occupational Radiation Dose Limits for Nuclear Facilities." *Nuclear News* 23, no. 2 (1980): 130.

Bair, W. J.; Richmond, C. R.; and Wachholz, B. W. *A Radiological Assessment of the Spatial Distribution of Radiation Dose from Inhaled Plutonium* (WASH-1320). Washington, D.C.: U.S. Atomic Energy Commission, 1974.

Bair, W. J., and Thompson, R. C. "Plutonium: Biomedical Research." *Science* 183 (1974): 715.

Brodine, V. *Radioactive Contamination.* New York: Harcourt, Brace, Jovanovich, Inc., 1975.

Comar, C. L. *Plutonium Facts and Inferences* (EA-43-SR). Palo Alto, Calif.: Electric Power Research Institute, 1976.

Eisenbud, M. "Radiation Standards and Public Health." *Nuclear Safety* 12 (1971):1.

Environmental Policy Association. "Radiation Standards and Public Health." *Proceedings of a Congressional Seminar on Low-Level Radiation.* Washington, D.C., 1978.

Federal Radiation Council. *Background Material for the Development of Radiation Protection Standards* (Report No. 1), 1960.

Hall, E. J. *Radiation and Life.* Elmsford, N.Y.: Pergamon Press, Inc., 1976.

International Atomic Energy Agency. *Proceedings of the Seminar on Radiological Safety Evaluation of Population Doses and Application of Radiological Safety Standards to Man and the Environment.* Vienna: International Atomic Energy Agency, 1975.

International Commission on Radiological Protection. "The Evaluation of Risks from Radiation." *Health Physics* 12 (1966): 239. See also, *Health Physics* 17 (1969): 389.

______. *Radiation Protection—Recommendations of the International Commission on Radiological Protection* (ICRP-26), 1977. See also, *Nuclear Safety* 20 (1979): 330.

McBride, J. P., *et al.* "Radiological Impact of Airborne Effluents of Coal and Nuclear Plants." *Science* 202 (1978): 1045.

Morgan, K. Z. "Adequacy of Present Radiation Standards." *The Environmental and Ecological Forum 1970-1971* (TID-25857), p. 104. Oak Ridge, Tenn.: U.S. Atomic Energy Commission, Division of Technical Information, 1972.

______. "Ionizing Radiation: Benefits Versus Risks." *Health Physics* 17 (1969): 539.

Morgan, K. Z., and Struxness, E. G. "Criteria for the Control of Radioactive Effluents." *Proceedings of the Symposium on the Environmental Aspects of Nuclear Power Stations*, p. 211. Vienna: International Atomic Energy Agency, 1971.

Morgan, K. Z., and Turner, J. E., eds. *Principles of Radiation Protection.* New York: John Wiley and Sons, Inc., Publishers, 1967.

National Bureau of Standards. "Maximum Permissible Body Burdens and Maximum Permissible Concentrations of Radionuclides in Air and Water for Occupational Exposure." *National Bureau of Standards Handbook 69.* Washington, D.C., 1959.

National Council on Radiation Protection and Measurements. *Alpha-Emitting Particles in Lungs* (Report No. 46), 1975.

______. *Basic Radiation Protection Criteria* (Report No. 39), 1971.

______. *Review of the Current State of Radiation Protection Philosophy* (No. 43), 1975.

Richmond, C. R., "Current Status of the Plutonium Hot-Particle Problem." *Nuclear Safety* 17 (1976): 464.

Taylor, L. S. *The Origin and Significance of Radiation Dose Limits for the Population* (WASH-1336). Washington, D.C.: U.S. Atomic Energy Commission, 1973.

______. "Principles of Radiation Protection—An Exercise in Judgment." *Proceedings of the Symposium on Radiation Safety and Protection in Industrial Application* (FDA No. 73-8012). Washington, D.C.: U.S. Department of Health, Education and Welfare, 1973.

______. "Radiation Protection Standards." *CRC Critical Reviews in Environmental Control* 2 (1971): 81, 147.

U.S., *Code of Federal Regulations*, Title 10 (Energy): Part 20. Standards for Protection Against Radiation. Washington, D.C.

U.S., *Code of Federal Regulations*, Title 40 (Protection of the Environment). Chap. 1. Environmental Protection Agency, Subchap. F (Radiation Protection Programs): Part 190. Environmental Radiation Protection Standards for Nuclear Power Operations.

U.S., Department of Health, Education, and Welfare. "Report of the Interagency Task Force on the Health Effects of Ionizing Radiation." Washington, D.C., 1979.

U.S., Environmental Protection Agency. *Health Effects of Alpha-Emitting Particles in the Respiratory Tract* (EPA-520/4-76-013). Washington, D.C., 1976.

6

The Radiation Environment: Natural and Man-Made

NATURAL BACKGROUND RADIATIONS

Introduction

Environmental radiation is made up partly of the natural background radiations, to which human beings have always been exposed, and partly of man-made radiations arising from various human activities. Exposure to man-made radiations can occur, in particular, from medical and dental diagnosis and treatment and from nuclear power operations. It will be seen in the course of this chapter that, of these two radiation sources, medical and dental applications of x rays and radioisotopes are by far the more important.

The natural background radiation dose provides a useful reference level of exposure because the human race has evolved and developed in its presence. For example, when the Federal Radiation Council proposed its radiation protection standards for the general population, the natural background dose was treated as a starting point (p. 139). Comparison with the natural background (and its variations) also provides a basis for judging the significance of man-made radiations, especially those associated with the use of nuclear energy in electric power generation.

Sources of the natural background radiations are both internal to and external to the human body and to other living organisms. The external sources are cosmic rays and radioactive species in the ground, in the air, and in building materials. Natural internal sources are radioactive substances present in nature that have entered the body by ingestion of food and drink and by inhalation. In this connection, it should be noted that the elements uranium and thorium and their radioactive decay products are widely distributed in soil and rocks, although generally in proportions of only a few parts per million. Radioactive isotopes of such common elements as potassium and carbon also occur naturally.

External Radiation Sources

The primary cosmic rays consist of positively charged atomic particles (i.e., positive ions), mainly hydrogen nuclei, of high energy that come from outer space and, to a lesser extent, from the sun. When the primary particles enter the upper atmosphere, they interact with oxygen and nitrogen atoms in the air, and, as a result, various secondary particles (high-energy electrons, neutrons, etc.) and radiations (gamma rays) are produced. These particles and radiations are capable of causing ionization, either directly or indirectly. Thus, cosmic rays are an important source of ionizing radiation at and near the earth's surface.

The intensity of the ionizing radiation due to cosmic rays depends on the altitude and, to a lesser extent, on the geomagnetic latitude. Because of atmospheric absorption, the cosmic-ray dose at sea level is about half that at an altitude of 1640 metres (m) (5000 feet). The latitude effect is caused by the earth's magnetic field, which deflects the electrically charged particles in the cosmic rays. As a result, the cosmic radiation increases by 10 to 20 percent, depending on the altitude, in going from the geomagnetic equator to 50-degree geomagnetic (north and south) latitude. At still higher latitudes, the change is negligible. In general, therefore, the cosmic-ray contribution to the natural background radiation is largest at high altitudes and high latitudes. Thus, the cosmic-ray dose equivalent in the United States ranges from about 38 millirems (mrems) per year in Florida to about twice that amount in Wyoming.[a]

External radiation exposure also arises from natural radioactive species present in the soil; these are mainly potassium-40 and uranium and thorium and their decay products. The doses received from gamma rays emitted from the various radionuclides depend on the nature of the minerals in the ground. The radiation levels are higher than average where granite and other minerals containing uranium or thorium are present. For example, the annual doses from the ground are as low as 15 to 35 mrems for the Atlantic and Gulf Coastal plains and as high as 75 to 140 mrems on the Colorado Plateau. Some external radiation may also be received from radioactive elements in building materials. In structures made of stone, concrete, or brick, the radiation dose is generally higher than in nearby wooden buildings.

Internal Radiation Sources

The major source of natural internal radiation is potassium-40. The element potassium is essential to man and to other animals and plants. Plants derive their potassium from the soil, and hence they always contain a definite proportion of the radioactive isotope potassium-40.[b] From plants, the potassium-40 finds its way, through food chains, into man. The potassium content (and hence the content of potassium-40) varies to some extent with age and sex, but it does not depend greatly on the place where the individual lives.

Another internal source of natural radiation is the gas radon (Chapter 8). The isotopes radon-220 and radon-222 are decay products of thorium-232 and uranium-238, respectively; thus, they are always present in the atmosphere. Radon gas entering the lungs is dissolved to some extent in the blood and hence is carried to all parts of the body. The remaining radon decays in the lungs where the solid decay products may be deposited. Several of these products emit alpha

[a]Doses from natural background radiation are variable; hence, somewhat different values are often recorded in the literature. For consistency, data in this section are taken (when available) from the 1972 report of the NAS-NRC Advisory Committee on the Biological Effects of Ionizing Radiations (BEIR Report), referred to on page 118.

[b]The abundance of potassium-40 in potassium in nature is 0.0118 percent.

particles, which have large quality factors (p. 123). The resulting lung dose can vary over a wide range, since it depends on the quantities of thorium and uranium in the ground, the structural material of the buildings in which people live and work, and on the efficiency of the ventilation.

One of the radioactive isotopes of carbon—namely, carbon-14—also constitutes an internal source of radiation. The interaction of neutrons in (secondary) cosmic rays with nitrogen nuclei in the atmosphere results in the formation of this isotope. In the process of photosynthesis, green plants take up carbon dioxide from the atmosphere and produce carbohydrates, such as sugars and starches. These compounds always include a small proportion of radioactive carbon-14, which enters the human body by way of food. Thus, carbon-14 is present in all carbon-containing compounds in the body (e.g., proteins, carbohydrates, fats, enzymes, and nucleic acids). The ratio of carbon-14 to carbon-12, the common stable form of carbon, in the human body, as well as in plants and animals, is about the same as it is in the atmosphere. The amount of carbon-14 in the atmosphere, and hence in all living organisms, has been increased by the testing of thermonuclear weapons. This is a man-made source of radiation that is discussed later in this chapter.

Whole-Body and Gonad Doses

The total whole-body dose from all sources of natural background radiations clearly depends on the location and on the nature of building materials. A building may contribute to the radiation level, but it may also serve as partial shielding from external sources. An approximate breakdown of the *average* whole-body doses contributed by natural radiations in the United States is given in Table 6-I. Actual annual background doses range from about 80 mrems at sea level to more than 200 mrems at higher altitudes and in areas where the ground contains minerals with a larger than average content of uranium and thorium. In some parts of India and Brazil, where the thorium-rich mineral monazite is present in the soil, the natural background doses are several (eight or more) times as large as in the United States.

TABLE 6-I

AVERAGE WHOLE-BODY DOSES FROM NATURAL RADIATIONS AT SEA LEVEL IN THE UNITED STATES

External Sources	mrems/yr
Cosmic rays	44
Uranium, thorium (and their decay products), and potassium-40 in soil and building materials	40
Internal Sources	
Potassium-40, radon (and its decay products), and carbon-14 in the body	18
	102

The background radiation dose to the gonads, which is the significant quantity from the genetic standpoint, is less than the whole-body dose because of partial shielding of external radiation by overlying tissue. The average gonad dose to people in the United States, regardless of age, is estimated to be between 80 and 90 mrems/year. This is also the *genetically significant dose* for the U.S. population.[c]

Localized Radiation Doses

Inhaled radon gas makes only a small contribution to the lung dose, but the dose from the deposited decay products can be significant, though variable.[d] On the average, the alpha-emitting radon decay products contribute about 100 mrems/year, averaged over the whole lung, compared with a total lung dose of 180 mrems/year from all natural radiation sources. The local dose to the segmental bronchioles (i.e., the smaller branches of the bronchial tubes) is estimated to be 450 mrems/year; the significance of this high value is uncertain.

Part of the uranium and thorium and their decay products that have entered the body by way of food and drink is deposited in the skeleton, where it tends to remain. This is particularly true for the isotopes of radium and also for the alpha-emitting decay products polonium-210. As a result of the accumulation of the natural radioactive material, especially the alpha-particle emitters, in the skeleton, the surface of the bone receives a dose of 60 mrems/year; this represents about 50 percent of the total bone surface dose from all natural sources.

Technologically Enhanced Natural Radiation

Technologically enhanced natural radiation arises from natural radiation sources that are normally sequestered in the ground but are released by an industrial activity. One example is radon-222 in natural gas, which is widely used as a fuel in the home and in industry. The concentration of radon in natural gas in the ground is variable, and partial decay occurs during transmission by pipeline. Nevertheless, the amount of radon remaining at the point of consumption is quite significant; it is estimated to be from 20 to 50 $\times 10^{-9}$ microcurie (μCi) per millilitre (ml) of gas.[e] When the gas is burned, the radon is unaffected, and its presence in the air could lead to radiation exposure of the lung. The dose received by an individual depends on whether the gas appliance is vented or not, but the tracheobronchial (or lung) dose could be as high as 50

[c]A strict definition of the genetically significant dose is the product of the average annual gonad dose, D_i, received by each person in age group i, the average child expectancy, P_i, of that age group, and the number of individuals, N_i, in the group (i.e., $D_i \times P_i \times N_i$) summed over all age groups and divided by the sum of $P_i \times N_i$ for all groups. If everyone in the population receives the same (average) gonad dose, as is the case for the background radiation, the genetically significant dose is equal to the average gonad dose.

[d]The data in this and some following sections are taken from *Natural Background Radiation in the United States*, Report No. 45, National Council on Radiation Protection and Measurements, Washington, D.C. (1975).

[e]The maximum permissible concentration of radon-222 in air for continuous breathing is given in Title 10 of the *Code of Federal Regulations* (10CFR20) as 3×10^{-9} μCi/ml.

mrems/year to a person near an unvented kitchen range. The average annual dose, however, is probably not more than a few millirems.

Radon gas is released continuously from many building materials of mineral origin (e.g., concrete, brick, and gypsum), most of which contain small amounts of uranium and its decay products, including radium. This represents another technological enhancement of natural radiation that can affect many people. The concentration of radon in indoor air depends on the degree of ventilation, but it is often in excess of the recommended maximum permissible concentration for continuous breathing. Particularly high radon concentrations, leading to lung doses of 1000 mrems/year (or more), have been measured in buildings where shale or phosphate rock, which contain larger than average amounts of uranium, has been used in structural materials.

The residues or solid *tailings* remaining after the extraction of uranium from its ores contain essentially all the radium originally present in the ore. The decay of the radium leads to the continuous emission of radon from the tailings. A source of radiation that would normally remain in the ground is thus released and can lead to significant radiation exposure of people in the vicinity. The radiation from uranium tailings is considered later in this chapter in connection with nuclear power operations.

MAN-MADE RADIATIONS

Medical and Dental Radiation Sources

Medical diagnostic radiology is the most important source of man-made radiation to which a large proportion of the population is exposed. In addition, smaller exposures result from dental x rays and and the use of radiopharmaceuticals in medical diagnosis.

The frequency of radiographic examinations in the United States has been increasing steadily over the years; according to U.S. Public Health Service data, during 1970 there were some 112 million x-ray visits (in a population of approximately 200 million). In that year, the abdominal dose from medical diagnostics, averaged over the population of the United States, was estimated to be about 70 mrems per person per year. In addition, there is some radiation exposure from the use of x rays in medical treatment. Attention may be called to the fact that the average individual whole-body dose from the natural background and from the healing arts, both of which are excluded from the radiation protection standards, is roughly 170 mrems/year. This is the same as the average maximum acceptable level for members of the general population from all other radiation sources (Table 5-VI).

The use of radioactive pharmaceutical materials for diagnostic purposes, such as thyroid function tests and kidney scans, has been growing steadily. The average whole-body dose from radiopharmaceuticals in the United States in the year 1970 is given in the BEIR report as roughly 1 mrem/year. Assuming no technical changes and the same growth rate pattern as in the past, the average dose was expected to increase by a factor of at least 10 by the year 1980.

The average gonad dose from medical radiation is roughly the same as the average abdominal dose—namely, about 70 mrems/year. However, the genetically significant dose, as defined in the footnote on page 148, is less, chiefly because the actual gonad dose and the average child expectancy vary with the age group. Although the uses of medical radiation have been increasing in recent years, the corresponding genetically significant dose, which was about 30 mrems/year in the mid-1970s, has been decreasing. This has resulted from the realization that all radiation exposures should be minimized, especially exposure of the gonads of individuals who might be expected to have children at any future time.

Global Fallout Radiation Sources

The atmospheric testing of nuclear weapons has resulted in a worldwide distribution of radioactive material referred to as *global fallout.* The fallout reached a maximum after the United States and the USSR had completed extensive atmospheric tests of high-yield thermonuclear devices before the end of 1962. With the signing in 1963 of a treaty by the United States, the United Kingdom, and the USSR banning nuclear weapons tests in the atmosphere, the global fallout decreased. Subsequently, however, there has been an increase of several percent in the weapons residues in the atmosphere as a result of tests conducted by France and the People's Republic of China. The inventory of global fallout is, however, significantly less than it was in 1963.

Radiation doses from fallout can be both external and internal. The external exposure results from gamma-ray emitters deposited on the ground and suspended in the atmosphere. The external dose has decreased substantially over the years because of the decay of the radionuclides of short half-life. The chief source of external radiation from fallout presently is cesium-137 (half-life 30 years).

Internal radiation exposures arise mainly from strontium-90, cesium-137, and carbon-14. The radioactive strontium and cesium isotopes are direct fission products, whereas the carbon-14 is formed in the same way as it is in nature (p. 141), except that the neutrons are generated in thermonuclear (fusion) reactions. The three species mentioned are important because they are produced in significant amounts and also because they have long radioactive half-lives. Strontium-90 is readily deposited in the skeleton, from which it is removed very slowly. On the other hand, cesium-137 and carbon-14 become more or less uniformly distributed throughout the soft body tissue.

Radiation doses from weapons fallout vary from one place to another; average whole-body doses per person in the United States for the years 1963, 1965, and 1969 are given in Table 6-II. The bone dose from strontium-90 is roughly 10 times the whole-body dose from this nuclide, whereas the gonad (or genetically significant) dose is about the same as the total whole-body dose. Some increase in the whole-body and bone doses is expected in the future, mainly from strontium-90.

Substantial amounts of plutonium, possibly some 8 metric tons (MT), have entered the environment as a result of weapons tests in the atmosphere. Part of the plutonium is nuclear explosive material that has escaped fission, and part is the result of neutron capture by uranium-238 of neutrons generated in thermonuclear reactions (p. 17). Because of the worldwide distribution of fallout, the amount of plutonium in the human body is small, but its activity (in frac-

TABLE 6-II
ANNUAL DOSES FROM GLOBAL FALLOUT
(mrems/person)

	1963	1965	1969
External	5.9	1.8	0.9
Internal			
Strontium-90	0.9	1.9	2.1
Cesium-137	4.3	2.3	0.4
Carbon-14	0.3	0.7	0.6
Total	13.0	6.7	4.0

tions of a curie) has been measured. The average plutonium burdens per individual from weapons residues, as determined in 1970-71 in approximately 100 postmortem examinations of members of the general population in the United States are recorded in Table 6-III. The corresponding bone and lung doses are estimated to be about 1 and 0.3 mrem/year, respectively. These are very small in comparison to the bone and lung doses from natural sources of radiation (p. 148).

Miscellaneous Radiation Sources

Because of the increase in cosmic-ray intensity with increasing altitude, a person flying in a jet aircraft from coast to coast across the United States receives an additional radiation dose of almost 5 mrems. (This is strictly natural radiation, but it is usually considered with miscellaneous sources of man-made radiations because the exposure is a consequence of man's action.) Taking into account the number of passenger-miles flown per year in the United States, the average dose over the whole population from air transport is estimated to be about 1 mrem per person per year.

Luminous paints used for watch and instrument dials contain a trace of radioactive material; in the past this was an isotope of radium, but in recent years tritium (the radioactive isotope of hydrogen) and promethium-147 have been used to an increasing extent. The annual whole-body dose to the population of the United States from luminous paint, almost exclusively on watch dials, was estimated to be less than 1.5 mrems in the mid-1970s. This dose is expected to decrease in future years as tritium and promethium-147 replace radium.

TABLE 6-III
AVERAGE HUMAN BODY BURDENS OF PLUTONIUM FROM GLOBAL FALLOUT

Organ	Burden (10^{-6} μCi)
Bone	1.4
Lung	0.3
Liver	1.4
Kidney	0.1
Lymph	0.03
Total	3.23

Color television receivers represent a relatively minor source of man-made radiation. There is a decreasing trend in radiation emissions from these receivers, but the number of sets in use is increasing. The average dose per person in the United States is estimated to remain constant at about 0.1 mrem/year.

Occupational Radiation Exposure

About 0.4 percent of the people in the United States are engaged in occupations in which they are exposed to man-made radiation through the use of x ray machines or various radioactive materials. The majority of these workers are involved in medical, dental, and pharmaceutical activities and in industry other than nuclear power operations. From data collected in the early 1970s, the U.S. Environmental Protection Agency (EPA) estimated the average annual occupational dose per worker to be 210 mrems.

RADIATION EXPOSURES FROM NUCLEAR POWER PLANTS

Introduction

The contribution of nuclear power generation to the average population dose is by no means as important as some of the other man-made radiation sources described above. It is also very much less than the dose from natural radiation background. The radiation exposures associated with nuclear power operations are treated here at some length, however, since this book is concerned with the environmental effects of these operations.

Controlled amounts of radioactive effluents may be released at all stages of the nuclear fuel cycle, from the mining of uranium ores, through the use of enriched uranium as the nuclear reactor fuel, to the final disposal of the waste products (p. 257). There are now, and always will be, more nuclear reactor plants, where electric power is generated, than any other nuclear facilities. Furthermore, there has been a substantial amount of experience with the environmental aspects of these plants. Consequently, the sources of man-made radiation exposures from nuclear power plants will be considered first.

Radiation Exposure Pathways

Radioactive effluents from a nuclear power plant fall mainly into two categories: airborne effluents to the atmosphere and liquid effluents to an adjacent water body. The most important airborne species are tritium and radioactive isotopes of the noble gases (i.e., krypton and xenon) and of iodine. The liquid effluent contains fission products and corrosion and erosion products that have become radioactive as a result of neutron capture.

There are many ways, as seen in Fig 6-1, whereby a person living in the general vicinity of a nuclear plant could be exposed to radiations from the effluents. The more important sources of exposure are summarized below:

External

Submersion in air
Immersion in water (swimming)
Radiation from particles deposited on the ground
Direct radiation from the plant

Internal

Inhalation of air
Ingestion of food (e.g., leafy vegetables, fish, etc.) and water
Drinking milk from grazing cows

The principal radiation exposure pathways from nuclear power plant effluents, as identified by the EPA from calculations and environmental measurements, are shown in Table 6-IV. The radioisotopes of iodine, principally iodine-131 and iodine-133, are important because of their relatively high yield in uranium fission and the marked affinity of the thyroid gland for iodine. This element is deposited in the gland as the compound thyroxine, which is

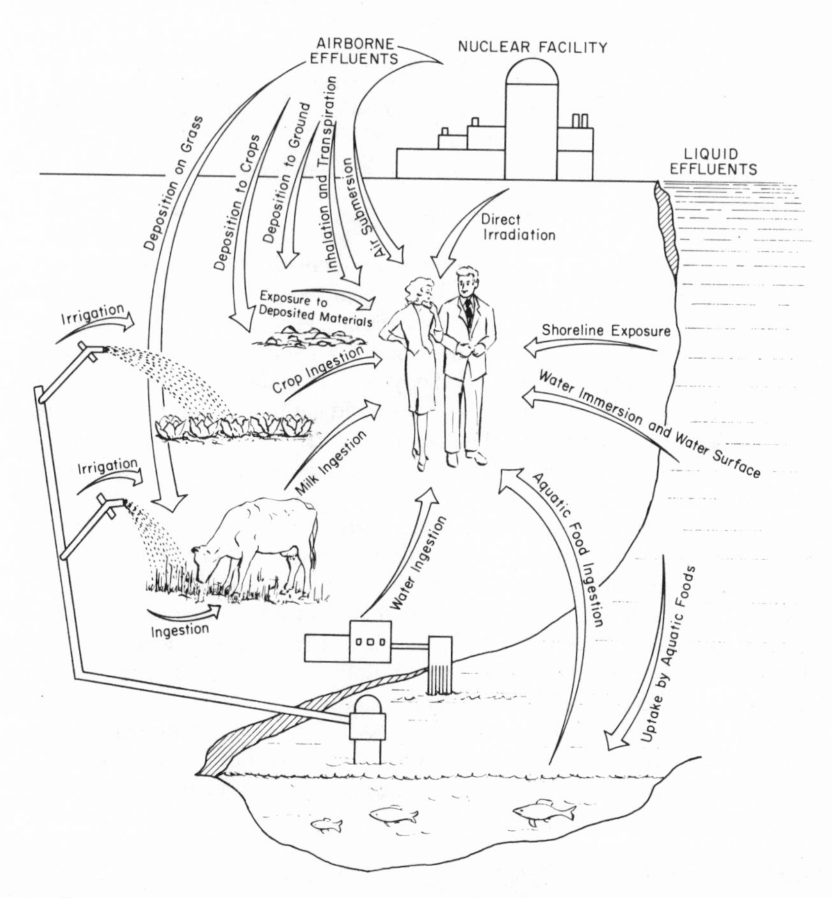

FIG. 6-1. Radiation exposure pathways to man (adapted from WASH-1209, U.S. Atomic Energy Commission, Washington, D.C. [1973]).

TABLE 6-IV
PRINCIPAL RADIATION EXPOSURE PATHWAYS FROM NUCLEAR POWER PLANT EFFLUENTS

Radionuclide	Effluent	Exposure Pathways	Critical Organ
Iodine	Airborne	Ground deposition (external)	Whole Body
		Air inhalation	Thyroid
		Grass → cow → milk	Thyroid
		Leafy vegetables	Thyroid
	Liquid	Drinking water	Thyroid
		Fish (and shellfish) consumption	Thyroid
Tritium	Airborne	Submersion (external)	Skin
		Air inhalation	Whole body
	Liquid	Drinking water	Whole body
		Food consumption	Whole body
Cesium	Airborne	Ground deposition (external)	Whole body
		Grass → cow → milk	Whole body
		Grass → cattle → meat	Whole body
		Inhalation	Whole body
	Liquid	Sediments (external)	Whole body
		Drinking water	Whole body
		Fish consumption	Whole body
Metals (iron, cobalt, nickel, zinc, manganese)	Liquid	Drinking water	GI tract
		Fish consumption	GI tract
Direct radiation from plant		External	Whole body

eliminated slowly. As far as radioiodines are concerned, the thyroid is the critical organ, and the young child is the critical receptor, as seen shortly.

The radioisotopes of cesium, mainly cesium-134 and cesium-137, are also produced in significant amounts in fission. This element becomes fairly uniformly distributed throughout the body, and essentially all the organs, including the gonads, are exposed to both beta and gamma radiations. The chief ways in which cesium can enter the body are by drinking water and by eating fish from the general vicinity of the liquid effluent discharge. The concentration of cesium in fish is discussed later.

Several radioactive isotopes of the noble gases occur in the airborne effluent, and their presence in the atmosphere can lead to an external radiation exposure. The internal (lung) dose from inhalation of these gases is small. Exposure from the short-lived gases, such as xenon-133 and krypton-88, can be minimized by providing adequate holdup in the plant before release to the atmosphere (Chapter 9). However, the possible accumulation in the atmosphere of long-lived krypton-85 may become significant (p. 173).

Tritium, the radioactive isotope of hydrogen, is present in both gaseous and liquid effluents. It can enter the body in various ways, but the tritium dose is usually small. Like krypton-85, tritium may accumulate in the environment (p. 168 *et seq.*).

Exposure from Radioiodines

Some of the radioiodines released from a nuclear plant are present in the airborne effluent. In the course of time, the iodine deposits on various surfaces including leafy vegetables and other food products and grass. There are then three main pathways for the entry of radioiodines into the human body and, hence, into the thyroid gland. These are by (a) inhalation of air, (b) eating leafy vegetables, etc., and (c) consumption of milk from grazing cows and goats.

The radioiodines in the airborne effluents from a nuclear power plant occur partly as elemental iodine and partly as an organic compound, chiefly methyl iodide.[f] In the elemental form, the iodine deposits fairly readily on vegetation and on the ground, the average deposition velocity being roughly 1 centimeter per second (cm/s) (120 feet/hour). The deposition velocity of methyl iodide is less by a factor of at least 1000.

Consequently, when the radioiodines in the effluent are mainly in the elemental form, the principal exposure pathway is from deposited iodine, as described below, but for organic iodide in the effluent, the principal pathway is by direct inhalation. The metabolic behavior of organic iodides in the human body is not well known, but for calculating thyroid doses it is assumed to be the same as that of elemental iodine. This is probably a conservative (i.e., cautious) assumption.

For elemental iodine, the major radiation exposure pathway is likely to be the consumption of milk if there are grazing animals (dairy cows or goats) in the vicinity of the nuclear plant. A cow or goat grazes a large area every day and, as a result, can ingest a substantial quantity of deposited radioiodine. Part of this iodine enters the thyroid gland of the animal and about an equal amount appears in the milk. The cow (or goat) thus has the ability to concentrate in its milk the radioiodine that has been deposited thinly on grass over a substantial area.

The use of this milk results in the accumulation of radioactive (and other) iodine in the thyroid glands of humans. The radiation dose to the thyroid is a maximum for children below about 10 years of age, because of the large quantity of milk consumed and the small mass of the thyroid gland in which the radiation from radioiodines is absorbed. For example, a four-year-old child drinks, on the average, 700 millilitres (ml) (1½ pints) of milk per day (compared with 230 ml for an adult), whereas the mass of the thyroid is about 3 grams (g) (compared to 20 g for an adult). Hence, the critical receptor for the radioiodines from the airborne effluent of a nuclear power plant is usually taken to be the thyroid of the young child. The critical path is then represented by

Iodine in air ⟶ grass ⟶ cow (or goat) ⟶ milk ⟶ child's thyroid.

The Environmental Reports submitted with the applications for a construction permit and an operating license and the Final Environmental Statements prepared by the U.S. Nuclear Regulatory Commission (NRC) staff (see Chapter 3) include calculations of the thyroid (and other) doses expected from normal plant operations. The plant design must be such as to comply with the numerical guides for keeping radioactive materials in the effluent at a level as low as is reasonably achievable. (These guides are described on p. 159 *et seq.*).

[f]Organic iodides are formed in reactor systems by the interaction of iodine, under the influence of radiation, with organic (carbon) compounds, such as traces of lubricating oils and greases, or even with methane gas in air, or with carbon in steel in the presence of water or steam.

Furthermore, when an operating license is issued, the Technical Specifications accompanying the license place a limit on the activity (in curies) of the radioiodines that may be discharged with the airborne effluent. Measurements made by independent health authorities have shown repeatedly that the radioiodine activity in milk from dairies in the general vicinity of nuclear power plants is well below the acceptable level of 10CFR20.

In October 1957, an accident, of a type that could not possibly occur in most power reactors in the United States, resulted in the release of relatively large amounts of radioiodines from a reactor installation at Windscale in England.[g] The milk from cows in the vicinity was confiscated because the activity of iodine-131 was eight times as great as the maximum acceptable level. The concentrations of the other radionuclides, including these of strontium and cesium, were regarded as safe. If the milk had been converted into cheese and allowed to age for three months, the iodine-131 (half-life eight days) remaining would have been insignificant. In vegetables, eggs, meat, and drinking water, the activity of iodine-131 and other radioisotopes was below the prescribed limit.

Radiation Exposure from Fish Consumption

For people living close to water bodies into which liquid effluents from a nuclear power plant are discharged, an important, and possibly major, pathway for radiation exposure could be the consumption of fish (including shellfish). It is well known that fish are able to concentrate certain elements from the environment, either directly from water or through their food sources. The amount of a particular element present in a given mass, say one pound, of the fish is greater than that in a pound of the water in which the fish lives. The ratio of these amounts when a steady state is reached is called the *concentration factor*—or, preferably, the *bioaccumulation factor* (an abbreviation for *biological accumulation factor*)—for the given element in the fish. If radioactive isotopes of the element are present in the water, then the bioaccumulation factor is the same as for the stable isotopes of that element that are already present in the water. Sometimes an element that normally occurs in insignificant amounts in fish (and other organisms), tends to become concentrated in the fish because of its chemical similarity to a biologically essential element. Radioisotopes of cesium are of special interest in this respect because cesium is chemically similar to the essential element potassium.

The bioaccumulation factor for a given element in fish depends on various circumstances. One is the particular chemical form (e.g., soluble or insoluble) in which the element is present in the water. Although the products of fission and of neutron capture are largely in solution in the liquid discharge, some of the elements present may be converted into insoluble oxides by reaction with water, especially if the latter is more alkaline than the plant effluent.

The normal composition of the water also has an influence on the bioaccumulation factor. For example, the factor for a given aquatic species is often

[g]The accident at Windscale was caused by a too rapid, controlled release of energy stored in graphite used as the moderator. Measures were subsequently taken to prevent a recurrence. In the United States very few reactors have graphite moderators, and the operating temperatures are generally high enough to preclude storage of energy in the graphite.

quite different in fresh water than in seawater. If the water contains a substantial quantity of potassium, less cesium will be taken up than if there were little potassium in the water.

A further important consideration is the food habits of fish, and these vary with the time of the year. The aquatic food chain is quite complicated, as will be seen in Chapter 12, and the relative amounts of the several food sources vary with the climatic conditions, the season of the year, and the fish species. Consequently, the bioaccumulation factor is a highly variable quantity. For example, the reported factor for the important element cesium in fish ranges from about 100 to almost 10 000.

There is clearly no single bioaccumulation factor for a given element that is applicable to all fish species at all times; furthermore, there is no certainty that a steady-state condition is attained when radioactive discharges into a body of water are made intermittently, as they usually are. Nevertheless, numbers such as those in Table 6-V, based on steady-state measurements, are commonly used to make estimates of the amounts of various radionuclides present in fish living in water into which liquid effluents are discharged. From these amounts, the radiation doses that may be received by people eating fish caught in this water can be calculated.

To verify the acceptability of the liquid discharges from nuclear power plants, several radiological surveillance studies have been made of fish from water bodies adjacent to these plants. In no case was a significant level of radioactivity found in the fish. The levels were far less than would have prevented people from eating the fish even in much larger than normal quantities. The same was true in two special cases described below.

When the Humboldt Bay (boiling water reactor [BWR]) nuclear power plant in California started operation, the tubes in the feedwater heaters were made of a zinc alloy. As a result, the liquid effluent from the plant contained appreciable, although not hazardous, amounts of radioactive zinc-65. It was known that oyster flesh has an exceptionally large bioaccumulation factor for zinc; consequently, oysters were deliberately cultivated (but not for consumption) in the effluent as it left the plant site. Even when the zinc-65 content of the oyster flesh reached a maximum, the radioactivity would not have constituted a hazard. In the unlikely event that an individual had eaten a substantial meal of these oysters every day, the radiation dose received would have been smaller than is presently regarded as acceptable for members of the general public

TABLE 6-V
STEADY-STATE BIOACCUMULATION FACTORS
IN EDIBLE PORTIONS

ELEMENT	SEA WATER		FRESH WATER	
	Fish	Shellfish	Fish	Shellfish
Cobalt	500	1000	500	1500
Strontium	1	6	40	70
Iodine	10	50	1	25
Cesium	30	20	1000	1000
Tritium	1	1	1	1

(Chapter 5). When the zinc alloy tubes in the feedwater heaters were replaced with stainless-steel tubes, the zinc-65 content of the oysters returned to normal.

Another situation of interest occurred in connection with the Indian Point (pressurized water reactor [PWR]) nuclear power plant No. 1 on the Hudson River in New York. During 1967 and 1968, the level of radioactivity, although well within the limits prescribed by the plant's operating license, was higher than normal because leaks had developed in the steam generator. Analysis of the fish caught in the Hudson River in the vicinity of the plant during the years 1967 through 1969 showed that a person eating one pound of such fish per week would receive a whole-body dose of less than 0.1 mrem/year.

Direct Radiation Exposure

The external radiation dose from immersion of the human body in air or water containing radioactive effluent from a nuclear power plant must be kept small compared to background. But exposure to radiation coming directly from the plant—the last item in Table 6-IV—could be important in some circumstances. Gamma radiation (or "gamma shine") is emitted from waste storage and treatment vessels, from spent-fuel storage tanks, and from other equipment containing radioactive materials. A more significant potential source of direct radiation is the turbine building associated with a BWR.

In BWRs, steam is produced directly from the water in the reactor vessel (Chapter 2). As a result of the action of neutrons on the oxygen in water, radioactive nitrogen-16, a gamma-ray emitter, is present in the steam. Because of its very short half-life, about 7 seconds, much of the nitrogen-16 decays while the steam is in the turbine, accompanied by the emission of gamma rays. This contribution to BWR gamma shine is called "turbine shine." It is not significant in a PWR plant because the steam is generated in a separate vessel.

Where gamma shine might be important, the radiation dose at the plant boundary can be reduced to a few millirems per year by proper shielding procedures within the plant. Such shielding is required in any event for the protection of plant workers. Since the gamma rays come from a fixed location, the dose from gamma shine falls off more rapidly with distance from the plant than do the possible doses from the mobile liquid and gaseous effluents. Hence, the direct radiation doses received by people living beyond a mile or two from the plant are very small.

Radiation Levels As Low As Reasonably Achievable

The U.S. Atomic Energy Commission (AEC), now the NRC, had always recommended that nuclear plants be operated at all times such that the radiation exposure to people outside the plant boundary be kept below the dose limits of the Federal Radiation Council. Current NRC regulations require that the licensee of a nuclear plant "make every reasonable effort to maintain radiation exposures, and release of radioactive materials in effluents to unrestricted areas, as low as is reasonably achievable. The term 'as low as is reasonably achievable' means as low as is reasonably achievable taking into account the state of technology, and the economics of improvements in relation to benefits to the public health and safety, and other societal and socioeconomic considerations, and in relation to the utilization of atomic energy in the public interest" (10CFR20, Paragraph 20.1 [c]).

Furthermore, an application for a permit to construct a nuclear power reactor must "include a description of the preliminary design of equipment to be installed to maintain control over radioactive materials in gaseous and liquid effluents produced during normal operations, including expected operational occurrences . . . . The applicant shall also identify the design objectives, and the means to be employed for keeping levels of radioactive material in effluents to unrestricted areas as low as is reasonably achievable" (10CFR50, Paragraph 50.34a).

After extensive public hearings, including the presentation of oral and written testimony, and the preparation of Draft and Final Environmental Statements by the AEC regulatory staff, numerical guides were issued in April 1975 (by the NRC) for meeting the criteria of radiation exposures as low as is reasonably achievable (10CFR50, Appendix I). These guides refer specifically to light-water-cooled reactors (LWRs)—that is, to PWRs and BWRs. Reactors of this type constitute the great majority of those now installed or under construction in the United States for the generation of electric power. As a result of operating experience with LWRs in recent years, numerical guides have been formulated for radioactive materials in effluents that are compatible with current technology and are practicable from the economic standpoint. Specific guides have not yet been issued for other power reactors or for spent-fuel processing plants. Nevertheless, the general requirement that radiation levels be maintained as low as is reasonably achievable is applicable to all licensed facilities.

Design Objectives for Radiation Levels

The four numerical guides described below have been adopted for meeting the design objectives for radiations from LWR plant effluents. It should be understood that these guides are to be used in assessing the adequacy of a proposed nuclear plant design. The actual radiation doses to the public as a result of plant operation must comply with EPA standards for the uranium fuel cycle, as given in Chapter 5.

Liquid Effluents. The calculated total quantity of radioactive material (above background) to be released from each LWR to unrestricted areas (i.e., outside the plant site) in liquid effluents shall not result in an estimated annual dose (or dose commitment[h]) for any individual in such areas in excess of 3 mrems to the whole body or 10 mrems to any organ. In evaluating the dose (or dose commitment), it is assumed that an adjacent river provides drinking water and that rivers (and other water bodies) serve as sources of fish (including shellfish), unless evidence is provided to prove otherwise.

Gaseous Effluents. The calculated total quantity of all radioactive material (above background) to be released to the atmosphere from each LWR shall not result in an estimated annual *air dose* from gaseous effluents in excess of 10 millirads (mrads) for gamma radiation or 20 mrads for beta radiation at any

[h]The annual dose commitment is the annual dose that eventually would be attained as a result of the retention of radioactive material in the body or critical organ.

location near ground level that could be occupied by individuals in unrestricted areas. A smaller quantity of radioactive material may be specified, if necessary, to prevent an annual external whole-body dose in excess of 5 mrems to any individual in an unrestricted area. On the other hand, if assurance is provided that a larger quantity of radioactive material can be released without the estimated annual dose exceeding 5 mrems to the whole body or 15 mrems to the skin, such larger quantity may be deemed to meet the requirements.

Radioactive Iodine and Particulates. The calculated total quantity of all radioactive iodine and radioactive material in particulate form (above background) to be released in effluents to the atmosphere from each LWR shall not result in an estimated annual dose (or dose commitment) from all pathways of these materials in excess of 15 mrems to any organ of any individual in unrestricted areas. Since the main hazard from radioactive iodine arises from its accumulation in the thyroid glands of children (p. 155), there must be assurance that the annual thyroid dose (or dose commitment) to any child will not exceed 15 mrems.

In estimating the dose (or dose commitment) due to the intake of radioactive material through food pathways, the calculations may be made for locations where such pathways actually exist. Since the situation may change with time, the nuclear plant licensee is required to conduct monitoring and surveillance programs to identify any changes that may result in radiation doses exceeding the design objectives. If such changes were to occur, the licensee would be obligated to take appropriate action by controlling emissions or other aspects of the exposure pathway so as to assure a thyroid dose of not more than 15 mrems/year to any individual in unrestricted areas.

Decrease of Population Dose. Very few people will live close enough to a nuclear power plant to receive the maximum doses associated with design-objective guidelines. However, there may be a substantial population in an area within which the plant effluents have some influence on the radiation dose. An indication of the potential effect of the radiation is then given by the *population dose* within a distance of 50 miles from the nuclear plant. The annual population dose in the given area, expressed in person-rems, is the sum of the actual annual dose (or dose commitment) received by every person living in that area. Alternatively, the population dose is equal to the product of the number of people living in the area and the average annual radiation dose (or commitment) per individual in the population group. One of the purposes of the design objectives is to minimize the population dose to residents within a radius of 50 miles from the plant.

The design-objective guidelines require that the system for treating the radioactive wastes from an LWR, as described in Chapter 8, "shall include all equipment items of reasonably demonstrated technology which can for a favorable cost-benefit ratio effect a reduction in dose to the population reasonably expected to be within 50 miles of the reactor. As an interim measure and until establishment and adoption of better values (or other appropriate criteria) the values of $1000 per total body man-rem [i.e., per person-rem to the whole body] and $1000 per man-thyroid-rem [i.e., per person-rem to the thyroid] (or such lesser values as may be demonstrated to be suitable in a particular case) shall be accepted for use in this cost-benefit analysis."

TABLE 6-VI
CALCULATED ANNUAL INDIVIDUAL DOSES FROM NUCLEAR POWER PLANT EFFLUENTS
(mrems/yr)

LOCATION	PATHWAY	DOSE	
		Whole-Body	Thyroid
	Airborne Effluents		
Nearest land boundary (0.9 km [0.59 mile])[a]	External	0.15	0.15
	Inhalation		0.31
Nearest residents (1.1 km [0.68 mile])	External	0.46	0.46
	Inhalation		0.87
	Leafy vegetables		1.5
Nearest park (1.4 km [0.85 mile])[b]	External	0.018	0.018
	Inhalation		0.035
Nearest milk cow (4.8 km [3.0 miles])	Milk (child)		4.8
	Liquid Effluent		
Discharge region	Fish (including shellfish)	0.003	0.006
	Recreation	4.9×10^{-5}	
Municipal (water intake 9.2 km [5.7 miles])	Ingestion	3.9×10^{-5}	0.002
Nearest park (1.4 km [0.85 mile])	Recreation	1.3×10^{-4}	

[a]Assumes 40 hours/week occupancy.
[b]Assumes 500 hours/year occupancy.

The amount of radioactive material discharged from a nuclear power plant can be steadily decreased by adding more and more stages to the waste treatment system. However, each successive stage is less cost-effective than the preceding one in decreasing the population dose in person-rems. In other words, the cost per person-rem decrease in the population dose becomes greater as the dose is made smaller. Provisionally, the guide requires that additions be made to the radioactive waste treatment system up to the point at which the annualized cost reaches $1000 for a decrease of 1 person-rem/year in whole-body dose and $1000 for a decrease of 1 person-rem/year in the thyroid dose to the population living within 80 kilometres (km) (50 miles) of an LWR power plant.[i]

Compliance with Design Objectives

Table 6-VI, adapted from the NRC's Environmental Statement for a large nuclear power plant to be constructed near Lake Erie, gives the results of calculations showing that the installation is designed to comply with the

[i]The annualized cost includes annual operating and maintenance costs plus an appropriate share of the capital cost of the system.

TABLE 6-VII
CUMULATIVE ANNUAL POPULATION DOSES
FROM NUCLEAR POWER PLANT EFFLUENTS

Distance from Plant (miles)	Cumulative Population in 1980	Annual Population Dose (person-rems)	Annual Average Individual Dose (mrems)
1	250	0.04	0.16
2	1 510	0.13	0.086
5	9 750	0.27	0.028
10	85 700	0.64	0.0075
30	770 000	1.26	0.0016
50	2 750 000	1.82	0.00066

guidelines in 10CFR50, Appendix I, discussed above. It is evident that, as in most nuclear plants, the liquid effluent is a much smaller source of radiation exposure than the airborne effluent. The thyroid dose from leafy vegetables is based on an annual consumption of 18 kg (40 lb) by an adult; it would be smaller for young children. In calculating the dose to the thyroid of the young child from milk, it is assumed that the cows graze for six months of the year and the child drinks 1 litre (2.2 pints) of milk daily.

Estimates of the cumulative population (whole-body) doses from the radioactive effluents within various distances from the nuclear power plant are given in Table 6-VII. In addition, a dose of 14 person-rems/year is expected to the population within 80 km (50 miles) from the plant as a result of transportation of nuclear fuel and wastes to and from the plant. The occupational dose to people working in the plant is estimated to be 1000 person-rems/year at most. These numbers may be compared with a population dose of approximately 400 000 person-rems within an 80 km (50-mile) radius from the natural background radiation.

NUCLEAR FUEL PRODUCTION AND REPROCESSING

Radiation Exposures from Nuclear Fuel Production

The production of nuclear fuel material in the form of enriched uranium dioxide (Chapter 2) involves the following stages:

1. mining the uranium ore
2. milling the ore to concentrate the uranium
3. conversion of the uranium concentrate to uranium hexafluoride
4. enrichment of the hexafluoride in uranium-235 by gaseous diffusion or centrifuge
5. conversion of the enriched hexafluoride into uranium dioxide fuel material.

Radioactive effluents associated with stages 3, 4, and 5 are relatively small; in fact, in these stages chemicals in the effluent are more important. Consequently, the discussion here of radiation exposures from nuclear fuel production is restricted to mining and milling operations. Since effluents from these operations are basically similar, they are considered together.

Radiation Doses from Mines and Mills

The major radioactive effluents from uranium mills and mines are airborne; they consist of the gas radon-222 and dust particles containing uranium-238 and its decay products. The latter include thorium-230 and radium-226, the grandparent and parent, respectively, of radon-222.

It has been estimated that for a given quantity of uranium ore, almost four times as much radon-222 is released to the atmosphere in mining than in milling (p. 164). However, there are more mines than mills processing the mined ore. In the 1970s, there were some 30 open-pit and 120 underground uranium mines in the United States, producing roughly equal amounts of ore, but only about 20 operating mills. Hence, the background radiation dose from radon-222 received by an individual is generally larger near a mill than at the same distance from a mine. The dose from dust particles is also larger near a mill.

The main radioactive exposure pathways to people living in the vicinity of a uranium mine or mill are as follows:

External
- Submersion in air (containing radon gas)
- Airborne dusts and tailings
- Radiation from deposited particles

Internal
- Inhalation of radon gas and dust (mainly lung dose)
- Ingestion of food products (mainly bone dose)

Most uranium mines and mills are located in arid and semiarid areas, where the only local food source is likely to be beef.[j]

Since the radiation dose to the individual is expected to be larger near a mill than a mine, the former will be considered. This dose depends on many variables, including meteorological conditions (e.g., rainfall, wind speed and direction), details of the milling process (e.g., method of extracting the uranium, types of dust filters), and the proportions of wet and dry areas in the tailings pile (p. 222).

The radiation doses in Table 6-VIII were calculated for an individual living either 0.8 or 1.6 km (0.5 or 1.0 mile) from an average installation milling 1800 MT (2000 short tons) of ore daily. (One metric ton is 1000 kg [2205 lb].)The bone dose is due mainly to radium and polonium-210 (a radon decay product); the lung dose arises from radon and its decay products. There is no radioiodine in the mill (or mine) effluent, and so the thyroid dose is essentially the same as (or less than) the whole-body dose.

Many of the basic parameters used in calculating the values in the table are conservative and are likely to maximize the estimated doses. The individual is supposed to spend all his time in the mill vicinity, with no allowance for shielding or for time spent elsewhere. Half the diet is assumed to consist of local-

[j]Uranium is now mined and milled mainly in New Mexico and Wyoming, with smaller amounts in Colorado, Texas, Utah, and Washington. There is also some mining, but no milling, in Arizona.

TABLE 6-VIII
CALCULATED INDIVIDUAL RADIATION DOSES
FROM A URANIUM MILL
(mrems/yr)

DISTANCE km	miles	WHOLE BODY	BONE	LUNG
0.8	0.5	25	250	90
1.6	1.0	5	50	18

ly produced meat; a smaller proportion of such meat would reduce the whole-body and bone doses but would have only a small effect on the lung dose. Furthermore, the mill is taken to be near the end of its normal lifetime when the tailings pile has its maximum size and there is an accumulation of deposited radioactive material on the ground. If these factors are taken into consideration, the average doses during the operating life of the mill are estimated to be about half the values in Table 6-VIII. It is possible that the doses could be reduced still further at a moderate cost by improving the mill's dust filters.

Uranium mills, like other licensed nuclear installations, are required to maintain radioactivity levels in plant effluents as low as is reasonably achievable. But no specific design guidelines, other than 10CFR20 (p. 140), have as yet been established by the NRC.

As already indicated, the radiation dose to an individual living near a uranium mine is expected to be less than for a mill. However, the overall effect of mining operations on the radiation background is estimated to be greater than that of milling. This can be seen in a general way by considering the total activities of the radon-222 released in mining and milling.

It has been estimated that mining a million (metric) tons of 0.1 percent uranium ore results in the release to the atmosphere of 15 000 Ci of radon-222. During the year 2000, assuming 350 LWRs are in operation in the United States, there should be a requirement for about 95 million tons of uranium ore. The total activity of the radon released in that year from mining would then be approximately 1.4×10^6 Ci. However, airborne emissions from milling this uranium ore would contain less than 0.4×10^6 Ci of radon. That is to say, the total amount of this gas released from active uranium mills is almost one-fourth of that from the mines supplying the ore.

In estimating the increase in the radiation background resulting from mining and milling uranium, allowance must also be made for the radon released by the tailings from inactive mills. Measures are being taken to reduce these emissions substantially, but if the tailings piles are assumed to remain uncovered, the total amount accumulated by the year 2000 would emit roughly 0.4×10^6 Ci of radon in that year (and in subsequent years).

Because radon-222 has a relatively short half-life (3.8 days), little of the activity will escape beyond the United States. Hence, it appears that 2.2×10^6 Ci of radon will be added to the nation's atmosphere in the year 2000 as a result of uranium mining and milling operations in that and preceding years. This can be compared with the estimates of 10^8 and 2.4×10^8 Ci of radon that are released to the atmosphere of the United States every year from uranium in the ground.

To be conservative, and assuming that the lower value is correct, it follows that uranium mining and milling should increase the natural background activity of radon-222 by little more than 2 percent by the year 2000.

On page 148, it was stated that the alpha-emitting decay products of radon-222 normally in the atmosphere contribute a lung dose of 100 mrems/year. On this basis, uranium mining and milling operations in support of 350 nuclear power reactors would be responsible for an average lung dose of only 2 mrems/year to people in the United States in the year 2000. The contribution to the total body dose would be less.

When a uranium mine is shut down, the release of radon to the atmosphere becomes small. However, the tailings that have accumulated during the whole period of operation of a mill continue to emit radon for many thousands of years. The contribution to the background radiation dose is small to individuals, except perhaps to those living near the tailings piles. Nevertheless, because of the persistence of the radon emission, the problem of tailings from inactive uranium mills is very important, as seen in Chapter 8.

Radiation Doses from Reprocessing Plants

Spent reactor fuel may be sent to a chemical reprocessing plant for the removal of fission products and the recovery of uranium and plutonium (Chapter 9). One commercial spent-fuel reprocessing plant in New York State operated between 1966 and 1972, but it has now been shut down. A plant of more advanced design has been partially completed in South Carolina, but its future is uncertain in view of President Carter's decision in 1977 to delay indefinitely the reprocessing of spent nuclear fuel. The discussion of the radiation doses that follows is based on the anticipated releases from the operation of a plant of this type. Note that even if fuel reprocessing were ever undertaken, the number of plants would be small, since one reprocessing plant would serve about 50 nuclear reactors of the type now in general use.

In modern fuel reprocessing plants, there are no radioactive liquid effluents; all the effluents are airborne. Among the substances present are the gases tritium and krypton-85, volatile radioiodines in both elemental and organic form, partially volatile ruthenium isotopes, and particulate matter. The particles contain several radioactive species, of which some of the more important are strontium-90, cesium-134, and cesium-137, and very small amounts of plutonium and other transuranium (TRU) elements (p. 132).

Although there are no liquid effluents from the projected fuel reprocessing plant, a nearby water body might represent a source of internal and external radiation exposure because of the deposition of airborne iodine and particulate matter. This source is small, however, in comparison with others, and the main radiation exposure pathways are those given below. The elements included in parentheses are those considered to be the most significant in each case.

External

Submersion in air (krypton)
Radiation from deposited particles (cesium, iodine)

Internal

Air inhalation (tritium)
Food and milk consumption (iodine, cesium, strontium, ruthenium)

TABLE 6-IX
CALCULATED INDIVIDUAL RADIATION DOSES FROM
A SPENT-FUEL REPROCESSING PLANT
(mrems/yr)

DISTANCE		WHOLE BODY	CHILD'S THYROID[a]	BONE	GI TRACT
km	Miles				
0.8	0.5	4.4	90	12	56
2.4	1.5	2.9	65	7.5	36

[a]Estimated from calculated data, based on a daily consumption of 1 litre of milk.

The calculated radiation doses from a typical reprocessing facility, with an annual capacity of 1500 MT (1650 short tons) of spent fuel, located in the southeastern United States are given in Table 6-IX for an individual located 0.8 or 2.4 km (0.5 to 1.5 miles) from the plant. It is assumed that at least a year has elapsed between removal of the fuel from the reactor and its reprocessing; this would permit iodine-131 to decay almost completely. Because of the lack of reprocessing facilities, there may well be a delay of several years before the spent fuel is reprocessed.

Conservative conditions on which the calculations are based are (a) that the exposed individual lives downwind in the predominant wind direction and (b) that all food and milk originates at the distance specified (i.e., 0.8 or 2.4 km). It is highly unlikely that the second condition would be satisfied in a real situation; in practice, therefore, the organ doses, especially the thyroid dose, would be substantially lower than given in the table. The whole-body dose, a considerable part of which arises from tritium and krypton, would not be greatly affected. For a population of about 550 000 within a radius of 89 km (55 miles) from the plant, the annual whole-body population dose is calculated to be about 170 person-rems.

It may be possible to lower the thyroid dose by improving iodine removal from the airborne effluent. The dose to the gastrointestinal (GI) tract is largely due to ruthenium, and this could be decreased by scrubbing the gases with an alkaline solution or passage through a silica gel bed before discharge. Plutonium and other TRU elements make a small contribution to the lung and bone doses; the amounts in the effluent could be reduced by a deep sand filter.

ACCUMULATION OF TRITIUM, KRYPTON-85, AND CARBON-14

Release of Gaseous and Volatile Substances

In principle, it should be possible to retain in solid form within restricted areas almost all the nonvolatile radioactive products formed as a result of nuclear power generation. In fact, most nuclear power and fuel reprocessing facilities are designed to realize this possibility. The situation is different, however, for long-lived products that are either gaseous or readily volatile. It is difficult to avoid some release of such substances to the environment. From the long-range standpoint, the most important gaseous or volatile species are tritium (in water and water vapor) and radioisotopes of krypton, xenon, iodine, and carbon (as carbon dioxide gas).

The rates at which radioisotopes in gaseous or liquid form are released to the air or to a body of water are, of course, controlled so that they are below (generally well below) the limits specified in the NRC regulations. The volatile effluents are, therefore, not a significant hazard at present. But questions have been raised concerning the accumulation in the environment of certain radionuclides as a result of the increasing use of nuclear power.

The isotopes of xenon have relatively short half-lives and decay rapidly. The same is true for most of the radioisotopes of iodine, except iodine-129; in any event, much of the iodine in the effluent gas can be removed (and retained) in solid absorbers. Hence, of the elements mentioned above, only tritium, krypton (as krypton-85), and carbon (as carbon-14) need be considered as the main potential long-range radioactive hazards from nuclear plant effluents.

Tritium and krypton-85 have moderately long half-lives—namely, 12.3 and 10.8 years, respectively—and carbon-14 has a half-life of 5730 years; hence, these nuclides will tend to accumulate in the course of time. If their rate of release should ever become constant, the total amounts present in the environment (i.e., in the atmosphere and in water bodies) will eventually attain steady-state (or equilibrium) values. The rate of production (or release) will then be exactly balanced by the rate at which the radionuclides undergo radioactive decay. It can be shown that the quantity (or activity) of a given species in the steady state is equal to the constant rate of release per year multiplied by 1.44 times the half-life in years. Theoretically, an infinite time is required for a completely steady state to be attained, but a close approximation (over 99 percent) is reached after seven half-lives.

Part of the tritium emitted from nuclear power and spent-fuel reprocessing plants is in the gaseous form (as HT) and part is present as tritiated water (HTO). Some of the tritium, either as gas or as water vapor, is discharged to the atmosphere; some, as tritiated water, to a river or stream. Tritium gas is fairly innocuous, but it is soon oxidized in the atmosphere to produce tritiated water. For all practical purposes, therefore, it may be supposed that all tritium release associated with the generation of nuclear power is present as tritiated water. The tritiated water will eventually find its way into the oceans and other water bodies, but some will always be present in atmospheric water vapor and droplets. All the krypton-85, on the the other hand, is discharged into the air, and since it is inert chemically and almost insoluble in water, it will accumulate in the atmosphere.

In recent years it has become apparent that radioactive carbon-14 produced during reactor operation could become a significant problem over the course of time. The carbon-14 would be released in gaseous form, mainly carbon dioxide and possibly simple hydrocarbons (carbon-hydrogen compounds). The carbon-14 in the gaseous effluent from a nuclear power plant (i.e., about 8 Ci/year from an LWR with an electrical capacity of 1000 MW) would not of itself be important, but since this nuclide has such a long half-life, the quantity in the environment will increase steadily from the continued operation of nuclear power plants.

Carbon-14 is formed in various neutron reactions. One such reaction is that of neutrons with ordinary nitrogen (nitrogen-14); this element is present as a trace impurity in the uranium dioxide fuel material and also in the air dissolved in the coolant water of an LWR. The reaction is the same as that by which carbon-14 is formed in nature (p. 147). Another important way whereby carbon-14 can be generated in a water-cooled reactor is by the interaction of neutrons with oxygen-17, a natural isotope of oxygen, which is always present in water (and in air). High-temperature gas-cooled reactors (p. 28) contain large amounts of carbon (as graphite). All natural carbon contains 1.1 percent of carbon-13, which captures neutrons to yield carbon-14.[k]

The following discussion is concerned with radiation exposures that may result from long-range accumulations of tritium, krypton-85, and carbon-14 in the environment. Such accumulations would arise from the fairly general use of nuclear power and spent-fuel reprocessing plants in several parts of the world over a period of years.[l] The estimates given below are based on an assumed extensive distribution of tritium, krypton-85, and carbon-14 in the atmosphere.

Long-Range Accumulation of Tritium

Ever since the earth has had an atmosphere containing oxygen and nitrogen, there has been tritium present in nature. It is produced continuously by the interaction of cosmic rays with the nuclei of oxygen and nitrogen atoms. It also decays continuously so that a steady state has been reached for natural tritium. There is some uncertainty concerning the amount of tritium arising from cosmic rays, but an average of the best estimates of the world inventory is about 50 megacuries (MCi) (1 MCi = 1 million Ci).

About 90 percent of the tritium from cosmic rays is in the ocean and other terrestrial waters (i.e., in the hydrosphere). Nearly all of the remainder is in the stratosphere, where the tritium is actually produced by the cosmic-ray reactions. From the stratosphere, the tritium gradually descends into the lower part of the atmosphere by natural diffusion. It is then brought down as tritiated water by rain or snow to the earth's surface, and accumulates in the hydrosphere. As a result of natural circulation—namely, evaporation of water, cloud formation, and precipitation—tritium is fairly uniformly distributed wherever water is present, including plants and animals.

The testing of thermonuclear weapons in the atmosphere, particularly since 1954, brought about a considerable increase in the amount of tritium on the earth. The total quantity introduced in this manner before the Nuclear Test Ban Treaty was signed (p. 150) has been estimated to be approximately 1700 MCi. Smaller quantities of tritium have been added from the atmospheric testing of nuclear explosives by China and France since 1963, but, on the whole, there has been a general decline due to radioactive decay. Nevertheless, provided there is no extensive weapons testing in the interim, by the year 2000 there will still remain at least 300 MCi of tritium from this source.

[k]The three neutron reactions leading to the formation of carbon-14 are:

$^{14}N(n, p)\,^{14}C$; i.e., nitrogen-14 + neutron → carbon-14 + proton

$^{17}O(n, \alpha)\,^{14}C$; i.e., oxygen-17 + neutron → carbon-14 + alpha particle

$^{13}C(n, \gamma)\,^{14}C$; i.e., carbon-13 + neutron → carbon-14 + gamma ray.

[l]Although spent-fuel reprocessing has been suspended indefinitely in the United States, it will probably continue in other countries.

TABLE 6-X
ESTIMATED WORLDWIDE NUCLEAR POWER GENERATION CAPACITY

Year	Power Capacity (in 1000 MWe)	Uranium Fuel (percent)	Plutonium Fuel (percent)
1980	250	85	15
1990	800	35	65
2000	2000	17	83

The amounts of tritium that will be produced by the generation of electric power from nuclear reactor plants up to the year 2000 have been estimated by the EPA (Report ORP/CSD 72-1). The data upon which the estimates are based are as follows. For 1000 megawatt (electric)-years (MWe-years) of reactor operation, as defined on page 143, the amount of tritium produced is 24 kilocuries (kCi) (1 kCi = 1000 Ci) with uranium as fuel and 38 kCi with plutonium as fuel. Projected values for worldwide nuclear power generation capacity in the years 1980, 1990, and 2000, with the proportions obtained from uranium and plutonium, are given in Table 6-X. For the ensuing calculations, the plant factor is taken to be 70 percent (p. 4).

By combining these data, and allowing for the natural decay of tritium, the worldwide accumulation of this radioactive species from nuclear power operations can be calculated. The results in megacuries are shown in Fig. 6-2. According to this estimate, the world inventory of tritium in the year 2000 from nuclear power operations will be roughly 450 MCi; this is considerably below the level reached in the early 1960s from weapons testing. It seems highly unlikely at present that plutonium will be utilized as fuel, in both LWRs and fast breeder reactors, to the extent indicated in Table 6-X. Hence, the values in the figure are too high. Nevertheless, they will be used as a basis for estimating the radiation doses that might be received from the release to the environment of all the tritium produced in nuclear power plants.

Radiation Doses from Tritium. The whole-body radiation dose from tritium depends on the amount of tritium in the body; this is related to the tritium content of the food (including milk) and water consumed. Since tritium (as tritiated water) in food and water is ultimately derived from the hydrosphere (mainly surface waters and groundwaters), the radiation dose from tritium depends on the concentration of this radionuclide in the hydrosphere. If a steady state were reached, such as is the case for tritium produced by cosmic rays, the ratio of tritiated water to ordinary water in the body would be the same as in the hydrosphere. The annual whole-body dose per person arising from tritium would then be given approximately by

$$D \text{ (mrems/yr)} = 90\,000 \times T ,$$

where T is the number of microcuries of tritium per millilitre of water.

In estimating the tritium concentration in water arising from nuclear power operations, allowance must be made for the fact that most of this tritium will be released in mid-latitudes (30 to 50 degrees) in the Northern Hemisphere. Since considerable time is required for the tritium to become distributed throughout

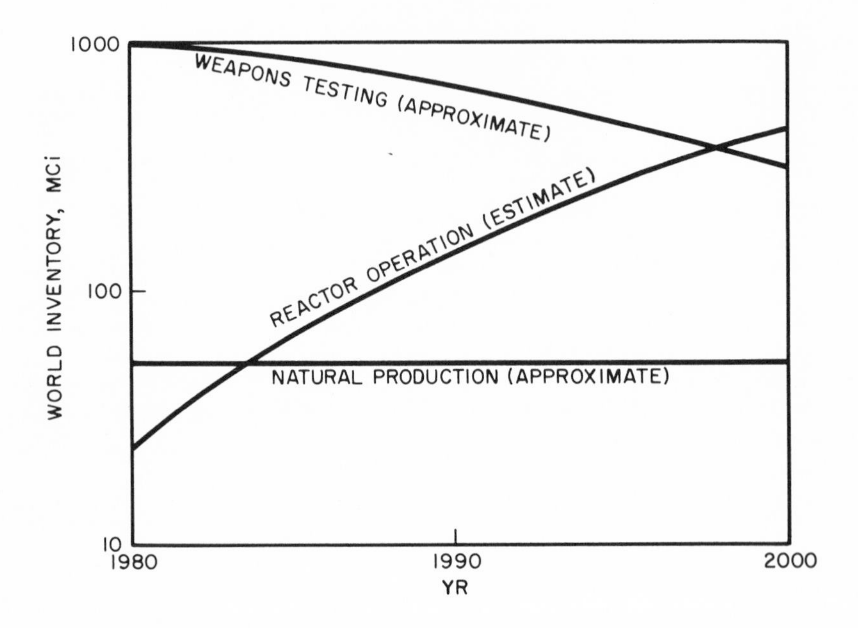

FIG. 6-2. World inventory of tritium from major sources (based on U.S. Environmental Protection Agency data).

the hydrosphere, it is assumed that 50 percent remains within the waters between 30 and 50 degrees north latitude. The volume of these waters to a depth of 40 m (about 130 feet) is estimated to be about 1.5×10^{15} m^3. If half the 450 MCi of reactor tritium (i.e., 225 MCi) is dissolved in this volume of water, the concentration would be 1.5×10^{-7} μCi/ml.[m]

If the proportion of tritium in body fluids is assumed to be the same as in the local hydrosphere, the annual dose per person from reactor tritium in the northern mid-latitudes (e.g., in the United States) in the year 2000 would be about 0.015 mrem, based on the equation given above. This is smaller than the radiation dose (almost 0.06 mrem/year) received by people in the United States in 1964 from tritium resulting from nuclear weapons tests. For purposes of comparison, it may be recalled that the average whole-body dose from natural background radiation is about 100 mrems, of which 18 mrems arises from sources within the body (Table 6-I).

Of course, some people will receive larger doses of radiation than the estimated average values, while others will receive smaller doses. But from the standpoint of the biological effects of radiation, especially genetic effects, the important quantity is the number of person-rems for a given population (p. 35). This is obtained by multiplying the number of people exposed by the average radiation dose per individual. Hence, the average dose does have significance.

The average dose calculated above, small as it is, may be an overestimate because the amount of tritium released to the environment as a result of nuclear

[m]This represents a ratio of 1 tritium atom (or 1 molecule of tritiated water) to 10^{16} hydrogen atoms (or molecules of ordinary water).

power generation may be less than assumed. As already mentioned, plutonium may be used as fuel to a smaller extent than expected. Furthermore, a technique is being developed for decreasing the tritium released from spent-fuel reprocessing plants, especially those treating breeder reactor (plutonium) fuels. After the fuel rods are sheared, the pieces would be heated in dry air or oxygen, thereby oxidizing the tritium to fully tritiated water (T_2O) vapor. After condensation, the small volume of tritiated water, essentially free from ordinary water, would be stored as liquid or solidified with cement or other material.

Biological Effects of Tritium. Since tritium is an isotope of hydrogen, tritiated water, in which part of the hydrogen has been replaced by tritium, has essentially the same chemical properties as ordinary water. Hence, tritium entering the body in food and water is quickly distributed among all the body fluids. A very small proportion of the ordinary hydrogen atoms in the body are thus replaced by tritium atoms. Consequently, all body tissues inevitably contain minute amounts of compounds of tritium corresponding to the normal hydrogen compounds.

When tritium nuclei undergo radioactive decay, they emit beta particles of very low energy (maximum 0.019 MeV, overall average roughly 0.006 MeV),[n] and no gamma rays. The product of radioactive decay of tritium is helium-3, a stable (nonradioactive) isotope of the inert element helium. Some questions have been raised concerning the biological hazard arising from the absorption of a certain amount of energy from the beta particles emitted by tritium. The general opinion is that the quality factor (as defined on page 122) for these particles is not greatly different from unity. The energy absorption (in rads) is then a good measure of the biological effectiveness (in rems).

The suggestion has been made that tritium might become concentrated in some body organ, just as iodine is in the thyroid gland and strontium (because of its chemical similarity to calcium) is in the skeleton. It should be pointed out, however, that the circumstances are different. When iodine concentrates in the thyroid or strontium in bone, there is no discrimination between the different isotopes, radioactive or stable, of the respective elements. All isotopes of iodine, for example, are retained to the same extent relative to the quantities present. Since tritium is an isotope of hydrogen, it would be expected to behave just like ordinary hydrogen. Wherever hydrogen atoms normally appear in the body, in fluids and tissue, there will be some tritium atoms. But there are presently about a hundred quadrillion (10^{17}) times as many ordinary hydrogen atoms as tritium atoms in the human body, and this ratio is probably much the same in all biological compounds.

Although tritium and ordinary hydrogen exhibit the same chemical reactions, tritium often interacts somewhat more slowly. It is conceivable, therefore, that if a tritium atom became incorporated in a compound (or tissue) in the body, perhaps from a food source, it might be removed less rapidly than would an ordinary hydrogen atom. This could lead to a higher ratio of tritium to hydrogen in the compound (or tissue) under consideration than in the body as a whole. Experiments with small animals in the laboratory have given no indication of such an effect; in fact, the reverse has often been found to be true.

[n]MeV is the abbreviation for million electron volts, a unit of energy defined in the footnote on page 215. The average maximum energy of the beta particles from the fission products is about 1.2 MeV, and the overall average is 0.4 MeV.

A few cases have been reported in which the ratio of tritium to ordinary hydrogen in some organs of mammals in the wild state was higher than in the body fluids. These results seem to imply a concentration (or bioaccumulation) of tritium in those organs. Since the animals had access to water containing larger proportions of tritium (from nuclear explosions) than found in the body fluids, the significance of the observations is uncertain. In any event, the concentration factors found were less than about 1.5.

Possible Genetic Effects of Tritium. There has been concern expressed that if tritium replaces ordinary hydrogen in deoxyribonucleic acid (DNA), the genetic material of the nuclei of living cells, the effects would be particularly serious. Any changes in DNA of the germ cells caused by extraneous factors, such as radiation (e.g., from tritium), heat, or certain chemicals, can cause mutations that may appear in later generations (Chapter 7). Laboratory studies have been made in which animals were deliberately fed a constituent of DNA (e.g., the compound thymidine) containing tritium in place of ordinary hydrogen. The conditions are, however, quite different from those that would arise in nature. There is no evidence of a natural tendency for the ratio of tritium to ordinary hydrogen in DNA to exceed that in the body fluid, and this would be essentially the same as in the food and drink.

It is true that the nucleus of the living cell is more sensitive to radiation than is the remainder of the cell contents, the cytoplasm. Since any tritium present is probably distributed in a uniform manner throughout the cell, the amount of radiation energy deposited in the nucleus will be proportional to its volume. Thus, only 3 to 10 percent of the beta particles from tritium in the cell would actually affect the nucleus.

The potential radiological effects of tritium in the body may be seen in perspective by comparing them with those of carbon-14 present in nature. All living organisms, plants and animals alike, contain a small proportion of the radioactive iosotope carbon-14, which, like tritium, is produced in the atmosphere by cosmic rays and by nuclear explosions. There is no evidence that carbon-14 behaves any differently in the body than does the common stable isotope, carbon-12, nor does it tend to concentrate in any paticular compound or organ.

The internal radiation dose from natural carbon-14 and the associated genetic consequences are much greater than would be expected from the tritium accumulated as a result of nuclear reactor operations. The total body (and gonad) dose from natural (cosmic-ray) carbon-14 is about 1 mrem/yr; this had increased to almost 2 mrems by 1964 as a result of nuclear weapons testing in the atmosphere, but has subsequently declined somewhat. As seen above, the radiation dose from the tritium that is expected to accumulate by the year 2000, as a result of the generation of nuclear power, is only a small fraction of a millirem per year. Thus, the genetic (and other) effects of tritium should be insignificant in comparison with those of the carbon-14; the latter has undoubtedly been present in man during the whole period of his existence.

Some experiments with bacteria and fruitflies have led to the conclusion that if a tritium atom is incorporated at a specific location in a pyrimidine molecule, which is one of the components of DNA, the genetic effect is greater than would be expected from the radiation dose alone. The difference is not large—about 25 percent—and it has not yet been demonstrated in higher organisms. Even if the increase did occur in mammals, the probability that tritium atoms in the body will enter the specific position in the pyrimidine molecule is remote. In the first place, the ratio of tritium atoms to ordinary hydrogen atoms in the human body is extremely small and will always be so. Furthermore, there are many different possible locations in a DNA molecule in which a tritium atom may occur. In only one of these would the reported increase in the genetic effect be observed.

The general conclusion has been reached that the biological effects of a given internal radiation dose from the beta particles emitted by tritium are the same as for an equal dose of gamma rays or x rays. There is no clear evidence that any additional significance or potential hazard is associated with the fact that the dose is derived from tritium rather than from other sources of beta particles or gamma rays.

Long-Range Accumulation of Krypton-85

About 500 kCi of krypton-85 are generated per 1000 MWe-year of reactor operation. Almost 99.6 percent of krypton-85 nuclei decay with the emission of beta particles with a maximum energy of about 0.67 MeV (overall average about 0.22 MeV) but no gamma rays. In the remaining roughly 0.4 percent of the decays, the available energy is divided between a beta particle (maximum energy 0.16 MeV (overall average 0.05 MeV), and a gamma ray (0.51-MeV energy). Although the amount of krypton-85 produced is greater than that of tritium and the energy of the beta particles is larger, the difference in health hazard is not as great as might appear; the reason for this will be explained shortly.

Because of its moderately long half-life (10.8 years), krypton-85 accumulates in the atmosphere as it is released from nuclear power and fuel-reprocessing plants. If krypton-85 were discharged at a constant rate, a steady-state amount of $10.8 \times 1.44 = 15.6$ times the annual release rate would be approached in about 50 years (p. 137). The rate of discharge, in the United States at least, is to be limited by EPA regulations to 50 kCi per 1000 MWe-year after January 1, 1983 (p. 143). The steady-state amount consequently will be only one-tenth of that which would result if all the krypton-85 formed in fission were released.

Krypton-85 discharged through a stack mixes with the stable krypton and other atmospheric gases in the surrounding air. It is then transported by wind and other forms of atmospheric motion over wide areas. In this way, the concentration of krypton-85 in the air decreases as the distance from the discharge point increases. The element krypton is essentially inert chemically, and it has a low solubility in water; hence, almost all the krypton-85 released remains within the atmosphere. The quantities in soil and water and in plants and animals are negligible.

Biological Effects of Krypton-85. Krypton-85 can enter the body by inhalation of air containing this nuclide. Because of its low solubility in water, the concentration of krypton in body fluids is very small, although it appears to be somewhat more soluble in fat. Since krypton has only very weak chemical activity, it is not involved in any metabolic processes and is not incorporated into any biological molecules. Consequently, there is no krypton (stable or radioactive) in any body tissue, except for the small amount that may have dissolved in fluid or fat. In this respect, krypton-85 behaves quite differently from tritium.

As a result of the low solubility in water and its unreactive nature, the total amount of krypton-85 in the body at any time is essentially only that in the lungs. Hence, the lining of the lungs receives a certain dose of radiation, but this is smaller than the skin dose arising from immersion of the body in a medium (air) containing a small concentration of radioactive krypton-85. The skin is relatively insensitive to radiation and there is evidence that radiation injury of the skin is, at least partially, reparable.

Since the beta particles from krypton-85 cannot penetrate the outer layers of the skin, they contribute very little to the whole-body radiation dose. The latter arises mainly from the gamma rays produced inside and outside the body in about 0.4 percent of the decays of krypton-85. The dose to the whole body or to the gonads, which would be capable of causing genetic mutations, is estimated to be only about 2 or 3 percent of the skin dose. The internal dose to the lungs, which is largely due to beta particles, would be roughly 4 percent of the skin dose.

Calculations have been made of the annual radiation dose to be expected from krypton-85 that has accumulated in the atmosphere as a result of the worldwide operation of nuclear power plants. Estimates of the average annual doses to the skin, lungs, and whole body (gonads) of a person living in the United States in the years 1980, 1990, and 2000 are given in Table 6-XI. They are based on the predicted nuclear power generation in Table 6-X and the assumptions that all the krypton-85 produced in fission is discharged to the atmosphere and that 75 percent remains in the northern hemisphere, where most nuclear plants are located. If, as is possible, the krypton-85 is more uniformly distributed throughout the atmosphere, the doses would be lower than in the table. The limitation proposed by the EPA on the release of krypton-85 would also result in a considerable decrease in the average estimated doses after 1983.

Removal of Krypton-85. The accumulation of krypton-85 will not be of great concern, especially with regard to its genetic effects, for many years. To comply with EPA regulations limiting the discharge of krypton-85 to the atmosphere, methods are being developed for removing this nuclide from gaseous effluents. This removal is particularly important for spent-fuel reprocessing plants, where fission product gases are released in the largest amounts.

TABLE 6-XI

ESTIMATED AVERAGE ANNUAL DOSES IN THE UNITED STATES FROM WORLDWIDE PRODUCTION OF KRYPTON-85 ASSUMING COMPLETE RELEASE*

(mrems/person)

Year	Skin	Lung	Whole Body
1980	0.07	0.004	0.002
1990	0.2	0.012	0.006
2000	0.9	0.05	0.025

*Recalculated from ORP/CSD 72-1 (U.S. Environmental Protection Agency, Washington, D.C.) using conversion factors in EPA-520/4-73-002 (U.S. Environmental Protection Agency, Washington, D.C.).

Krypton can be separated from other gases by passage through certain rubber-like membranes or by the use of solvents of the Freon type. Another possibility is to liquefy the gases and separate the krypton by distillation at low temperatures. The krypton gas would then be absorbed on charcoal or other suitable solid or compressed into tanks. The containers could be buried at the plant site or at an approved repository where they would be kept under surveillance. Since the half-life of krypton-85 is not extremely long, compared with that of some other fission products, the surveillance time could be limited to 100 years or so.

Long-Range Accumulation of Carbon-14

The carbon-14 formed by the action of neutrons on the oxygen-17 in the water coolant of an LWR will be released at the nuclear power plant, whereas most of the remainder, present in the fuel, will be released during the reprocessing operation. It has been estimated that the total amount of carbon-14 produced would be about 60 Ci per 1000 MWe-year. About the same rate of release is expected for a fast breeder reactor plant, although essentially all the carbon-14 would be formed in the fuel and in the breeder blanket (p. 25). The carbon-14 would then be released at the reprocessing plant. For high-temperature gas-cooled reactors, carbon-14 formation is estimated to average roughly 300 Ci per 1000 MWe-year. Much of this would probably be released to the atmosphere when the graphite-based fuel elements are burned as the first step in fuel reprocessing.

Regardless of the form of the carbon-14 when it is produced in a reactor, it will eventually be present in the atmosphere as carbon dioxide. Part of this gas will slowly dissolve in the oceans and other surface waters, but the remainder will be available for uptake by green plants. Carbon-14 can become distributed throughout the body as a result of the consumption of food, all of which originates from green plants. Although carbon-14 emits a beta particle of low energy (maximum 0.16 MeV, average about 0.05 MeV) and no gamma rays, it can, like tritium, represent an important internal source of radiation. However, since carbon-14 has a much longer half-life (5730 years), a steady state would not be approached for thousands of years, assuming continued release to the atmosphere.

Under certain assumptions, the amount of carbon-14 discharged from nuclear operations that would remain in the lower atmosphere (troposphere) and be available to plants has been estimated. Furthermore, it is assumed that the ratio of carbon-14 to normal carbon in the body would closely follow the ratio in the atmosphere. Then, by using the accepted relationship that 6×10^{-6} μCi of carbon-14 distributed uniformly in the human body would produce a whole-body dose of 1 mrad (or approximately 1 mrem) per year, one may calculate the estimated annual doses per person as shown in Table 6-XII. The worldwide nuclear power generation on which these estimates are based are those in Table 6-X.

Two other assumptions were made in deriving the estimates in Table 6-XII: namely, that the carbon-14 is distributed evenly throughout the whole troposphere and that all the carbon-14 generated in reactor operations is released to the environment. If the carbon-14 remains preferentially in the northern

TABLE 6-XII
ESTIMATED AVERAGE ANNUAL DOSES FROM WORLDWIDE PRODUCTION AND DISTRIBUTION OF CARBON-14 ASSUMING COMPLETE RELEASE
(mrems/yr/person)

Year	Whole-Body Dose
1980	0.01
1990	0.04
2000	0.14

hemisphere, rather than being distributed uniformly worldwide, the annual doses to people in the United States would be about 50 percent larger than in the table. On the other hand, a decrease in the amounts of carbon-14 escaping to the atmosphere would reduce the dose estimates in Table 6-XII.

If the various assumptions on which Table 6-XII are based should prove valid, the annual doses from carbon-14 produced by nuclear operations would be less than 0.2 mrem in the year 2000. This can be compared to 1 mrem/year whole-body dose from carbon-14 in nature resulting from the action of neutrons in cosmic rays. Furthermore, as seen on page 151, 0.7 mrem/year was received in 1965 from carbon-14 generated in nuclear weapons tests; this dose has been decreasing, however, since that time because of the gradual transfer of carbon dioxide from the atmosphere to the oceans.

The matter of the potential accumulation of carbon-14 from nuclear power operations is being studied by the EPA. It is possible that a limit will be set on the amount that may be released, just as it has for krypton-85. The removal of carbon dioxide from gaseous effluents should not prove to be too difficult. For example, it could be condensed at low temperature with the krypton-85 and then separated by distillation. The carbon-14 dioxide would then be converted into a solid compound, such as calcium carbonate, which would have to be kept segregated for thousands of years.

Environmental Radiation Ambient Monitoring System

The Environmental Radiation Ambient Monitoring System (ERAMS), administered by the EPA's Office of Radiation Programs, was initiated in July 1973 by the consolidation and redirection of several radiation monitoring networks operated by the Bureau of Radiological Health prior to the formation of the EPA. The previous programs had been oriented primarily to measurements of radioactivity in the fallout from nuclear weapons tests. In the ERAMS, the emphasis is toward identifying trends in the environmental accumulation of long-lived radionuclides, such as those arising from nuclear power operations. Sampling stations are distributed over the United States, including Alaska and Hawaii, and the U.S. territories.

At regular intervals measurements are made of tritium (in rain, surface water, drinking water, and milk), of krypton-85 (in the atmosphere), and of carbon-14 (in milk). Other nuclides included in the monitoring system are strontium-90 (in water, milk, and bone) and isotopes of uranium and plutonium (in the air). In addition, gross (or overall) measurements of radioactivity are made on solid particles deposited from the atmosphere (drinking water and milk). From 1960 until December 1974, the results obtained from radiation surveillance programs were published monthly in *Radiation Data and Reports* (and its predecessors). Currently, ERAMS data are available in quarterly EPA summaries and an annual topical report.

ESTIMATED ANNUAL DOSES IN THE YEAR 2000

Nuclear Power Operations

Two independent studies have been made of the average dose to each individual in the whole or a large part of the United States that would be expected in the year 2000 from radioactive sources arising from nuclear power operations. In that year, possibly 50 percent of the total electric power used in this country might be generated in nuclear plants, with either LWRs or fast breeder reactors (Chapter 2). One of these studies was made by the AEC [o] and the other by the EPA.[p]

The AEC study, which was the more detailed of the two, was concerned with an area of more than 780 000 km^2 (300 000 square miles) in the watersheds of the upper Mississippi River (above its confluence with the Ohio River). The area includes the entire state of Iowa, most of Illinois, Minnesota, and Wisconsin, and part of Kansas, Nebraska, and South Dakota. The population is roughly 9 percent of the United States total, and the area accounts for about 10 percent of the power generated and consumed in the whole country. These proportions are expected to remain essentially unchanged through the year 2000.

The main assumptions on which the dose estimates were based are the following:

1. Half of the nuclear power plants would have LWRs, and the other half would have liquid-metal-cooled fast breeder reactors.
2. All plants would have advanced waste treatment equipment needed to conform with the guidelines for maintaining environmental radiation levels as low as is reasonably achievable.
3. Noble gases (mainly krypton-85) from breeder reactors only and from all fuel reprocessing plants would not be discharged.

The results of the study show that, on the average throughout the region, the whole-body radiation dose to a representative individual would be increased by about 0.2 mrem/year by the year 2000. This does not include potential contributions from abnormal or accidental releases of radioactivity. Also excluded were contributions from fabrication of new and spent fuel and from radioactive wastes; when averaged over the population of the region, these should be minor. People living close to a nuclear plant site would receive larger than average doses, but the estimates indicate that 99 percent of the population would receive

[o] *The Potential Radiological Implication of Nuclear Facilities in a Large Region of the U.S.A. in the Year 2000*, WASH-1209, U.S. Atomic Energy Commission, Washington, D.C. (1972).

[p] *Estimates of Ionizing Radiation Doses in the United States 1960—2000*, ORP/CSD 72-1, U.S. Environmental Protection Agency, Washington, D.C. (1972).

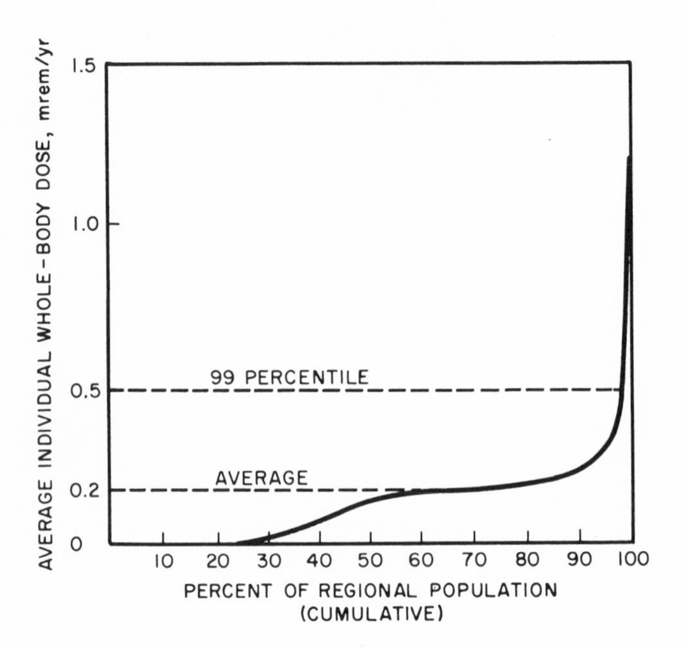

FIG. 6-3. Average individual radiation doses from predicted nuclear power operations in the Upper Mississippi River Basin estimated for the year 2000 (data from WASH-1209, U.S. Atomic Energy Commission, Washington, D.C. [1973]).

whole-body doses of less than 0.5 mrem/year from the major sources related to the generation of nuclear power (Fig. 6-3).

The primary purpose of the EPA report was "to provide . . . estimates of future doses to the United States population and major contributors to those doses that may assist in the formulation of general and specific radiation protection guidance." To this end, an assessment was made of the doses received by people in the United States from all radiation sources, natural and man-made, from 1960 and 1970 with predictions through the year 2000. The report gives data in terms of population doses (p. 35) from various sources; the average annual dose per individual is obtained upon dividing the annual population dose by the number of people in the U.S. population.

In estimating the expected average doses from nuclear power operations, the EPA assumed that noble gases are not stored, but are all released to the atmosphere. The whole-body dose from gamma-ray emitters in the gaseous effluents was assumed to be 5 mrems/year at each reactor (LWR or breeder) site boundary. Allowance was made, however, for larger doses in the vicinity of spent-fuel reprocessing plants. Based on actual experience, the doses from liquid effluents were assumed to be small.

The results of the EPA study are summarized in Table 6-XIII; the expected dose from carbon-14, which was not available when the report was prepared, has been included. The doses from tritium, krypton-85, and carbon-14 allow for the amounts of these radionuclides arising from worldwide nuclear power

TABLE 6-XIII
ESTIMATED AVERAGE WHOLE-BODY DOSE TO INDIVIDUALS IN THE UNITED STATES POPULATION IN THE YEAR 2000 FROM NUCLEAR POWER OPERATIONS

Source	(mrem/yr)
Reactors	0.175
Fuel reprocessing	0.20
Tritium	0.015
Krypton-85	0.04
Carbon-14	0.14
Total	0.57

operations, as already described in this chapter. From the AEC and EPA studies, which are based on somewhat different assumptions, it would be reasonable to conclude that the average increase in the individual radiation dose to people in the United States, as a result of nuclear power generation, would not be likely to exceed 0.6 mrem/year by the end of this century.

At the time the foregoing estimates were made, it was accepted that all spent fuel from reactors in the United States would be reprocessed, but the future of this operation is now in doubt. On the other hand, both the AEC and EPA assumed that the contributions of uranium mining and milling would be very small. This assumption was based on the fact that at distances of more than half a mile or so from an open-pit uranium mine or from a mill, the radioactivity level was found to be not significantly different from that at a greater distance in the same general area.

More recently, however, attempts have been made to estimate the amounts of radon-222 and other radioactive material released from uranium mines and mills (p. 163). From these estimates, it appears that these releases would probably increase the average whole-body radiation dose to people in the United States from nuclear operations to about 1 mrem/year. An additional lung dose of 1 mrem/year might also be expected from radon-222 and its alpha-emitting decay products.

Natural and Man-Made Radiation Doses

To place the radiation dose from nuclear power in perspective, it should be compared with the average doses from other man-made sources as well as from natural sources. The doses from the natural background and from medical, global fallout, miscellaneous, and occupation sources were discussed in the early sections of this chapter. The data for the average annual whole-body dose for the period 1960-2000 are summarized in Fig. 6-4.[q] The corresponding figure in the EPA report, referred to above, included a small dose from nuclear weapons tests conducted at the Nevada Test Site prior to September 1962. It has been omitted from Fig. 6-4 since it is no longer significant. However, the dose from

[q]Note that the vertical (annual dose per person) scale is logarithmic; the divisions (i.e., 0.1, 1, 10, and 100) represent increases by a factor of 10 in each case. The purpose is to permit representation of a large range of dose values on a single figure.

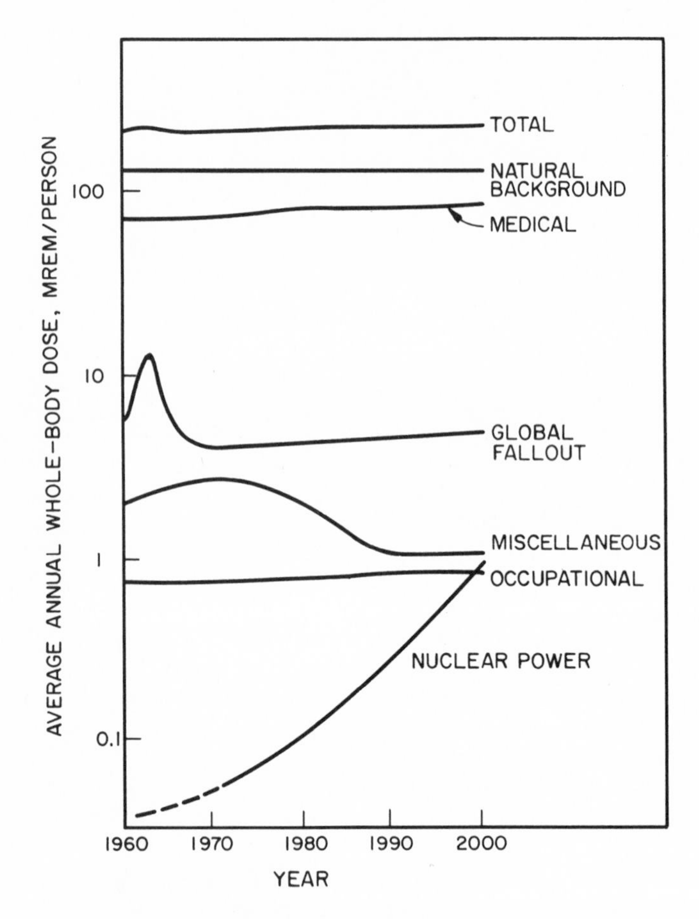

FIG. 6-4. Summary of estimated average whole-body radiation doses in the U.S. (adapted from ORP/CSD 72-1, U.S. Environmental Protection Agency, Washington, D.C. [1972]).

nuclear power has been increased to allow for radioactive releases in the mining and milling of uranium.

The estimates of radiation doses arising from nuclear power operations described above were based on amounts of radioactive materials released to the environment. The doses are thus applicable to members of the general public outside the operational areas. People working within such areas will receive additional radiation doses from a variety of sources. Allowance for occupational exposure might add another 1 mrem/year, averaged over the whole population of the United States, by the year 2000 (Fig. 6-4). The additional dose of about 2 mrems/year from nuclear power operations may be compared with the total of some 200 mrems/year from natural background and man-made sources. The contribution from nuclear power is less than the normal variations occurring in the background radiation from time to time and from place to place.

Bibliography—Chapter 6

Bond, V. P. "Evaluation of Potential Hazards from Tritiated Water." *Proceedings of the Symposium on the Environmental Aspects of Nuclear Power Stations*, p. 287. Vienna: International Atomic Energy Agency, 1971.

Buchanan, J. R. "Nuclear Power and Radiation in Perspective" (ORNL-NSIC-100). Selections from *Nuclear Safety*. Oak Ridge, Tenn.: Nuclear Safety Information Center, 1974.

Eisenbud, M. *Environmental Radioactivity*. New York: McGraw-Hill Book Company, 1973.

Eisenbud, M., and Petrow, H. G. "Radioactivity in the Atmospheric Effluents of Power Plants That Use Fossil Fuels." *Science* 144 (1964): 288.

Elwood, J. W. "Ecological Aspects of Tritium Behavior in the Environment." *Nuclear Safety* 12 (1971): 326.

International Atomic Energy Agency. *Proceedings of the Symposium on the Radiological Impacts of Releases from Nuclear Facilities into Aquatic Environments*. Vienna: International Atomic Energy Agency, 1976.

International Commission on Radiological Protection. *Implications of Commission Recommendations That Doses Be Kept As Low As Readily Achievable* (ICRP-22). New York: Pergamon Press, Inc., 1973.

Kahn, B. *et al. Radiological Surveillance Studies at a Boiling Water Nuclear Power Reactor* (BRH/DER-70-1). Washington, D.C.: U.S. Bureau of Radiological Health, 1970.

Killough, G. G., and Till, J. E. "Scenarios of ^{14}C Releases from the World Nuclear Power Industry from 1975 to 2020 and the Estimated Radiological Impact." *Nuclear Safety* 19 (1978): 602.

Klement, A. W., Jr., *et al. Estimates of Ionizing Radiation Doses in the United States, 1960-2000* (ORP/CSD-72-1). Washington, D.C.: U.S. Environmental Protection Agency, 1972.

Magno, P. J.; Nelson, C. B.; and Ellett, W. H. "A Consideration of the Significance of Carbon-14 Discharges from the Nuclear Power Industry." *Proceedings of the 13th Cleaning Conference* (CONF-740807), p. 1047. Washington, D.C.: U.S. Atomic Energy Commission, 1975.

Martin, J. E., *et al.* "Radioactivity from Fossil-Fuel and Nuclear Power Plants." *Proceedings of the Symposium on the Environmental Aspects of Nuclear Power Stations*, p. 325. Vienna: International Atomic Energy Agency, 1972.

McBride, J. P., *et al.* "Radiological Impact of Airborne Effluents from Coal and Nuclear Plants." *Science* 202 (1978): 1045.

National Council on Radiation Protection and Measurements. *Environmental Radiation Measurements* (Report No. 50). Washington, D.C., 1976.

______. *Krypton-85 in the Atmosphere—Accumulation, Biological Significance, and Control Technology* (Report No. 44). Washington, D.C., 1975.

______. *Natural Background Radiation in the United States* (Report No. 45). Washington, D.C., 1975.

Reinig, W. C., ed. *Proceedings of the Symposium on Environmental Surveillance in the Vicinity of Nuclear Facilities*. Springfield, Ill.: Charles C. Thomas, Publisher, 1971.

Rohwer, P. R., and Wilcox, W. H. "Radiological Aspects of Environmental Tritium." *Nuclear Safety* 17 (1976): 216.

Rowe, W. D.; Galpin, F. L.; and Peterson, H. T. "EPA's Environmental Radiation-Assessment Program." *Nuclear Safety* 16 (1975): 667.

Schaefer, H. J. "Radiation Exposure in Air Travel." *Science* 173 (1971): 780.

Travis, C. C., *et al.* "Natural and Technologically Enhanced Sources of Radon-222." *Nuclear Safety* 20 (1979): 722.

United Nations, Scientific Committee on the Effects of Atomic Radiation. *Ionizing Radiation, Levels (Vol. I) and Effects (Vol. II)*. New York: United Nations, 1972.

U.S., Atomic Energy Commission. *The Potential Radiological Implications of Nuclear Facilities in a Large Region in the U.S.A. in the Year 2000* (WASH-1209). Washington, D.C., 1973. See also, *Nuclear Safety* 15 (1974): 56.

U.S., *Code of Federal Regulations*, Title 10 (Energy). Washington, D.C.
Part 50. Appendix I. (Numerical Guides . . . to Meet the Criterion 'As Low As Is Reasonably Achievable . . .').

U.S., Environmental Protection Agency. *Environmental Analysis of the Uranium Fuel Cycle* (EPA-520/9-73-003 B, C, and D). Washington, D.C., 1973.

______. *Environmental Radiation Protection Requirements for Normal Operations of Activities in the Uranium Fuel Cycle.* Washington, D.C., 1975.

______. *Environmental Radiation Dose Commitment: An Application to the Nuclear Power Industry* (EPA-520/4-73-002). Washington, D.C., 1974.

______. *Radiological Quality of the Environment* (EPA-520/1-76-010). Washington, D.C., 1976. See also, *Nuclear Safety* 18 (1977): 215 and 20 (1979): 342.

______. *Radiological Surveillance Study at the Haddam Neck PWR Nuclear Power Station* (EPA-520/3-74-007). Washington, D.C., 1975.

Veluri, V. R.; Boone, F. W.; and Palms, J. M. "The Environmental Impact of ^{14}C Released by a Nuclear Fuel-Reprocessing Plant." *Nuclear Safety* 17 (1976): 560.

Yeates, D. B.; Goldin, A.S.; and Moeller, D. W. "Natural Radiation in the Urban Environment." *Nuclear Safety* 13 (1972): 275.

7

Biological Effects of Radiation

INTRODUCTION

Somatic Cells and Germ Cells

Injury to living organisms by radiation undoubtedly arises from damage to the cells in the body. Some cells that are affected in this way may be killed and may be replaced by new cells, just as is the case for many mechanical, burn, and similar injuries. In other cases, however, the cell damage may have permanent or irreversible effects.

All living cells are basically alike in the respect that they have a central region (or nucleus) containing a number of somewhat elongated units called *chromosomes*. Each chromosome is made up of a large number of segments, known as *genes*, that carry the hereditary (i.e., genetic) characteristics of the individual organism. The great majority of body cells, referred to as somatic cells, contain 23 pairs of chromosomes; the two chromosomes in each pair are similar but not identical.

When a somatic cell reaches a certain stage of growth, it divides into two cells. This is achieved in the following manner. Each member of the 23 pairs of chromosomes reproduces (or replicates) itself, so that two identical sets of 23 chromosomes are formed. The original somatic cell then divides into two cells, each of which contains one set of 23 pairs of chromosomes. In other words, each of the two somatic cells formed is a replica of the original cell. In this way a living organism can grow or generate new body cells to take the place of the billions of cells that are continuously being lost either by injury or in normal life functions.

In the gonads (or sex glands), where the germ cells—sperm cells in the male and egg cells in the female—are produced, a different type of cell division occurs. After a series of stages, somatic cells containing 23 pairs of chromosomes give rise to germ cells with 23 single chromosomes. In the process of sexual reproduction, a sperm cell and an egg cell unite, and the resulting fertilized cell (zygote) contains 23 pairs of chromosomes; one set of 23 chromosomes is derived from the male parent and the other, different set from the female parent. The successive replication of these (somatic) cells then leads to the production of an embryo and eventually of a new individual. The information required by the organism for the development of its own particular characteristics is contained in the genes present in the chromosomes of the sperm and egg cells of the parents.

Somatic and Genetic Mutations

Changes in chromosomes are called *mutations*. Those that occur naturally, without any action by man, are known as spontaneous mutations. In addition, mutations can be induced by various mutagenic agents, including radiation, heat, and many chemicals. The mutated chromosome will then replicate itself in the process of cell division. Thus, if a somatic cell survives the action of the mutagenic agent, the organism will continue to produce mutant forms of the cell. The living system in which the mutation has occurred may thus suffer an injury that may or may not be reparable. Some of the delayed effects of radiation, which are described below, may possibly arise from somatic mutations.

Somatic mutations may affect the individual in whom the mutation has occurred, but the modified chromosome will not be passed on to succeeding generations. If a mutation occurs in the gonad cells, however, the modified chromosome might appear in germ cells formed in the gonad and might be carried over into the offspring. Such a mutation can result in the development of new characteristics that may appear in subsequent generations. Changes in germ cells produced by so-called spontaneous mutations have played an essential role in the evolution of living organisms.

Because of the differences in cell types, the biological effects of radiation may be conveniently discussed under two general headings: somatic effects and genetic effects. Somatic effects are not inherited, but genetic effects, resulting from the absorption of ionizing radiation by the gonads, may be passed on from one generation to the next.

SOMATIC EFFECTS OF RADIATION

Early Somatic Effects

The somatic effects of radiation are, in general, either early or delayed. Early somatic effects are observed only when fairly large radiation doses, usually in excess of 50 rems (i.e., 50 000 mrems), are received over the whole body within a short time, usually a day or less.[a] These are called *acute radiation exposures*. Such acute exposures cannot arise during normal operations associated with nuclear power production, but they are conceivable as a consequence of a serious reactor accident, as discussed in Chapter 4. The early biological effects of acute radiation exposures are, therefore, described briefly below.

Because of natural biological variability, people (and animals) are not all equally affected by a given radiation dose. However, certain conclusions, as outlined in Table 7-I, have been reached on a statistical basis. Much of the information on early somatic effects has been obtained from a study of the Japanese victims of the atomic bombs in August 1945. There have been a few cases of

[a] As explained in Chapter 5, dose equivalents are commonly referred to simply as doses in rems.

TABLE 7-I
PROBABLE EFFECTS OF ACUTE WHOLE-BODY RADIATION DOSES

Acute Dose[a] (rems)	Probable Observable Effect
5-75	Chromosome aberrations; temporary depression of white blood cells in some individuals; no other observable effects.
75-200	Vomiting in 5 to 50 percent of exposed individuals within a few hours, with fatigue and loss of appetite; moderate blood changes; recovery from most symptoms within a few weeks.
200-600	Vomiting within two hours or less for doses of 300 rems or more; severe blood changes with hemorrhage and increased susceptibility to infection, particularly at the higher doses; loss of hair after two weeks for doses over 300 rems; recovery within from one month to a year for most individuals at the lower end of the dose range; only about 20 percent may survive at the upper end of the range.
600-1000	Vomiting within one hour; severe blood changes, hemorrhage, infection, and loss of hair; from 80 to 100 percent of exposed individuals succumb within two months; survivors convalesce over a long period.

[a]The doses are those in soft tissue near the body surface; the doses in the center of the body interior would be about 70 percent of the tabulated values.

laboratory accidents where workers received fairly large doses of radiation (50 rems or more) in a short time and exhibited the early effects. Data have also been obtained from hospital patients receiving such doses of radiation for therapeutic purposes.

The lower dose indicated in Table 7-I at which early effects of radiation are observed—namely, 5 rems—refers to an increase in the frequency of chromosome aberrations in white blood cells detected by special laboratory techniques. The biological significance, if any, of these changes is unknown at present. A (temporary) decrease in the number of white blood cells is barely detectable in a group of people who have received about 25 rems whole-body dose. Only when the dose is roughly 50 rems can this effect be readily observed in some exposed individuals by conventional laboratory procedures. Obvious clinical symptoms—specifically, nausea and vomiting—appear in a small fraction of individuals when the acute dose exceeds 75 to 100 rems.

The effects described above apply when the whole or a major portion of the body is exposed to the respective doses of radiation. The absorption of the same doses by a small region of the body will cause very much less injury. For example, radiation doses of several thousand rems have been applied locally for the destruction of malignant growths. The patients sometimes become nauseated, but other symptoms appear to be minor.

Delayed Somatic Effects

Delayed somatic effects of radiation from external sources have been observed in radiologists (particularly those who practiced during the first few decades of this century), in patients who received radiation treatment for various diseases, and in the survivors of the atomic bombings in Japan. The most apparent effects in these cases have been an increased incidence of leukemia and other types of cancer, including tumors of the thyroid gland, the lung, and the breast (in women). In addition, underground uranium miners have exhibited a marked increase in lung cancers, and painters of self-luminous instrument dials an increase in bone cancers, as a result of the deposition of radionuclides within the body. The delay time between the receipt of a substantial dose of radiation and the appearance of cancer is called the *latent period*. The average latent period is about 15 years, but it may range from a year or two for leukemia to 30 years or more for bone cancer.

The cancers induced by radiation do not differ from those that occur normally in human beings. Consequently, evidence for radiation-induced cancer of a given type can be obtained only by comparing the incidence in a group of people exposed to radiation with the normal incidence of that type of cancer in an equivalent unexposed group. However, the "normal" incidence is an uncertain and variable quantity. The average incidence of cancer in a large population, such as that of the United States, is fairly well known, but this incidence will not necessarily apply to a smaller (and often special) group. For example, the expected normal frequency of cancer in people who have required extensive x-ray therapy or diagnosis may well differ considerably from the average for the whole country. The same may be true for the wartime population of the bombed Japanese cities of Hiroshima and Nagasaki.

As a general rule, except possibly for very large doses, the incidence of radiation-induced cancer increases with the dose received. Consequently in a group of only a few thousand people, the dose received must be fairly large, usually at least 50 to 100 rems, for the incidence of a particular type of cancer to be clearly greater than expected, taking into consideration statistical uncertainty in the normal incidence in a group of moderate size. If the dose is small, however, the small increase (if any) in the incidence of cancer could be detected only by studying a very large number of exposed individuals, for which the normal incidence would be reasonably well known.

The situation may be illustrated in connection with the problem of immediate interest—namely, to estimate the cancer incidence that might result from exposure to very small doses of radiation from nuclear power operations. A possible approach, at first sight, would be to consider the natural background radiation described in Chapter 6. Most of the people in the United States, especially those living in coastal regions, receive an average dose of about 100 mrems/year from this radiation (i.e., a lifetime dose of roughly 7 rems). On the other hand, there are some 3 million people residing in the Rocky Mountain states whose annual background radiation dose is at least 200 mrems (lifetime dose 14 rems or more).

Despite the difference of about 7 rems in the lifetime radiation doses of the two population groups, epidemiological studies have not revealed any significant difference in cancer incidence. Clearly, if the additional radiation dose of 7 rems results in an increase in cancer incidence—and it is by no means certain that it does—a much larger population group than 3 million would have to be studied.

There seems little prospect that the incidence of cancer resulting from very small radiation doses, such as those expected from nuclear power operations or even the larger natural background radiation doses, can ever be observed directly. For the results to be significant, groups containing billions of individuals would have to be studied. The only practical way to obtain the required information is to examine the effects of fairly large doses of radiation in population groups of reasonable size and to infer from the results what might be expected for much smaller doses. This procedure involves various assumptions described in due course. First, however, it may be instructive to review the available information on the delayed somatic effects on human beings of radiation doses generally exceeding 50 rems.

Leukemia

Leukemia appears to be the most readily detectable of the delayed somatic effects of ionizing radiation. It is generally regarded as a form of cancer associated with a marked increase in the number of white cells (leukocytes) in the circulating blood. As far back as 1911, there were indications that more radiologists died from leukemia than would have been expected. Later studies have confirmed these indications. Among radiologists who practiced in the early decades of this century, the incidence of leukemia was two or three times as great as in a comparable group of physicians who were not occupationally exposed to radiation. It has been estimated that these radiologists received annual doses of roughly 15 to 30 rems or lifetime doses of several hundred rems. In recent years, radiation doses have been decreased to 5 rems (or less) per year, and deaths from leukemia among radiologists have been about the same as among other physicians.

An increased occurrence of leukemia has also been observed among the Japanese who received a single large dose of ionizing radiation from the atomic explosions in 1945. The increase was first detected in 1948, and a peak was reached in the period 1951-52. Since then there has been a general annual decline, with some variations from year to year. The latent period has ranged from a little over 2 years up to about 25 years or possibly more in a few cases. In Nagasaki, where the radiation consisted almost entirely of gamma rays, 20 leukemia deaths (against 6 expected) were recorded between 1950 and 1970 in a population of some 7500 who received estimated doses of 10 to 600 rems (estimated average about 120 rems). In Hiroshima, the radiation contained a substantial proportion of neutrons, which are more effective biologically than gamma rays. About 16 500 people received doses in excess of 10 rems (probable average at least 100 rems) and there were 61 deaths from leukemia between 1950 and 1970, compared with 13 expected. Individuals who were less than 10 years of age at the time of the explosions were about twice as susceptible as those who were older.

A study has been made of some 14 500 people (mostly men) in the United Kingdom who were given radiation treatment for an arthritic disease known as ankylosing spondylitis. The total dose to the spine, received over a period ranging from a month to a few years, has been estimated to vary from roughly 250 to 2250 rems for different individuals, with an average of about 370 rems. An increase in leukemia incidence was first observed after a latent period of about 2 years. The annual incidence increased to a peak in the later years and then declined until it was about the same as normal. By this time, 60 leukemia deaths were recorded, whereas about 7 would have been expected in an average population of the same size. A better comparison would have been with sufferers from ankylosing spondylitis who were not treated with radiation, but the number of such people was too small to provide data of statistical significance.

An increase of leukemia has also been found in other people who have received fairly large doses of x-rays for diagnostic and therapeutic purposes. There are, however, some contradictions that have not been explained in connection with radiation treatments in the pelvic region. Four independent studies have shown leukemia excess in women who received x-rays for the relief of certain gynecological conditions. On the other hand, no increased incidence of leukemia was observed in four groups of patients who were treated with radiation for cervical cancer, although the dose range was apparently much the same in both cases.

The greater sensitivity of young children than adults to radiation-induced leukemia in Japan was mentioned earlier. A similar sensitivity has been found among children who were exposed to radiation for medical reasons. Particular instances in which this increased sensitivity has been reported are a group of 1450 infants treated with x-rays for enlarged thymus glands and 2000 young children who received x-ray treatment for ringworm of the scalp. Despite the relatively small numbers, there seems little doubt that the incidence of leukemia in these two groups was greater than would have been expected among adults receiving the same doses of radiation.

Several studies have been made of leukemia occurrence among children whose mothers had received radiation doses to the abdomen for medical purposes during pregnancy. The results, however, are not decisive; in some of the studies the leukemia incidence appeared to be above normal, but in others the incidence was essentially normal. No increase in leukemia has been found in children born to Japanese mothers who received much larger doses of atomic bomb radiation while pregnant. Despite the inconclusive nature of the evidence, it is generally assumed, when making estimates of the effects of radiation (p. 191), that exposure before birth increases the risk of leukemia (and other kinds of cancer) during the first 10 years of life.

Bone Cancer

Bone cancer is not very common in man, but its incidence is increased by radiation. Except for a few instances associated with x-ray therapy, most cases of radiation-induced bone cancer have resulted from the deposition of radionuclides in the skeleton. A group of 775 painters of self-luminous instrument dials, who worked between 1915 and 1935, have been followed for periods of 36 to 56 years. These people had substantial amounts of radium-226 (an alpha-particle emitter) deposited in their skeletons. About one in ten developed some form of bone cancer, but the absorbed dose in all cases was estimated to be more than 500 rads; this is considered to represent a dose equivalent of about 5000 rems. For larger doses, the incidence of bone cancer was found to increase rapidly with the dose. The latent period appears to range from about 10 years to possibly 30 or 40 years.

Intravenous injection of radium-224, which also emits alpha particles, has been used for the treatment of tuberculosis and ankylosing spondylitis. Of a group of 900 patients who received this treatment, more than 50 developed bone cancer after latent periods ranging from 4 to 20 years. Those who were less than

20 years of age when they had the injections were about twice as sensitive as older individuals.

Four cases of bone cancer, compared with less than one expected normally, were observed among the approximately 14 500 people who were given external radiation (x-ray) therapy for ankylosing spondylitis. One of these four, however, had also received radium-224 injection. Since the patients already suffered from a disorder of the musculoskeletal system, the significance of the small number of cases of bone cancer is uncertain. After a latent period of more than 20 years, there was an indication of an increase in bone cancer among the atomic bomb survivors in Japan who had received external radiation doses exceeding about 400 rems. Here again, there are too few instances for reliable analysis.

Lung Cancer

As with bone cancer, most of the evidence for radiation-induced lung cancer comes from cases in which alpha-emitting radionuclides are present within the body. It has been realized for many years that men working in underground uranium, fluorspar, and some other mines are prone to develop lung cancer. Minerals containing uranium emit radioactive radon-222, which decays rapidly to form solid radioactive products. These solids, which emit alpha particles, are deposited in the lungs, where they may induce malignancies. Before adequate precautions were taken to reduce the radon concentration in the atmosphere (Chapter 8), many uranium miners accumulated lung doses of several thousand rems.

Among the ankylosing spondylitis patients who received (external) x-ray treatment, 96 developed lung cancer over a period of several years; in a similar-sized group of the general population, about 54 cases would have been expected. The average dose to the lining of the lungs was estimated to be about 400 rems. The possible effects of such factors as smoking, the disease itself, and other forms of medication on the development of lung cancer among the spondylitis patients have not been investigated. There are indications of an increase in lung cancer in the Japanese survivors, but there are uncertainties in interpreting the data.

Breast Cancer

Breast cancer can be induced in women exposed to sufficiently large doses of radiation. Before 1950, people suffering from tuberculosis were often subjected to frequent fluoroscopic (x-ray) examinations of the chest. In a group of women who received radiation doses in the range of 50 to 7000 rems from these examinations, the incidence of breast cancer was significantly greater than expected. The time delay between the beginning of the examinations and the detection of the cancers was from 8 to 24 years. A similar increase in breast cancer was observed in women who had been treated with x-rays for acute inflammation of the breast following childbirth. The doses were estimated to be from 50 to 450 rems, and the minimum latent period was about 10 years.

Prior to 1965, there was little evidence of a significant increase of breast cancer among women who survived the atomic bombings of Hiroshima and Nagasaki in 1945. In subsequent years, however, the number of cases of breast cancer was larger than in a normal Japanese population. The minimum latent period was apparently about 15 years.

Thyroid Nodules

The thyroid glands of adults are comparatively resistant to ionizing radiation, but the thyroids of young people exposed before about 20 years of age are much more sensitive. An increase in the incidence of thyroid nodules has been observed for such individuals among

1. the Japanese survivors of the atomic explosions
2. the inhabitants of the Marshall Islands who had substantial deposits of radioactive iodine in their thyroid glands, following ingestion of fallout from a nuclear weapons test
3. children in the United States who had been treated with x rays (at doses down to 20 rems and possibly less) for various throat disorders.

Most of the radiation-induced thyroid nodules are benign (i.e., not malignant) and can be readily removed by surgery. The latent period is about 6 years at least, but it is often longer.

Other Types of Cancer

The mortality data for the period 1950 to 1970 of the atomic bomb survivors in Japan indicate some increase in the incidence of other types of cancer in addition to those mentioned above. The difference between the observed incidence and that expected in a similar group not exposed to radiation was detectable only for doses in excess of 200 rems. In the ankylosing spondylitis patients, there was no evidence of an increase in cancer of parts of the body that receive only small doses, such as the kidney, liver, and urinary system. An increase was observed, however, in the highly irradiated regions, including the pharynx, stomach, and pancreas, as well as the organs and tissues already considered.

Life Shortening

It was once thought that exposure to ionizing radiation could cause a *nonspecific life shortening*. That is to say, there might be a general decrease in life expectancy apart from that arising from specific causes, such as an increased incidence of cancer. A review of the mortality data for about 5000 Japanese atomic bomb survivors who died between 1950 and 1970 has, however, provided no evidence for such nonspecific life shortening resulting from exposure to radiation.

Some clarification of the situation may be provided by a report published in 1972 of the mortality rates of mice that received various whole-body doses of gamma rays. Provided the conditions were such that the life shortening was less than 15 percent of the average life expectancy, the decrease arises specifically from the increase of various forms of cancer. If the radiation dose is large enough to cause a decrease of more than 15 percent in the average life, there is apparently also a nonspecific effect. Possibly something similar may occur in humans.

Mention may be made of an experiment with mice exposed to 110 mrems of gamma rays for 8 hours each day throughout their lifetime of 2 to 2 1/2 years.

These mice lived longer, on the average, than the unexposed controls did (or those receiving substantially larger daily radiation doses). This result does not appear to have been refuted, although it has not been explained satisfactorily.

RADIATION RISK ESTIMATES

Basic Assumptions

Information obtained from studies of delayed somatic effects, such as those described in the preceding sections, have been used, together with certain assumptions, to estimate the potential risks that might be associated with exposure to low levels of ionizing radiation. The radiation levels of interest are so low that, if there are any delayed effects, there is no presently known way in which they can be observed. The assumptions may be conveniently stated as follows:

1. There is no *threshold* dose below which radiation has no delayed biological effect. In other words, any dose, no matter how small, is associated with some risk.
2. The risk associated with a given dose of radiation is proportional to that dose. That is to say, if the dose is increased (or decreased) by a certain factor, the risk is increased (or decreased) by the same factor.
3. The risk associated with a given dose is the same regardless of whether the radiation is received in a very short time (e.g., from an atomic explosion), over a period of weeks or months (e.g., for medical purposes), or over a lifetime (e.g., from nuclear power generation).

The significance and validity of these assumptions are now examined.

Except possibly for exposures of fetuses during pregnancy (p. 188), neither early nor delayed clinical effects of radiation have been observed at doses below about 20 rems. This does not necessarily mean that there are no such effects; they may exist, but are not detectable in a population group of reasonable size (e.g., a few million people). Consequently, the conservative approach is to assume that there is no threshold below which radiation has no biological consequences.

If the observed response (e.g., incidence of a given type of cancer) to a particular dose of radiation is plotted on a graph against that dose, the result is called a *dose-response curve.*[b] According to the second assumption, that the risk is proportional to the radiation dose, the dose-response curve is a straight line. Hence, it is sometimes referred to as the *linear hypothesis*. If there is no threshold dose, as required by the first assumption, the straight line extends to the origin of the graph (i.e., the zero point).

For the occurrence of thyroid nodules among individuals exposed below 10 years of age, the evidence for a linear dose-response curve, possibly extending to the origin, appears to be fairly good. But in most studies, because of uncertainties in the radiation dose and the relatively small numbers of exposed individuals, the response curve may or may not be linear. A typical situation is il-

[b]The word *curve* is used here quite generally; it could be a straight line.

lustrated in Fig. 7-1, which applies to the incidence of radiation-induced leukemia between 1950 and 1966 among the atomic bomb survivors in Nagasaki. By allowing for the possibility of substantial uncertainties that are known to exist, the points can be regarded as falling on a straight line. However, an elongated (i.e., flattened) S-shaped curve, implying a very low (but not zero) incidence for small doses, appears possible. In fact, a later analysis of the data has shown that, for the gamma-ray component of the radiation, the dose-response curve is more likely to be S-shaped than linear. Risk estimates (for small doses) based on the simple linear hypothesis would then be too large.

The radiation dose rate is the dose received in a specified time. If a given dose is received in a short time—say, a minute or a day—the dose rate will be higher than if the same dose is received in a longer time—say, months or years. The dose rates for atomic weapons radiation can be very high, and they are generally fairly high when radiation is used for medical purposes. For exposures to natural background radiation or to the effluents from nuclear power facilities, the dose rates are very low. The third assumption implies that the risk associated with a given dose of radiation is independent of the rate at which the dose is delivered. It does not allow for the possibility of some repair of radiation injury at low dose rates.

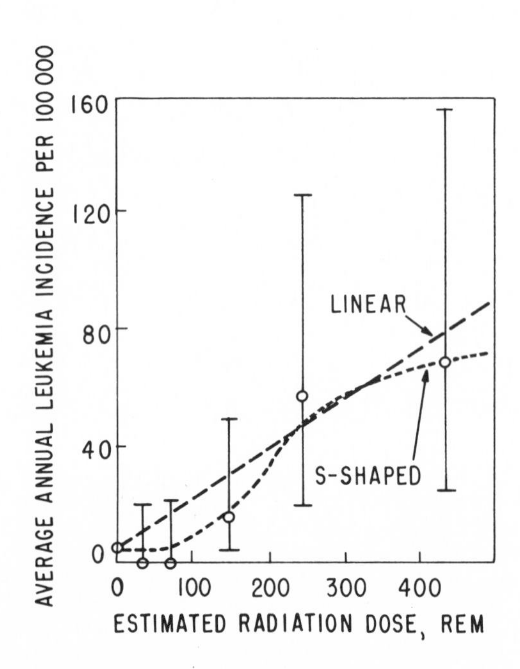

FIG. 7-1. Average annual incidence of leukemia calculated for a population of 100 000 among the atomic bomb survivors in Nagasaki during the period from 1950 to 1966. The value corresponding to zero radiation dose is the normal annual incidence of the disease in Japan. The dashed line indicates a possible linear relationship that lies within the 95 percent confidence limits (vertical lines passing through the points). In this particular case, an elongated (or flattened) S-shaped curve (dotted line) is equally probable.

There is no doubt that the early (or acute) somatic effects of radiation, such as vomiting, erythema, loss of hair, and decrease in blood cell count, are not observed at low dose rates. For the delayed effects, however, the evidence from human experience is uncertain. But several experimental studies with small animals with the use of low-LET (linear energy transfer) radiation, such as beta particles and gamma rays (p. 122), have indicated that the increase in cancer is less when a given dose is delivered over a long time period than over a short time. In other words, it is possible that partial repair occurs at low dose rates.

The results of a study of the incidence of myeloid (bone marrow) leukemia in male mice are shown in Fig. 7-2. (Myeloid leukemia is the type commonly induced by radiation in humans.) Various total amounts of gamma rays were administered either as a single dose (high dose rate) or in small doses at daily intervals, referred to as a *fractionated dose* (low dose rate). The leukemia incidence in the mice was definitely smaller at the low dose rate. Incidentally, it appears to be well established that genetic effects in mice are less at low dose rates than at high dose rates (p. 202). The assumption that the incidence of delayed radiation-induced somatic effects is independent of the dose rate thus results in overestimates of the risk.

It is seen from Fig. 7-2 that for large single doses of radiation—more than 350 rems in this case—the incidence of leukemia in mice decreases as the dose is increased. A similar decrease has been reported at high doses and dose rates in studies with other small animals. This decrease is attributed to the killing, at these high doses and dose rates, of susceptible cells that, if only injured but not killed, would have contributed to the development of leukemia. If this killing effect is significant, the use of data obtained at high doses and dose rates might tend to underestimate the incidence of leukemia at low doses and dose rates.

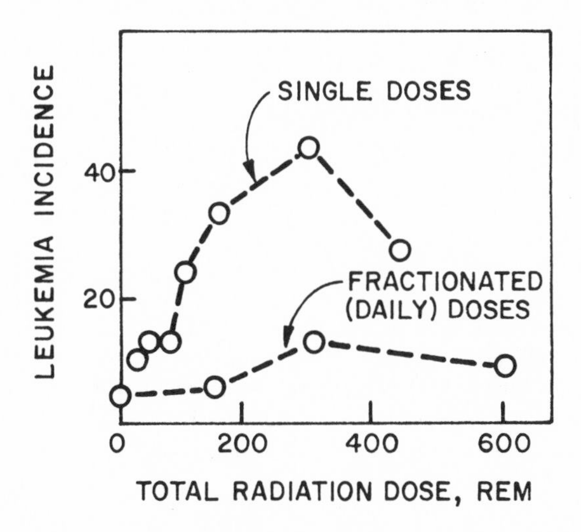

FIG. 7-2. Incidence of bone-marrow leukemia in male mice after receiving either one dose or a succession of smaller daily doses of gamma radiation. The decreased incidence for a large single dose is discussed in the text. (Adapted from A. C. Upton *et al. Radiation Research* 41 (1970): 467.)

In summary, there is no clear-cut evidence that can determine whether the linear extrapolation (with no threshold) of observations made at high radiation doses and dose rates correctly gives the effects to be expected at low doses and dose rates, or if it underestimates or overestimates them. In fact, each of these possibilities may be valid in a specific situation. The U.S. Environmental Protection Agency (EPA) has decided that the "linear, nonthreshold model [should be used] for deriving [radiation] standards to protect the public health." It will be seen later, however, that this model may be conservative, as indeed it should be when protection of the public is of concern.

Absolute and Relative Risks

The risk associated with a given dose of radiation may be expressed either as an *absolute risk* or a *relative risk*. The absolute risk is a measure of the excess incidence of a particular type of cancer in a large exposed population group over the expected (or normal) incidence in a similar unexposed group. The relative risk, on the other hand, expresses this excess incidence as a fraction (or percentage) of the normal incidence. The procedures for calculating the respective risks illustrate the difference between them.

A group of Japanese, who were more than 10 years old when exposed to radiation from the atomic explosion over Nagasaki in 1945, were observed from 1950 though 1971 for the incidence of leukemia. During the 20-year period, the group averaged 5300 persons, and they had received an average dose of 113 rems per individual. The population dose (p. 35) was thus $5300 \times 113 = 6.0 \times 10^5$ person-rems. The total incidence of leukemia in the group was 12 cases over the 20 years, compared with the normal expectation in Japan of 5.3 cases in an unexposed group of the same size. The excess incidence of leukemia was thus $(12 - 5.3)/20 = 0.335$ cases/year. The absolute risk of radiation-induced leukemia is then $0.335/(6.0 \times 10^5) = 0.56 \times 10^{-6}$/year per person-rem. For convenience, this number is multiplied by a million to give

$$\text{Absolute risk} = 0.56 \text{ cases/year per million person-rems.}$$

This is the expected average number of cases per year, for a period of 20 years, resulting from a total population dose of 1 000 000 person-rems.

The relative risk, in percent per rem, is defined by

$$\text{Relative risk} = \frac{\text{Absolute risk}}{\text{Normal incidence}} \times 100\% \text{ per rem.}$$

The normal incidence of leukemia in Nagasaki, as seen above, is 5.3 cases in 20 years—that is, $5.3/20 = 0.265$ cases/year, in a group of 5300 persons. The expected incidence per year per million persons is then $(0.265/5300) \times 10^6 = 50$ cases. Since the absolute risk has been found to be 0.56 cases/year per million person-rem, it follows that

$$\text{Relative risk} = \frac{0.56}{50} \times 100 = 1.1\% \text{ per rem.}$$

The relative risk is sometimes expressed in terms of the *doubling dose*; this is the average individual dose in a large population group that is expected to double the normal incidence of a particular type of cancer. In other words, the excess cases of cancer induced by the doubling dose is the same as the normal incidence in an unexposed population; that is to say, the increased incidence would be 100 percent. Since the relative risk gives the percent increase per rem, it follows that

$$\text{Doubling dose} = \frac{100}{\text{Relative risk (percent per rem)}} \text{ rems.}$$

If the relative risk is expressed as a fraction, rather than as a percentage, the doubling dose is equal to 1/relative risk.

Some of the data used in calculating the absolute and relative risks for radiation-induced leukemia in individuals exposed at 10 years of age or older are given in Table 7-II. The information in this table, as well as that in the next section, is taken from the BEIR Advisory Committee report entitled *The Effects on Populations of Exposure to Low Levels of Ionizing Radiation*, referred to on page 118.

A revised BEIR report was completed in May 1979, but publication was held up until 1981 by the National Academy of Sciences because of a difference of opinion among the committee members concerning the expected effects of very low radiation levels. It has been stated, however, that the revised report largely confirms the general conclusion reached in the earlier report. Numbers derived from new studies are said to be not substantially different from those in the 1972 BEIR report that are used here.

The absolute risk concept implies that, for a given value of the absolute risk, the increased incidence of a particular type of cancer would depend only on the radiation dose. On the other hand, a given value of the relative risk means that the increased incidence would also be proportional to the normal incidence. For example, the normal incidence of stomach cancer is greater in Japan than in the United States; the relative risk concept would then require a certain dose of radiation to induce more cases of stomach cancer in the former country than in the latter. Radiation biologists are not sure which of the two concepts is the cor-

TABLE 7-II

RISK ESTIMATES FOR LEUKEMIA FOR RADIATION EXPOSURE AT AGE 10 YEARS OR MORE

GROUP STUDIED	AVERAGE PERSON-REMS	YEARS OBSERVED	LEUKEMIA CASES		ABSOLUTE RISK[a]	RELATIVE RISK[b]
			Observed	Normal		
Nagasaki atomic bomb survivors	6.0×10^5	20	12	5.3	0.56	1.1
Spondylitis patients	2.1×10^6	25	52	5.5	0.88	2.3
Gynecological patients	1.6×10^5	24	6	1.3	1.2	2.7

[a]Absolute risk per year per million person-rems.

[b]Relative risk in percent per rem.

rect way to express the risks that might arise from radiation exposure. Consequently, results of calculations based on both concepts are given in the next section.

Since no other information is available (or even attainable), the risk estimates, such as those in Table 7-II, refer to people who have received fairly large doses of radiation at high or moderately high dose rates. According to the linear extrapolation hypothesis with no threshold, these absolute and relative risk estimates (per rem) would be the same at all doses and dose rates down to the very smallest values. Another consequence of this hypothesis is that the risk in a large population group would be proportional to the total population dose in person-rems regardless of how the doses were distributed among the members of the group.

Total Risk Estimates

For the purpose of making estimates of the total risk of radiation-induced cancer in a large population group, the BEIR Committee allowed for the possibly greater sensitivity before birth and below the age of 10 years than in later life. For a lifetime exposure (e.g., from natural background radiation or nuclear plant effluents), three periods were considered: (a) before birth, (b) from birth through 9 years of age, and (c) from 10 years onward. For the last two groups, the minimum latent period—that is, the time after exposure before any excess cancer deaths occur—is assumed to be 2 years for leukemia and 15 years for all other cancer types. Following the earliest radiation-induced deaths, the incidence is assumed to continue at the same annual rate for a number of years; this has been called the *plateau region*. The plateau is taken to be 25 years for leukemia and 30 years for other cancers. The possibility is also considered that the plateau region for other cancers may persist the whole lifetime. For exposure prior to birth, there is assumed to be no latent period, and the plateau region is 10 years for all cancer types.

The foregoing assumptions are summarized in Table 7-III. In addition, the last two columns give what are thought to be conservative absolute and relative risk estimates for radiation-induced leukemia and other cancer deaths as derived in the manner described in the preceding section.

In the BEIR report, the population of the United States is divided into 13 age groups, from zero to 85 + years of age. Then, with the use of the data in Table 7-III and the 1967 statistics for leukemia and other cancer deaths, the expected number of additional radiation-induced deaths in each age group is calculated. The results for the appropriate age groups are then added to give the values in Table 7-IV for a population dose of 1 million person-rems *each year* from conception to death.[c] It should be understood that the estimated number of annual deaths for the various exposure periods are not those occurring during these respective periods, but later in life averaged over the plateau region. The data in Table 7-IV are based on a plateau region of 30 years.

[c]The values in the BEIR report (Table 3-I) are for the 1967 U.S. population close to 200 million, receiving an average individual dose of 100 mrems/year; this corresponds to an annual population dose of 20 million person-rems. The data in Table 7-IV were obtained by dividing the BEIR values by 20.

TABLE 7-III

ASSUMED VALUES IN CALCULATING RADIATION-INDUCED CANCER INCIDENCE

Exposure Period	Cancer Type	Minimum Latent Period (years)	Plateau Region (years)	Death Risk	
				Absolute (per year per million person-rems)	Relative (percent increase per rem)
Before birth	Leukemia	0	10	25	50
	All others	0	10	25	50
Age 0-9 years	Leukemia	2	25	2.0	5.0
	All others	15	30	1.0	2.0
10 years and over	Leukemia	2	25	1.0	2.0
	All others	15	30	5.0[a]	0.2

[a]Made up as follows: breast, 1.5 (allowing for virtual absence in males); lung, 1.3; gastrointestinal tract, 1.0; bone, 0.2; all others, 1.0.

Suppose the plateau region for cancer other than leukemia, for exposure at 10 years of age and over, extends for the lifetime of the individual rather than for 30 years. For the absolute risk concept, the change from Table 7-IV would not be large; the estimated annual number of deaths would be 100, compared with 86 per million person-rems/year for the absolute risk concept. On the basis of the relative concept, however, the estimate would be 504 additional cancer deaths, compared with 159 in Table 7-IV. This marked increases arises mainly from a jump to 193 (compared to 35.8) in deaths from cancer other than leukemia resulting from exposure in the 0-9-year period, when the susceptibility to radiation is high.

The lifetime duration of the plateau region would imply that cancer initiated by radiation at an early age may appear at the same relative risk throughout life. At age 50 and beyond, which would require a plateau of more than 40 years, the natural incidence of cancer is very high. Consequently, the relative risk concept leads to prediction of a large number of extra deaths from radiation-induced cancers among older people.

After taking all factors into consideration, the BEIR Committee concluded that if an average population of 1 million persons (adults, pregnant women,

TABLE 7-IV

ESTIMATED ANNUAL NUMBER OF CANCER DEATHS FOR POPULATION DOSE OF 1 MILLION PERSON-REMS PER YEAR (30-YEAR PLATEAU)

Exposure Period	Absolute Risk Concept		Relative Risk Concept	
	Leukemia	Other Cancer	Leukemia	Other Cancer
Before birth	3.8	3.8	2.8	2.8
Age 0-9 years	8.2	3.6	4.7	35.8
10 years and over	13.8	53.1	29.4	83.2
Total	86		159	

children, etc.) were exposed to 1 rem of radiation per year, each year of their lifetime, the number of deaths from radiation-induced cancer would eventually be 150 to 200/year. If nonfatal cancers are included, the number could be higher by a factor of roughly 2, although this has not been established. In addition, there could be 100 to 200 cases per year of thyroid nodules, which are generally nonfatal. However, it must be emphasized that, as the BEIR report points out, *"these figures must not be taken to represent more than crude estimates of risk, based on the incomplete nature of the data at present available."*

The foregoing estimates are derived from the linear hypothesis, which assumes that the risks (per rem) are the same for low doses and dose rates as for high doses and dose rates. The BEIR Report notes that this assumption could well result in a substantial overestimate of the risk, and includes the following statement: "Use of a factor, if known for man, to take into account the influence of dose and dose rate on the dose-effect [i.e., dose-response] relationship might reduce these [risk] estimates appreciably."

Report No. 43 (January 1975) of the National Council on Radiation Protection and Measurements refers to a review of radiobiological data concerning life shortening in dogs and mice, induction of leukemia and bone cancer in mice, and mammary and thyroid tumors in rats. The conclusion reached in this review is that, for gamma rays, the factor referred to above has a mean value of 0.2 in small mammals. That is to say, a given low-LET radiation dose received at a low dose rate is only 0.2 times as effective as at high dose rates. The factor varies to some extent with the biological effect and the animal species studied, and possibly with the radiation levels at which comparisons are made. But if the overall, preferred value of 0.2 were also applicable to human beings, the cancer risk estimates given above (and below) would be five times too large.

Somatic Risks from Nuclear Power Operations

According to the estimates made by the EPA, modified in the manner described on page 179, the general population dose in the United States arising from nuclear power operations could be 0.3 million person-rems/year in the year 2000. This is based on an estimated population of 300 million and an average individual annual whole-body dose of about 1 mrem. If this population dose were to persist over the whole lifetime of everyone in the United States, then, based on the BEIR Advisory Committee conclusions given above, a maximum of 40 to 50 cancer deaths above normal would eventually be expected per year from whole-body radiation exposure associated with nuclear power operations. There could also be 10 to 15 additional lung cancers per year and some cases of malignant and nonmalignant thyroid nodules.[d]

Although an annual population dose of 0.3 million person-rems is expected by the year 2000, probably 50 more years will elapse before everyone in the United States will have received a lifetime exposure at this level. Hence, the foregoing estimates would not be reached until about the year 2050; in the intervening years, the numbers would be less.

[d]In the mid-1970s, the "normal" annual cancer mortality rate in the United States was about 1600 per million population; the total was thus about 350 000/year.

From a 1979 report of the Committee on Science and Public Policy of the National Academy of Sciences, it appears that for every 1000 megawatt (MW)-years of electricity produced by nuclear reactors, 0.2 additional cancer death may eventually be expected among the general population (i.e., excluding those occupationally involved). Although it seems unlikely, suppose that by the end of the twentieth century 300 000 MW-years of electricity are generated annually by nuclear means in the United States. There should then be some 60 additional cancer deaths per annum during the early years of the next century, in agreement with the number derived above.

A fundamental assumption made in estimating the additional cancer incidence from nuclear power operations should be borne in mind. It is that the number of radiation-induced cancers would always be the same for a given population dose in person-rems. That is to say, the number of cases in a population of 300 million receiving an average whole-body dose of 0.001 rem (1 mrem)/year would be the same as in a population of 3000 receiving an average dose of 100 rems (100 000 mrems) per year. There seems to be no way at present to determine whether this is true or not.

In any event, the potential effects of the effluents from nuclear power operations should be compared with those from other types of power generation. The sulfur and nitrogen oxides, hydrocarbons, particulate matter, etc.—not to mention the small amounts of radioactive material present in the discharge from a fossil-fuel plant—undoubtedly have some adverse consequences. However, these matters have not yet received the attention given to the effects of ionizing radiations. Nevertheless, a number of comparisons suggest that, under normal operating conditions, the generation of a given amount of electricity from nuclear energy will result in fewer deaths than from the use of coal as the energy source.

GENETIC EFFECTS OF RADIATION

Types of Mutations

As mentioned earlier, ionizing radiation can cause changes (mutations) in the chromosomes present in the sex (or germ) cells; such changes may not affect the individual exposed to radiation, but various effects may appear in subsequent generations. Observations made on insects and mammals deliberately exposed to radiation in laboratory experiments suggest that the mutations induced by radiation, as well as by other mutagenic agents (e.g., various chemicals and heat), are similar to those that arise spontaneously.

Changes in the chromosomes, whether caused spontaneously or by a mutagenic agent, may take two forms: namely, gene (or point) mutations and chromosome abnormalities. In a gene mutation, the general structure and number of chromosomes in the germ cell are unchanged, but there is a change in one or more individual genes that may lead to the appearance of characteristics not present in either parent. Chromosome abnormalities are represented either by a deviation from the normal number of chromosomes in the germ cell (numerical aberrations) or by an overall change in the structure as a result of chromosome breakage followed by faulty recombination of the parts (structural rearrangements). Since most mutations in mammals occur in the genes, the subsequent treatment is concerned mainly with genetic mutations.

Mutations (or the genes carrying them) are generally described as being either *dominant* or *recessive*. If a particular gene mutation is dominant, the appropriate characteristics will appear in the offspring, even though the mutation has occurred in the germ cells (sperm or egg) of only one of the parents. On the other hand, a recessive mutation must be present in both sperm and egg if the characteristics are to be fully evident in the individual resulting from their union. A recessive mutation may be latent for many generations until the occasion arises for the combination of sperm and egg cells, both of which have that mutation.

The majority of mutations, both spontaneous and induced, are recessive. Nevertheless, it appears that a mutation is seldom completely recessive, and some effect is often observable in the next generation even if the particular mutation is derived from only one parent. Most mutations have deleterious effects and, as a result, do not tend to accumulate in the population. The few beneficial mutations that do arise tend to persist.

The harmful effects of a deleterious mutation may be comparatively minor, such as a susceptibility to disease or a decrease of a few months in life expectancy. Other mutations may have more serious consequences, such as death in the embryonic stage or infertility of the offspring. Thus, individuals bearing deleterious genes may be handicapped relative to the rest of the population, particularly in the respect that they tend to have fewer surviving children or they die earlier. It is apparent that such harmful mutations will eventually be eliminated from the population. A mutation that does great harm will be removed rapidly, since few (if any) individuals bearing such a mutation will survive to the age of reproduction. On the other hand, a slightly deleterious mutation may persist for several generations, possibly in recessive form.

The main interest of mutations induced by radiation is, of course, their potential effects on man. It is known that even small doses (5 to 10 rems) of radiation can cause chromosomal changes in human cells, especially those in the blood because they can be seen under the microscope. But there is little information concerning the effects of radiation on the hereditary characteristics of man.

The genetic material of most living organisms is fundamentally similar and inferences concerning the effects of radiation on man can be made from observations on other species. Thus, studies of bacteria, fruitflies, mice, and other organisms have provided useful information about how mutations occur and the consequences of these mutations. However, since data from a species reasonably close to man might be regarded as most relevant, another mammal, the mouse, has been used as the experimental animal for evaluating the genetic effect of radiation in man.

The effects of radiation on gene mutations and chromosome abnormalities in mice have been studied extensively. Investigation of the effects of radiation by observations on gene mutations, commonly referred to as the *specific locus method*, is preferred because the results are clearly observed in the offspring of the irradiated parents. The main drawback, which has militated against its more general use, is that it requires large numbers of animals. A single experiment, for example, may involve 10 000 mice or more, if significant results are to be obtained even with large doses of radiation.

Genetic Studies by the Specific Locus Method

Many genes exist in two forms that often produce contrasting effects (e.g., different eye colors in man); such pairs of genes are called *alleles*.[e] It is generally accepted that contrasting alleles arise from gene mutations. One member of a pair of alleles is usually dominant, whereas the other is recessive. If both chromosomes of a given pair in a fertilized egg (zygote) contain the same alleles, one derived from each parent, the zygote or the resulting individual is said to be *homozygous* for the given gene. But if the two chromosomes of a pair contain contrasting alleles of a particular gene, the result will be a *heterozygous* individual. For a recessive characteristic to be apparent, the zygote must be homozygous for the recessive allele. If it is heterozygous (or homozygous for the dominant allele), only the dominant effect will be evident.

The specific-locus method in genetic studies depends on the use of a special strain of mice that is homozygous for a number of recessive genes. A uniform stock of "wild-type" mice, which is homozygous for the corresponding dominant allele, is exposed to a known amount of radiation. These irradiated mice and a similar number of nonirradiated mice are mated to unexposed mice of the special strain known to be homozygous for the specific recessive allele. By comparing the frequency of the occurrence of the recessive characteristics in the progeny of the two sets of mice, it is possible to determine how many mutations of the particular gene have been induced by the given dose of radiation in the exposed animals. Altogether, in the United States and in the United Kingdom, twelve different gene loci (or allele pairs), with such readily observable characteristics as coat coloring and spotting, ear length, tail type, foot malformation, etc., have been examined for the genetic effects of radiation.

From the early studies of radiation-induced mutations in the fruitfly, it had been concluded that the mutation frequency depended only on the total radiation dose accumulated by the parents up to the time of conception and was independent of the period over which the particular dose had been received. In other words, the genetic effects in the fruitfly, and presumably in man, were apparently independent of the dose rate of the radiation. However, results first obtained in 1958 (and confirmed later) from specific locus studies made in the United States revealed the important fact that the frequency of mutations in mice is significantly dependent on the dose rate in certain ranges. The effects are somewhat different for male and female mice that have been exposed to ionizing radiation, and the results for the two sexes are described in turn below.

Results of Specific Locus Studies

Despite the use of several thousand mice at each dose and dose rate, a significant increase in point mutations was detectable only when dose and dose rate were fairly high. The lowest dose rate that could be used, with an expectation of

[e]There may be several alleles of a given gene, but only two are present in a cell, one on each chromosome of a pair.

observable effects, was 0.001 rem/minute, which is equivalent to about 500 000 mrems/year. This is 5000 times the average natural background level of 100 mrems/year. The total gonad dose prior to conception ranged from 86 to 1000 rems compared to an average of about 3 rems for the natural radiation background dose to man in a generation (30 years). It is apparent that the situation here is similar to that for delayed somatic effects. If radiation-induced mutations are to be observed, even in a moderately large population group, the doses and dose rates must be much higher than those of immediate interest.

In male mice exposed to x rays or gamma rays in the dose-rate range from 1000 down to 90 rems/minute,[f] the frequency of induced mutations per rem was found to be essentially constant. In this dose-rate range, a given dose of radiation to the gonads produces the same effect regardless of the dose rate. The mutation frequency is then dependent only on, and is proportional to, the total dose. (Later studies have shown that when the total dose is small, the frequency is less than the proportional value, but this is not important here.)

If the dose rate is less than 90 rems/minute, however, the mutation frequency per rem decreases steadily down to a dose rate of 0.8 rem/minute. At this point, the mutation frequency per rem becomes constant again and remains so down to 0.001 rem/minute, the lowest value at which the specific-locus studies were feasible. The major finding was that at the lower range of dose rates (0.8 rem/minute and below) the induced mutation frequency per rem in male mice was only about one-third that in the higher range (90 rems/minute and above).

In each dose-rate range, with the exception of small doses in the high dose-rate range, there is a proportionality (linear relationship) between the number of mutations and the total dose received. But the proportionality constant at the higher dose rates is three times as large as at the lower dose rates. Thus, a given dose of radiation will produce three times as many point mutations if the radiation is delivered at a high dose rate than if it is delivered at a lower dose rate. The situation is illustrated in Fig. 7-3, which shows the mutation frequency in a large population of male mice plotted against the total radiation dose. The results are seen to fall on (or close to) two straight lines, the upper one for the higher dose rates and the lower line for the lower dose rates. For zero radiation dose (above background) both lines extrapolate approximately to the spontaneous mutation rate in mice. As noted above, however, the observations fall below the upper straight line as it approaches the spontaneous rate (i.e., at low total doses).

In female mice, the dose rate effect is even more striking. At higher dose rates, greater than 90 rems/min, the mutation frequency per rem is greater than for males, but as the dose rate is decreased, the induced mutation frequency was found to decrease steadily. At a dose rate of 0.009 rem/min, the mutation frequency in females, for total doses up to 400 rems, was essentially the same as the spontaneous mutation rate. Thus, at sufficiently low dose rates, even large doses of radiation do not produce a significant number of additional genetic mutations in female mice.

[f]The original results are recorded in terms of exposures expressed in roentgens (R). For simplicity, the corresponding doses (and dose rates) are stated in rems, based on the assumption that a gonad exposure of 1 R is equivalent to a dose of 1 rem (p. 122).

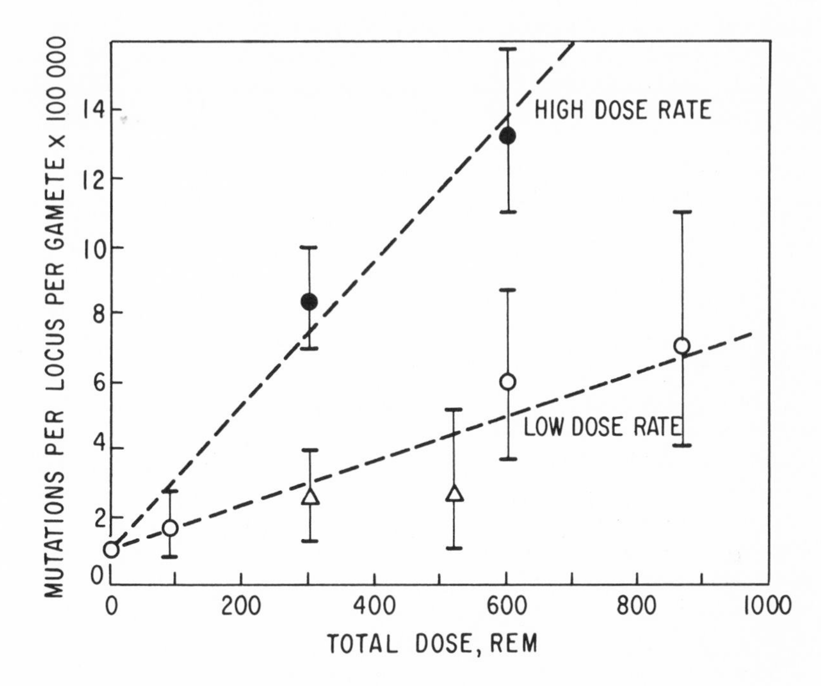

FIG. 7-3. Radiation-induced mutation in male mice. The high dose rate was 90 rem/min and the low dose rates were 0.001 rem/min (open circles) and 0.009 rem/min (triangles). The vertical lines passing through the points indicate 90 percent confidence intervals. (Adapted from W. L. Russell, "Environmental Effects of Producing Electric Power," Hearings before the Joint Committee on Atomic Energy, U.S. Congress, Washington, D.C. [1969].)

The difference in the genetic effects of radiation in male and female mice presumably arises from differences in the respective processes involved in the formation of the sperm (by the male) and the egg (by the female). The mechanisms, as outlined below, are the same in all mammals. In the male, the cells in the gonads, called spermatogonia, continually, throughout the reproductive lifetime, bud off cells by division; these cells undergo further divisions and pass through three stages of short duration before producing the mature germ cells (spermatozoa). In the female, the primitive germ cells (primary oocytes) are already formed before birth. They then go into an arrested state that remains without further division until a few hours before the mature egg (secondary oocyte) is released and is ready for fertilization. The most important stages for the induction of mutations are considered to be the spermatogonia in the male and the arrested oocytes in the female.

The recommendations for limiting the genetic (or gonad) dose of radiation for the general public, made in 1956 by the BEAR Committee of the National Academy of Sciences (p. 128), were based on the results of the specific-locus method as applied to male mice at high radiation dose rates. The frequency of radiation-induced mutations was taken to be the same in males and females. In humans, the frequency was assumed to be the same as in mice.

For the low radiation dose rates experienced in peaceful nuclear activities, however, the mutation frequency in males is lower than the value used by the BEAR Committee by a factor of 3, at least. Furthermore, an additional factor of 2 would be required to allow for the absence of radiation-induced mutations in females at low dose rates. Thus, the mutation frequencies upon which the recommendations of radiation protection standards for genetic effects in the general population were based in 1956 were probably too large by a factor of 6. Although the results described were obtained for mice, it is probable that the general conclusions are applicable to man in view of the similarity of the sexual processes in all mammals.

Genetic Risk Estimates

Firm projections of the absolute risk for radiation-induced mutations in man cannot be made because of the unavailability of reliable data. The doubling dose, based on the relative risk concept, can be used, however, although there is still a substantial degree of uncertainty.

From specific-locus experiments, it appears that at the lowest dose rate, which is still high in comparison with natural radiation levels, the average rate for radiation-induced (recessive) mutations in male mice is 5×10^{-8}/rem per gene. For females, on the other hand, the rate is too small to be measured. Hence, the average for the two sexes is taken to be 2.5×10^{-8}/rem per gene. (This average allows for the factor of 6 referred to above.) If the same three conservative assumptions are made as when estimating somatic risks (p. 191), the mutation rate would apply at all radiation doses and dose rates down to the very smallest. Since no corresponding data are available for human beings, the radiation-induced mutation rate for mice is assumed to apply to man.

According to the BEIR Advisory Committee's 1972 report, the rate of spontaneous (i.e., natural) mutations in humans is in the range of 5×10^{-7} to 5×10^{-6}/gene per generation. At the lower end of the range, the relative mutation risk per generation would be $(2.5 \times 10^{-8})/(5 \times 10^{-7}) = 0.05$/rem. Since the doubling dose (p. 195) is numerically equal to the reciprocal of the relative risk (expressed as a fraction), the doubling dose is here $1/0.05 = 20$ rems per generation. At the upper end of the range, the doubling dose is calculated to be 200 rems per generation. The doubling dose for radiation-induced mutations in mice is thus expected to be between 20 and 200 rems per generation. An independent estimate based on five studies with mice, which included specific-locus measurements and observations on chromosome aberrations, indicated an average doubling dose of about 180 rems per generation.

The only data on radiation-induced mutations in humans are available from the Japanese atomic bomb survivors. Children whose parents were exposed to the radiation before conception showed no significant difference in health, physical measurements, and death rates from children of unexposed parents. From this fact and an estimate of the average gonad dose, the conclusion has been drawn that the doubling dose must be at least 50 to 150 rems per generation.

On the basis of studies of genetic effects in human beings, the adverse consequences of mutations can be considered as falling into four categories:

1. dominant (single-gene) disease
2. chromosomal and recessive (single-gene) diseases
3. congenital anomalies (malformations)
4. constitutional and degenerative disease (e.g., diabetes, epilepsy, etc.).

The normal occurrences of these diseases per million live births is given in the second column of Table 7-V. The third column, adapted from the BEIR Committee's report, shows the estimated increases in the various diseases that would eventually result from an average gonad dose of 1 rem to each parent prior to conception (i.e., 1 rem per generation) for several generations.

The estimated increases in the table of genetically related diseases resulting from radiation exposure are *equilibrium values*. These values would be attained only after parents in several successive generations had received a dose of 1 rem per generation. The time required for equilibrium depends on the rate of elimination of the mutations by natural selection, but a rough estimate is that it would be about five generations, or 150 years, of exposure. After equilibrium is reached, there would be no further increase in the incidence of genetically related diseases provided the radiation level remained unchanged. For an exposure of 1 rem per generation, the occurrence of these diseases would then have increased, according to Table 7-V, from 60 000 to somewhere within the range of 60 060 to 61 500 per million live births.

The estimates in the various disease categories in Table 7-V were made in the following manner. The increased risk of dominant diseases, in the first category, is determined by the appropriate doubling dose, which is taken to be from 20 to 200 rems per generation in humans. These are the values derived above from the observed rate of recessive gene mutation in mice. There is evidence, however, that the rate of induction by radiation of dominant mutations in mice is considerably smaller. If this is the case, then the doubling dose for induced mutations of the latter type could well be more than 200 rems per generation. The conservative approach, as adopted by the BEIR Committee, is to assume that the doubling dose for recessive mutations is also applicable to dominant mutations.

If the doubling dose is taken to be 200 rems per generation, a generation dose of 1 rem will cause a fractional increase of 1/200 in the normal incidence of dominant diseases. The latter is given in Table 7-V as 10 000 per million live births; hence, the increase for a generation dose of 1 rem would be

TABLE 7-V
ESTIMATED INCREASE AT EQUILIBRIUM OF GENETICALLY RELATED DISEASES FOR 1 REM PER GENERATION IN 1 MILLION LIVE BIRTHS

Disease Category	Normal Incidence	Increase
Dominant diseases	10 000	50 to 500
Chromosomal and recessive diseases	10 000	Small
Congenital anomalies	25 000	10 to 1000
Constitutional and degenerative diseases	15 000	
Total	60 000	60 to 1500

10 000/200 = 50. Similarly, if the doubling dose is 20 rems per generation, the increase for 1 rem would be 10 000/20 = 500. This gives the range 50 to 500 in the first line of the third column.

Chromosomal and recessive diseases are somewhat less common than dominant diseases. Moreover, there are reasons for believing that recessive diseases are not likely to increase in proportion to the mutation rate. In any event, the incidence of recessive traits is expected to increase slowly over hundreds of generations. The effect of radiation exposure on the diseases in the second category is thus expected to be small.

Congenital anomalies and constitutional and degenerative diseases, included in the third and fourth categories, are not dependent on single-gene (or chromosome) mutations. These two categories are treated here as a single group. The BEIR Committee has defined a *mutational component* as the fraction of the incidence of these anomalies that is directly proportional to the mutation rate. The mutational component is thought to be between 1/20 and 1/2; there is thus an uncertainty factor of 10 in this respect multiplied by the same factor in the doubling dose. If the mutational component is 1/20 and the doubling dose is 200 rems per generation, the fractional increase for a generation dose of 1 rem will be 1/20 × 1/200 = 1/4000. Since the total normal incidence is 40 000, the increase would be 40 000/4000 = 10. At the other extreme, based on a mutational component of 1/2 and a doubling dose of 20 rems per generation, the increased incidence for 1 rem per generation would be (1/2 × 1/20) × 40 000 = 1000. Hence, the range on the last line of Table 7-V is 10 to 1000.

Apart from specific diseases, it is probable that, to a certain extent, unspecified ill health, both physical and mental, has a genetic basis. According to the BEIR Report, the mutational component may be taken to be approximately 1/5. Hence, for a doubling dose of 200 rems per generation, a radiation dose of 1 rem per generation would eventually result in an increase of 1/200 × 1/5 = 1/1000 (0.1 percent) in general ill health. For a doubling dose of 20 rems, the corresponding increase would be 1/20 × 1/5 = 1/100 (or 1 percent). Like the estimates in Table 7-V, these values would be reached only after several generations of exposure.

Genetic Risks from Nuclear Power Generation

Suppose that the average individual dose to the United States public from nuclear power operations is 1 mrem/year, as estimated (approximately) for the year 2000. If an average generation is taken to be 30 years, the generation dose would be 30 × 1.0 = 30 mrems, or 0.03 rem. Since a generation dose of 1 rem is estimated to increase the incidence of genetically related diseases by 60 to 1500 per million live births, the corresponding increase for a generation dose of 0.03 rem would be in the range from 0.03 × 60 = 1.8 (roughly 2) to 0.03 × 1500 = 45. At equilibrium, which would not be reached before the end of the next century, the total incidence of diseases with a genetic component would then be from 60 002 to 60 045 per million live births, if the normal (spontaneous) incidence is 60 000.

The increase in general unspecified ill health can be estimated in a similar manner. The increase for a generation dose of 1 rem is expected to be from 0.1 to

1 percent of the normal incidence. Hence, for a generation dose of 0.03 rem, the increase would be in the range of 0.003 to 0.03 percent. This means that for every million normal cases of genetically related ill health, an additional 30 to 300 cases would eventually arise from nuclear power operations.

The actual genetic risks might be smaller than the estimates given above because of the cautious approach used in deriving these estimates. In particular, the estimates depend on the value used for the radiation doubling dose. This has been obtained from experiments on mice made at doses and dose rates that are much higher than would be experienced by the general population from nuclear facilities or even from the natural background radiation. At the low doses and dose rates of interest, the doubling dose could be larger than 200 rems per generation; the actual genetic risks would then be less than the lowest of the ranges given above.

RADIATION EFFECTS ON AQUATIC ORGANISMS

Introduction

During normal operation, a nuclear power station discharges small amounts of radioactive materials into an adjacent water body. Although the concentrations of these substances that may be discharged are strictly limited, there is a possibility that the radiations may have an adverse effect on the aquatic ecosystem. Fish and lower organisms that form part of the food web living in the water body (Chapter 12) are exposed to radiation from both external and internal sources. As a result of the high bioaccumulation factors (p. 156) for some elements, the internal doses received by aquatic organisms at various stages of development could conceivably become significant.

Because fish are an important food source, the possible effects of radiation on the ecology of the Columbia River became a matter of concern when the first plutonium production reactors started operation at Hanford, Washington, in 1944. At that time, the Applied Fisheries Laboratory, later renamed the Laboratory of Radiation Ecology, was established at the University of Washington to study the effects of radiation on aquatic life forms. Many such radiobiological investigations have now been conducted at this and several other laboratories in the United States and in other countries. The results of some of these studies are described here.

Radiation Effects on Fish

The early stages of fish life are known to be more sensitive to radiation than are adults. Consequently, large numbers of eggs and alevins (i.e., newly hatched juveniles) of Columbia River salmon were exposed to gamma radiation from cobalt-60 at exposure rates varying from 0.5 to 2.8 R/day for periods of about 80 days after fertilization of the eggs. The exposed fish, together with a similar number from an unexposed control group, were then released to the ocean to undertake their normal migration. Upon their return to the Columbia River, the adult salmon were examined. So far, there has been no indication of injury as a result of total radiation exposures that range from 33 to over 200 R.

No significant increase in mortality has been detected, neither has there been any decrease in fecundity (i.e, the number of offspring per female). The radiation doses (and dose rates) in these studies were very much larger than would be experienced from nuclear power operations.

A laboratory approach to the study of radiation effects on the embryonic stage of fish is to make use of an aqueous medium containing radioactive isotopes. Exposures can then be both internal from the water imbibed by the eggs, and external from the surrounding medium. Fertilized eggs of carp and hybrid trout have been maintained in water to which various amounts of tritium were added. No significant change in hatchability was detected at tritium concentrations that were 67 000 to 520 000 times the maximum values permitted by Nuclear Regulatory Commission regulations. The total radiation exposure of the eggs ranged up to 400 R.

The development of oyster larvae has been examined in water containing radioactive chromium-51, zinc-65, or strontium-90 and its decay product yttrium-90; the decay products of chromium-51 and zinc-65 are not radioactive. The purpose of the experiment was to determine the concentrations at which radiation caused an increase in the frequency of occurrence of abnormal larvae—that is, larvae with incompletely developed shells. The frequency of this abnormality showed a significant increase only at concentrations between 100 000 and a million times the maximum levels acceptable for the effluent from a nuclear plant. At lower concentrations, no abnormal oyster larvae were detected. Similar results have been obtained in experiments on the mortality rates and production of abnormal larvae of several other fish species.

Somewhat different conclusions were reported by Russian scientists in 1966. They claimed that, for some marine fish, the occurrence of abnormal larvae was detectable in water containing as little as 10^{-7} microcurie/milliliter (μCi/ml) of strontium-90 (plus yttrium-90). This is a factor of 3 smaller than the accepted maximum permissible concentration in water suitable for regular consumption by members of the general public (Chapter 5).

Apart from the fact that investigations by others have not confirmed the Russian claims, the results are suspect for two reasons. First, about a millionfold increase in the concentration of strontium-90 (plus yttrium-90) was said to cause only a three-fold increase in larvae abnormalities and a five-fold increase in mortality. These relatively small increases in deleterious effects for such a large concentration increase are completely inconsistent with the results of many radiobiological investigations. Second, the lowest radioactivity level at which deleterious effects were said to be detectable is actually less than that of naturally occurring potassium-40 in seawater.

All the studies described above have referred to radiation exposures from an external gamma-ray source or from radioactive solutions of simple composition. Observations have also been made on fish living in actual or diluted reactor effluents. Freshly fertilized eggs of salmon and other fish were incubated, and the resulting juveniles were allowed to grow to the migratory stage in water containing radioactive effluent from the Hanford plutonium production reactors.[g]

[g]The liquid discharged at Hanford was the reactor cooling water in which the radioactivity originated mainly from neutron activation of impurities and of corrosion and erosion products.

Even when the radioactivity of the water was many times greater than that in the Columbia River, into which the liquid effluent was normally discharged, there was no significant change in the mortality or growth compared with fish in water containing no such effluent.

An extreme case of radiation exposure involved the small mosquito fish, so called because they exist largely on mosquito larvae. These fish were living in the shallow portion of White Oak Lake, Tennessee, into which radioactive effluents from the Oak Ridge National Laboratory had been discharged for several years. The sediments in the shallow areas contained relatively large quantities of cobalt-60, zinc-65, strontium-90, ruthenium-106, cesium-137, and other radionuclides. Over a hundred generations of mosquito fish have lived in the shallow water since the lake was first used for the discharge of radioactive effluents. The exposure from external radiation was estimated to be as high as 11 R/day.

The frequency of dead embryos and of abnormalities was greater in the irradiated fish than in a similar population in an adjacent stream containing no radioactive effluent. However, the fecundity was definitely greater in the females exposed to radiation. An analogous situation was found in connection with snails living in an area into which radioactivity seeps from a waste disposal site. The irradiated snails produced fewer egg capsules than did unirradiated controls, but the number of eggs per capsule was larger.

Radiation Effects on Algae

Primitive floating aquatic plants, including diatoms and other algae, form part of the food web of fish. Consequently, the effect of radiation on these plants has an important bearing on the survival of many fish species. Algae are known to accumulate various elements, especially cesium, manganese, and yttrium, if they occur in the water in which the plants live. The bioaccumulation factors of the elements mentioned are very large; hence, if radioisotopes are present, the aquatic plants may be exposed to fairly large internal doses of radiation. Should the algae suffer significantly, the fish population might be expected to decrease markedly.

A study has been made of the effect of cesium-137 on the cell division rate of a common marine diatom (*Nitzschia closterium*). The rate of cell division is important because it determines the quantity of the diatoms available as food for fish. In one experiment, the diatoms were started in a medium containing cesium-137 at a concentration equivalent to about a thousand times the maximum acceptable for drinking water. The concentration of cesium-137 within the diatoms was, however, several thousand-fold higher. After 26 weeks, cultures were transferred to a solution of cesium-137 with an activity 10 times greater, and the rate of cell division was followed for another 30 weeks. There was no evidence of injury to the individual cells or to the population of diatoms during the observation period of more than one year.

In another series of experiments, the development of marine algae *(Padina japoinia)* was studied in seawater containing tritium in excess of the normal concentration. Effects on the germination and growth of spores was observed only at tritium concentrations many thousand times greater than the maximum acceptable level in the liquid effluents from nuclear facilities (Chapter 5).

Genetic Effects

Radiation undoubtedly causes genetic mutations in fish, just as it does in insects, plants, and animals. In nature, the offspring with deleterious mutations are quickly eliminated, but those with beneficial mutations, which may include greater resistance to radiation, will continue to propagate. Provided there is no great decrease in mortality and fertility rates, as actually appears to be the case even at moderately high radiation dose rates, the fish population level will not change greatly although many individuals may be eliminated. Even in the absence of radiation above the background, there is a great overproduction of young fish. Only a small proportion survive, but this is adequate to maintain the population.

Bibliography—Chapter 7

American Nuclear Society, Public Policy Statement. "Comparative Risks of Different Methods of Generating Electricity." *Nuclear News* 22, no. 14 (1979): 193.

Archer, V. E. "Effects of Low-Level Radiation: A Critical Review." *Nuclear Safety* 21 (1980): 68.

Auerbach, S. "Ecological Considerations in Siting Nuclear Power Plants: The Problem of Long-Term Biotic Effects." *Nuclear Safety* 12 (1971): 25.

Blaylock, B. G., and Witherspoon, J. P. "Radiation Doses and Effects Estimated for Aquatic Biota Exposed to Radioactive Releases from LWR Fuel-Cycle Facilities." *Nuclear Safety* 17 (1976): 351.

Bodansky, D. "Electricity Generation Choices for the Near Term." *Science* 207 (1980): 721.

Brown, D. G.; Cragle, R. G.; and Noonan, T. R., eds. *Proceedings of the Symposium on Dose Rate in Mammalian Radiation Biology* (CONF-680410). Washington, D.C.: U.S. Atomic Energy Commission, 1968.

Gotchy, R. L. *Health Effects Attributable to Coal and Nuclear Fuel Cycle Alternatives* (NUREG-0332). Washington, D.C., 1977.

Grahn, D., *et al.* "Analysis of Survival and Cause of Death for Mice Under Single and Duration-of-Life Gamma Irradiation." *Life Sciences and Space Research* X (1972): 175.

Hull, A. P. "Radiation Perspective: Some Comparisons of the Environmental Risks from Nuclear- and Fossil-Fueled Power Plants." *Nuclear Safety* 12 (1971): 165.

International Atomic Energy Agency. *Proceedings of the Symposium on the Biological Effects of Low-Level Radiation Pertinent to the Protection of Man and His Environment.* Vienna: International Atomic Energy Agency, 1976.

______. *Proceedings of the Symposium on Late Biological Effects of Ionizing Radiation.* Vienna: International Atomic Energy Agency, 1978.

Jablon, S. "The Origin and Findings of the Atomic Bomb Casualty Commission." *Nuclear Safety* 14 (1973): 651.

Love, L. B., and Freeburg, L. C. "Health Effects of Electricity Generation from Coal, Oil, and Nuclear Fuels." *Nuclear Safety* 14 (1973): 409.

Luning, K. G., and Searle, A. G. "Estimation of the Genetic Risks from Ionizing Radiation." *Mutation Research* 12 (1971): 291.

Mays, C. W., and Lloyd, R. D. "Malignancy Risk to Humans from Total Body Gamma-Ray Irradiation." *Proceedings of the Third International Congress* (CONF-730907-Pl), p. 417. Washington, D.C.: International Radiation Protection Association, 1974.

Morgan, K. Z. "Adequacy of Present Radiation Standards." *Environmental and Ecological Forum 1970-1971* (TID-25857), p. 104. Oak Ridge, Tenn.: U.S. Atomic Energy Commission, Division of Technical Information, 1972.

National Academy of Sciences. *Risks Associated with Nuclear Power: A Critical Review of the Literature.* Washington, D.C., 1979.

National Academy of Sciences - National Research Council. "The Effects on Populations of Exposure to Low Levels of Ionizing Radiation." *Report of the Advisory Committee on the Biological Effects of Ionizing Radiations* (BEIR Report). Washington, D.C., 1972. Revised report (BEIR III) published 1981.

National Council on Radiation Protection and Measurements. *Review of the Current State of Radiation Protection Philosophy* (Report No. 43). Washington, D.C., 1975.

Rossi, H. H., and Kellerer, A. M. "Radiation Carcinogenesis at Low Doses." *Science* 175 (1972): 200.

Russell, W. L. "Factors Affecting the Radiation Induction of Mutations in the Mouse." *Biological Implications of the Nuclear Age* (CONF-690303). Washington, D.C.: U.S. Atomic Energy Commission, 1969.

______. "Genetic Effects of Radiation." *Proceedings of the Fourth United Nations Conference on the Peaceful Uses of Atomic Energy* 13 (1971): 487.

Sagan, L. A. "Human Costs of Nuclear Power." *Science* 177 (1972): 487.

Starr, C., and Greenfield, M. A. "Public Health Risks of Thermal Power Plants." *Nuclear Safety* 14 (1973): 267.

Storer, J. B., and Bond, V. P. "Evaluation of Long-Term Effects of Low-Level, Whole-Body External Radiation Exposures." *Proceedings of the Fourth United Nations Conference on the Peaceful Uses of Atomic Energy* 11 (1971): 3.

Taylor, L. S. "What Do We Know About Low-Level Radiation." *The Environmental and Ecological Forum 1970-1971* (TID-25857), p. 168. Oak Ridge, Tenn.: U.S. Atomic Energy Commission, Division of Technical Information, 1972.

Totter, J. R. "Some Observational Bases for Estimating Oncogenic Effects of Ionizing Radiation." *Nuclear Safety* 21 (1980): 83.

United Nations, Scientific Committee on the Effects of Atomic Radiation. *Ionizing Radiation Levels (Vol. I) and Effects (Vol. II).* New York: United Nations, 1972.

Upton. A. C., *et al.* "Late Effects of Fast Neutrons and Gamma Rays in Mice As Influenced by the Dose Rate of Irradiation." *Radiation Research* 41 (1970): 467.

U.S., Atomic Energy Commission. *Proceedings of the Symposium on the Biological Implicatons of the Nuclear Age* (CONF-690303). Washington, D.C., 1969.

U. S., Department of Health, Education, and Welfare. *Report of the Interagency Task Force on the Health Effects of Ionizing Radiation.* Washington, D.C., 1979.

Van Cleave, C. D. *Late Somatic Effects of Ionizing Radiation* (TID-24310). Oak Ridge, Tenn.: U.S. Atomic Energy Commission, Division of Technical Information, 1968.

8

Radioactivity in Reactor Fuel Production

RADIOACTIVITY IN URANIUM MINING

Radon and Its Daughter Products

The radiation hazard to uranium miners is a significant environmental aspect of the generation of nuclear power; the matter therefore merits consideration here. For more than a hundred years, there has been evidence that many men who had worked for 15 to 25 years in European mines where uranium was present were seriously affected by a disease of the respiratory tract. The symptoms of this disease increased with the working time and, on the average, the uranium miners died at an earlier age than men working in other mines. In 1913, it was realized that more than half the cases of pulmonary disease were actually cancers of the respiratory system, commonly referred to as lung cancer. The problem was not studied seriously, however, until the late 1930s. The general feeling at that time was that the harmful effect was due mainly to breathing air containing the radioactive gas radon and radioactive dust particles.

Radon arises in the following manner. Uranium-238, the main constituent (99.3 percent) of natural uranium, is the parent of a series of radioactive decay products ending in a stable lead isotope. Among these products is thorium-230, which decays into radium-226, the common isotope of radium. Radium-226 decays in turn to produce gaseous radon-222, which escapes to the atmosphere when uranium minerals are exposed. In underground uranium mines, and also in many other mines, because uranium is so widely distributed in the earth's crust, the radon tends to accumulate in the air unless it is continuously removed by ventilation.

Before 1951, the concentration of radon in the air breathed by a uranium miner was regarded as the best indication of the potential hazard. But in that year, studies made in the United States suggested that it was not so much the radon itself, but rather the radiations from the solid radioactive daughter products that were responsible for causing lung cancer. This view has since been confirmed by observations made with laboratory animals. In dusty mines, particulate matter containing the uranium-238 decay products thorium-230 and radium-226 contribute to the lung dose, but the daughter products of radon-222 are regarded as being of major importance.

TABLE 8-I
RADON AND ITS DECAY PRODUCTS

Nuclide	Half-life	Radiation
Radon (radon-222)	92 h	alpha
Radium-A (polonium-218)	3.05 min	alpha
Radium-B (lead-214)	26.8 min	beta, gamma
Radium-C (bismuth-214)	19.7 min	beta, gamma
Radium-C′ (polonium-214)	2.7×10^{-6} min	alpha
Radium-D (lead-210)	20 years	beta

Radon-222 has a half-life of 92 hours and its solid decay product, called radium-A (polonium-218),[a] has a half-life of about 3 minutes. In turn, radium-A decays to radium-B, then to radium-C, and radium-C′, all of which are solid elements with short radioactive half-lives, as indicated in Table 8-I. The decay product of radium-C′ is radium-D (lead-210), with the comparatively long half-life of about 20 years. The group of radionuclides (p. 9) consisting of radium-A, -B, -C, and -C′ is referred to as the *short-lived daughters* of radon. Because of the moderately short half-lives of radon itself and of its decay products, these daughters are found wherever radon gas is present.

Since the daughters of radon are solid elements, they tend to deposit on small dust particles and on minute droplets of moisture present in the air. If air containing radon and its decay products is breathed, the solid daughter nuclides are largely retained by the interior lining of the lungs. Here, the emission of short-range alpha particles from radium-A and radium-C′ can cause damage to the tissue that could lead to the development of lung cancer.

The effect of the beta particles and gamma rays from radium-B and -C is of little significance in comparison with that of the alpha particles. Furthermore, radium-D, the daughter of radium-C′, decays fairly slowly and emits beta particles. The radon gas itself does not remain in the lung, but is partly taken up by the blood and partly exhaled. Hence, the main hazard from the presence of radon in the air arises from its short-lived alpha-emitting daughters that are deposited in the lung.

If the parent element of a radioactive series has a much longer half-life than any of its decay (or daughter) products, an essential state of equilibrium is attained in the course of time. In this equilibrium state, equal numbers of atoms (or nuclei) of all the radioactive members of the series disintegrate in unit time. In other words, in view of the definition of the curie (p. 132), in a state of radioactive equilibrium, equal numbers (or fractions) of curies of the parent nuclide and of each of its radioactive daughters will be present.

The half-life of uranium-238 (4.5 billion years) is very long in comparison with that of the other nuclides in its decay series; hence, radon-222 and its daughters should be in radioactive equilibrium in uranium ores. This would be true, however, only if the radon gas could not escape; it would then be continuously produced from its immediate parent, radium-226, as fast as it

[a]The names radium-A, -B, -C, etc., were given to the successive solid decay products of radium-226 before the identity of the elements was established.

decayed. If the radon (or any daughter) is removed (e.g., by dispersal in the air), then equilibrium will no longer exist.

When uranium is extracted from its ores and purified for the fabrication of reactor fuel, nearly all the radium-226 and its immediate parent thorium-230 are removed and will not be regenerated in appreciable amounts for many years. Hence, uranium in manufactured products does not emit radon gas and so does not constitute a hazard in this respect.

Development of the "Working Level" Concept

When it was thought that the radon concentration in the atmosphere was the important criterion in determining the potential lung cancer hazard to underground uranium miners, standards were proposed on this basis without reference to the daughters. In 1953, after the importance of the latter was realized, the National Committee for Radiation Protection and Measurements (NCRP) in the United States recommended that 10 picocuries (pCi)[b] of radon in equilibrium with its short-lived daughters per litre of air be accepted as the maximum permissible concentration in operations involving uranium and its decay products. This particular concentration was chosen because calculations based on various assumptions indicated that it would represent a radiation dose to the lung of 0.3 rem per working week, which was then the maximum permissible dose for occupationally exposed individuals (Chapter 5).

At a conference held in 1953, in which representatives of the United States, the United Kingdom, and Canada participated, a recommendation was made that the maximum permissible concentration of radon in equilibrium with its short-lived daughter products be increased to 100 pCi/litre. Largely as a result of this recommendation, the Seven State Uranium Mining Conference on Health Hazards, held in Salt Lake City, Utah, in February 1956 agreed that this concentration in the mine atmosphere be taken as the level at which corrective action (e.g., increased ventilation) was desirable. The concentration of 100 pCi/litre of radon in equilibrium with its short-lived daughters was then referred to as a *working level* (WL). Although the so-called working level is no longer regarded as an acceptable maximum concentration, it has been redefined in such a way as to make it a useful practical unit for expressing the lung dose from the radon daughters.

Definition of the Working Level

Because the radon in a mine atmosphere is rarely, if ever, in equilibrium with its daughters, some difficulty was experienced in applying the working level concept as defined by the Salt Lake City conference. Consequently, in 1957, the U.S. Public Health Service (PHS) redefined the working level in a manner that was more precise from the practical standpoint. Instead of expressing the concentrations of radon and its daughters in picocuries, the essential quantity used was the alpha-particle energy, since this is probably the best available measure of the possible hazard to the lungs.

[b]The picocurie is one millionth of a millionth part (10^{-12}) of a curie, or one millionth part of a microcurie.

One WL was defined as any combination of the short-lived daughters of radon (i.e., radium-A, -B, -C, and -C′) in 1 litre of air that will result in the ultimate emission of 1.3×10^5 million electron volts (MeV) of potential alpha-particle energy.[c]

The definition of the WL as representing the ultimate emission of 1.3×10^5 MeV of alpha-particle energy per litre of air is actually based on the calculation of the potential alpha energy of the short-lived daughters *in equilibrium* with 100 pCi of radon. However, the definition of the WL refers to "any combination of short-lived daughters" of radon and does not specify equilibrium conditions. The essential quantity in the definition is thus the potential alpha energy, regardless of whether the system is in equilibrium or not. In a mine atmosphere, radioactive equilibrium between radon and its daughters does not exist, but the definition of the WL in terms of the alpha-particle energy is unaffected.

Protection Standards for Uranium Miners

The working level is a useful operational guide, but it does not directly indicate the actual exposure of the individual miner. For this purpose, the radon daughter concentration in the mine must be multiplied by the length of time the worker spends in a particular environment. The common unit used in this connection is the *working level month* (WLM). It is equal to the number of WLs in the mine air multiplied by the number of working months, where the working month is commonly defined as 170 hours of exposure.

At present, the U.S. Environmental Protection Agency (EPA) standard requires that radon and decay products in underground mines be controlled so that no individual receives an exposure of more than 4 WLM in any consecutive 12-month period. This limit is in accord with the observation that radiation-induced lung cancer is rarely found in nonsmokers receiving a lifetime exposure of less than 120 WLM. There are indications of a possible synergistic effect of cigarette smoking on the incidence of lung cancer among uranium miners, and the suggestion has been made that these miners be discouraged from smoking cigarettes.

The relationship between the WLM and the dose equivalent to the lung is uncertain. The 1972 Biological Effects of Ionizing Radiation (BEIR) Committee Report (p. 118) adopted a value of 0.5 rad/WLM for the absorbed dose to the sensitive lung tissue. The significant radiation from the radon daughters consists of alpha particles for which a quality factor of 10 is reasonable (p. 123). Hence, the lung dose equivalent of 1 WLM is roughly 5 rems. On this basis, the accepted standard maximum exposure of 4 WLM/year would represent a dose of 20 rems/year. This can be compared with the NCRP maximum permissible occupational lung dose of 15 rems/year (Table 5-V, last entry).

Control of Radon Daughter Concentrations

In the United States, uranium ores are mined in roughly equal amounts by open-pit mining and by underground mining. Open-pit mining is used when the

[c]The electron volt (eV) is a unit of energy equal to that imparted to an electron in passing through a potential difference of 1 V. It is equivalent to 1.6×10^{-19} joule, so that 1 MeV is equivalent to 1.6×10^{-13} joule.

ore body lies under relatively friable material at depths up to about 120 metres (m) (400 ft) or so. When the ore body is below this level, however, or when it is under rock strata, underground mining is necessary. In open-pit mining, the radon gas is continuously dispersed to the atmosphere and the concentration in the local air is much less than in underground mines.

In underground mines, the most practical way to decrease the radiation hazard to the workers is to remove the radon and its airborne decay products by an efficient ventilation system. Fans at the mine entrance continuously blow fresh air through long portable ducts into (or exhaust contaminated air from) the working areas. Appropriate design of passageways within the mine also contributes to good ventilation. Improvements in ventilation systems during recent years have resulted in a marked decrease in the radon daughter concentrations in uranium mine air. In mines at shallow depths and in those with horizontal single-level deposits, adequate ventilation can be achieved readily.

On the other hand, mines that are poorly laid out for ventilation, those with multilevel deposits, and those in which small uranium ore bodies occur at random may have difficulty in meeting the WL standards. The very high rates of air flow required in these mines could have serious drawbacks. Eye injuries, for example, may result from the larger sand particles carried by the air, and there may be an increased incidence of bronchitis and pneumonia. Since most of the airborne radon daughter products are attached to dust particles, which can be easily filtered out, the wearing of respirators by the miners is a possibility when adequate ventilation is difficult. According to a 1974 report of the U.S. Bureau of Mines, the radon concentration in uranium mine air can be reduced at least 60 percent by applying a polymer sealant to the mine walls.

Environmental Aspects of Uranium Mining

The area occupied by surface structures of an underground uranium mine may be only a few acres, but the underground passages may extend to a mile or more. Much of the surface area is taken up by the mine waste pile, the bulk of which could be returned to the pit if the operation were to be terminated. Groundwater entering underground workings is collected in sumps and is then pumped to the surface. The water contains small quantities of radioactive elements, including uranium and radium, but they are largely taken up by the soil. In some underground mining operations, it is economically feasible (or it may be necessary) to recover uranium from the wastewater before discharge. If effluents are to be discharged to an adjacent water body, state or interstate pollution control agency permits are required, in accordance with the Federal Water Pollution Control Act (FWPCA).

Open-pit mining is characterized by a large open excavation and piles of earth and rock overburden that have been removed. In addition, there are various structures, earth-moving equipment, and a network of yards and roads. During much of the operating life of the mine, overburden is used for backfilling mined-out areas, thereby keeping the overburden piles within reasonable bounds. Eventually, most of the land occupied by the mine is reclaimed. The piles of overburden remaining are graded and contoured to simulate the surrounding countryside and, if possible, reseeded with native plants and grasses.

The final pit areas are difficult to reclaim and may be graded and allowed to fill with water to form a lake, which might find recreational use. Groundwater is continuously pumped from the working area of the mine while ore is being removed. This water, which contains some radioactivity, is returned to the ground, or it may be discharged in the same way as in underground mines, provided the necessary permits have been obtained.

Substantial amounts of radon gas are released to the atmosphere in uranium mining. In underground mines, the gas is discharged with large volumes of ventilating air, and in surface mines it is released when the uranium ore body is exposed and broken up. This radon may be a hazard to people working on the mine site, and they must take appropriate precautions. As a result of dilution and natural dispersion in the air, however, the concentration of radon beyond the mine boundary is usually less than the accepted standard for the general population. From Chapter 6, recall that radon is always present in the atmosphere, arising from the decay of uranium-238 (and thorium-232) in the ground and in masonry building materials. The amount of radon produced in uranium mining is only a small fraction of that escaping from the ground. Nevertheless, questions have been raised concerning the possible health effects of additional radon in the atmosphere (p.224).

RADIOACTIVITY FROM URANIUM MILLS

Extraction of Uranium from the Ore

In a uranium mill, the ore, generally containing the equivalent of 0.1 to 0.5 percent (average about 0.2 percent) of the oxide U_3O_8, is first crushed into small pieces. Next, the pieces are ground in the presence of water to form a slurry. The slurry is treated (leached) with an acid or alkaline solution, depending on the

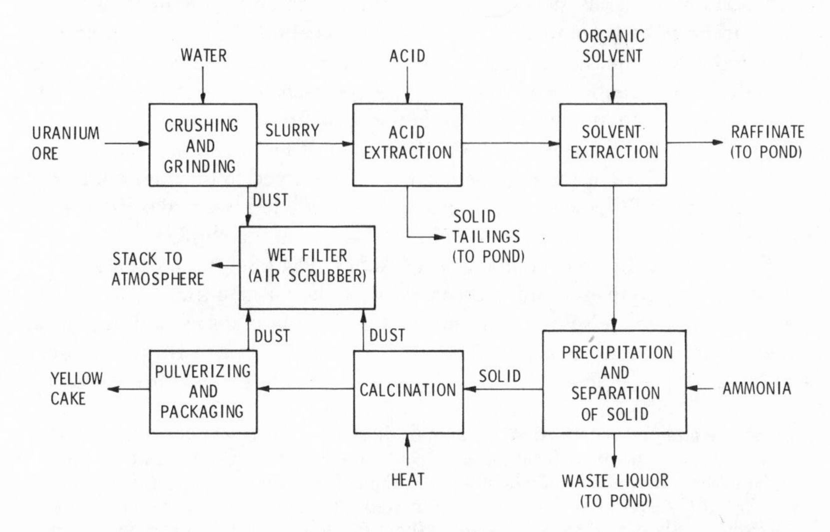

FIG. 8-1. Schematic outline of uranium milling process based on solvent extraction.

nature of the ore, which extracts most of the uranium together with small amounts of impurity elements, such as vanadium, molybdenum, and iron. The leach liquid containing the uranium is then separated from the residue, which constitutes the solid part of the plant *tailings*. Some of the thorium and nearly all the radium and lead isotopes originally present in the ore remain behind in the solid tailings. After washing with dilute acid or alkali to remove as much uranium as possible, the tailings in the form of a slurry are pumped to a tailings retention pond.

In one common procedure, based on acid leaching, the liquid and washings separated from the solid tailings are treated with an organic solvent to extract the uranium while leaving most of the impurities behind in an aqueous residue called the *raffinate*. Ultimately, the uranium is precipitated by means of ammonia; the solid is separated and calcined (heated) to yield the product known as *yellowcake*, which contains the equivalent of 70 to 90 percent of U_3O_8 (Fig. 8-1). The yellowcake is shipped to a uranium refining plant for further purification and conversion into uranium hexafluoride. The raffinate, which contains most of the thorium and radium extracted in the initial leaching, is sent to the tailings pond.

Licensing of Uranium Mills

To process or refine ore containing 0.05 weight percent or more of uranium after removal from a mine, a Source Material License is required from the U.S. Nuclear Regulatory Commission. This is, in effect, a license to operate a uranium mill. The procedure is described in Title 10 of the *Code of Federal Regulations*, Part 40 (10CFR40), entitled "Licensing of Source Material."[d]

An applicant for a license to possess and use source material for uranium milling must file an application with the NRC at least nine months before commencing construction of the mill. In accordance with the requirements of 10CFR51 and to satisfy the National Environmental Policy Act of 1969, the application must be accompanied by an Environmental Report similar to those described in Chapter 3 for reactor licensing, except that a single report should cover both construction and operation phases. Before granting a license, the NRC weighs economic, technical, and other benefits against the environmental costs. When the mill is operating, the licensee must submit annual reports specifying the quantities of radioactive materials discharged to the environment. In the future, the NRC is to require that a uranium mill licensee make provision for reducing the radon emission from tailings after the mill is closed down (p. 222).

By Section 274 of the Atomic Energy Act of 1954 (and its amendments), the NRC is permitted to enter into agreement with the states to transfer to them the authority to regulate source material. States entering into such agreements are called Agreement States. The four principal uranium-producing states—Colorado, New Mexico, Texas, and Washington—are Agreement

[d]*Source material* is defined as uranium or thorium or any combination thereof in any physical or chemical form or ores containing 0.05 weight percent or more of uranium, thorium, or any combination thereof, but not enriched in fissile species. The NRC's authority to control source material begins only after it has been removed from its place of deposit in nature; the NRC has no control over the mining of uranium.

States. Roughly half the operating uranium mills are licensed by Agreement States. The programs established by these states must be compatible with NRC programs for the regulation of source material and the control of radiation hazards.

Licensing requirements for uranium mills established by the Agreement States are similar to those of the NRC. The main difference is that in 1980 only the State of Washington required an environmental impact assessment. However, the NRC is offering technical help to the other Agreement States in connection with such assessments.

Guidance for the preparation of licenses is provided in NRC Regulatory Guide 3.5, entitled "Guide to the Contents of Applications for Uranium Milling Licenses." The following topics are covered:

1. Proposed Activities
2. Site Description (including geography, demography, meteorology, hydrology, geology, and seismology)
3. Facility Design and Construction (including mill process, major equipment, and safety and control instrumentation)
4. Waste Management System
5. Operations (including corporate organization, qualifications of radiation safety personnel, radiation safety and survey programs, and training of employees in radiation safety)
6. Accidents (including plans for coping with them)
7. Quality Assurance
8. Evaluation of Alternatives.

A diagram of the plant layout is required, showing, in particular, the areas where dust would be generated. The ventilation and dust control systems must be described in sufficient detail to permit their adequacy to be judged. A radiation survey program must be presented indicating the sampling locations and frequency, and the methods to be used in determining airborne radioactivity in the plant. A program for external radiation monitoring is required, and written radiological safety instructions must be prepared for issuance to mill employees.

The applicant for a source material license must also submit information on the method of dealing with liquid and solid wastes and on the geologic and hydrologic characteristics of the site. Means proposed for preventing or controlling the release of radioactive wastes to unrestricted areas must be described. There must be an environmental monitoring program for regularly sampling and testing all mill effluents, gaseous and liquid, that are to be discharged to the environment. If effluent is to be released to a river or lake, there must be prior approval as required by the FWPCA.

When an earth embankment (or dam) is used to form a pond for the retention of liquid and solid wastes, its retention capability and integrity must be established. Methods for inspection and maintenance of the dam must be developed. In order to provide assistance on these matters, the NRC has prepared a guide outlining basic design considerations to be used in the construction of embankments for retention systems and the information that should be submitted with the application for a license. Before a license is issued, a visit is made to the mill by a representative of the NRC to obtain first-hand knowledge of the location and of various aspects of the proposed operation. If the information submitted in the application and in the environmental impact statement and the results of the visit are all satisfactory, the applicant will be granted a license to possess and use source material.

The operation of the mill must be conducted in conformance with the radiation protection standards described in Chapter 5. Representatives of the NRC make inspections of uranium mills at approximately yearly intervals to verify that the operations are being carried out properly. Independent measurements are often made of air and effluent samples and of environmental radioactivity levels. If deficiencies are found, corrective action must be taken within a prescribed time. In cases involving violations of the regulations or in cases where there is an imminent hazard to public health and safety, the NRC has the authority to impose monetary fines and to suspend or revoke the mill's operating license.[e]

Dust Control in Uranium Mills

Because of the ample ventilation that is possible, the concentration of radon gas in a uranium mill is generally quite low, and the short-lived daughters are then not of significant concern. There is the possibility of external radiation exposure from gamma rays in some parts of a uranium mill, but simple precautions limit to a low level the potential of such exposure to the workers. The main radiological health problem in a mill is the dust produced in crushing and grinding the ore and raised during the drying, calcining, and packaging of the yellowcake. At all intermediate stages, the radioactive materials are either in a liquid solution or in the form of a slurry, and so there is no release of dust to the atmosphere.

As a general rule, ores are kept wet when being crushed or ground; this minimizes the formation of dust. Dust-producing operations are conducted, as far as possible, in enclosures and adequate ventilation serves to minimize the concentration of dust particles in the air. Respirators may be worn by the mill workers, if necessary, as a supplement to, but not as a substitute for, good ventilation.

Although airborne dusts constitute the main radiation sources from both the uranium ore and the yellowcake, the underlying factors are quite different. Radiation from the ore dust arises chiefly from alpha-particle emitters, especially thorium-230 and radium-226. The yellowcake, however, is much less radioactive because most of the thorium and radium isotopes are left in the solid tailings.

Environmental Aspects of Uranium Mills

When possible, uranium mills are located close to the mines to minimize transportation costs. Consequently, most—although not all—mills are in sparsely populated regions of the United States. A typical mill, processing 540 000 metric tons (MT) (600 000 short tons) of ore annually, might occupy some 120

[e]Radiation doses to people living in the general area of a uranium mill are discussed in Chapter 6.

hectares (300 acres), with about three-fourths taken up by the tailings retention pond. The normal average operating lifetime of a mill is roughly 15 years. Owners of uranium mills now being licensed are required to submit plans for restoration of the area when operation ceases.

Very little of the dust produced in uranium mill buildings reaches the environment because the air is passed through wet scrubbers before being discharged. If necessary, dry filters can serve as backup to the scrubbers. However, radon gas cannot be removed in this manner, and it is released from the mill buildings to the atmosphere. In addition, radon is continuously emitted from ore piles and from the tailings retention system, both of which are in the open. In most, if not all, uranium mills the tailings piles are kept wet while the plant is operating, thereby reducing the emission of radon and the dispersal of radioactive dust.

Waste liquids from a uranium mill, including the raffinate from the solvent-extraction process, have too high a concentration of thorium-230 and radium-226 to permit them to be discharged to an adjacent water body. They are consequently used to form a slurry with the solid tailings and are impounded in the retention pond. Such ponds are constructed to prevent escape into surface water systems and to minimize percolation into the ground. Natural evaporation of the water permits retention of all the discharged liquid. The accumulation of solid tailings in the course of mill operation serves to increase the capacity of the pond by raising the level of the containing dam.

If leakage should occur through the dam, the liquid would be collected and returned to the retention pond. To assess the effects of seepage of radioactivity into the ground, test wells are drilled in the vicinity of the tailings ponds, and water samples from these wells, and also from other wells and springs in the area, are analyzed for thorium-230, radium-226, and uranium. New uranium mills in the United States are being designed to reduce the levels of radioactivity almost to zero in all liquids discharged to the environment.

Uranium Mill Tailings Piles

One of the major environmental problems associated with uranium mills is concerned with the tailings piles.[f] A metric ton (1000 kg [2205 lb]) of uranium ore yields, on the average, about 1.8 kg (4 lb) of yellowcake. Hence, the quantity of solid tailings is essentially the same as the amount of ore milled; this may be several hundred thousand metric tons per year for a typical mill. The tailings (solid and liquid) contain essentially all the thorium-230 and radium-226 present in the ore, and so they produce radon-222 at about the same rate as the ore itself. As the radium-226 (half-life 1600 years) decays, it is regenerated by the decay of thorium-230 (half-life 80 000 years). The rate of formation of radon gas is determined by the slow decay of thorium-230; hence, it will remain almost constant for many thousand years. Furthermore, decay of uranium-238 (half-life 4.5 billion years) remaining in the mill tailings could result in the emission of radon at approximately one-tenth of the initial rate for a very long time.

[f]In some locations, uranium can be extracted directly from the ore without mining, by leaching with an aqueous solution of ammonium carbonate injected through holes drilled into the ore bed. Tailings piles are then eliminated.

Since the tailings pile in an operating mill is kept wet, very little radon and dust are released to the atmosphere. But when the mill is shut down, as several have been, the water evaporates, and the tailings pile dries out. The radon release then increases, and winds can carry the radioactive dust to considerable distances. Drying of parts of the pile during operation of the mill would have similar consequences, although on a smaller scale.

Stabilization of Inactive Tailings Piles

In 1963, complaints were made by the inhabitants of Durango, Colorado, concerning blowing dust from abandoned tailings piles. In response to these and similar complaints from other cities, the Atomic Energy Commission (AEC), as it then was, and the PHS, in cooperation with health officials from the states of Colorado and Utah, undertook studies of airborne radioactive particles and radon gas in the vicinity of inactive uranium mills. Although the results were not decisive, because of variations of meteorological and other conditions, they indicated that some action was desirable, and in February 1966 the Colorado Health Department proposed that tailings piles be treated (or "stabilized") to minimize the dispersal of radioactive dust.

The recommendation concerning stabilization was based on a successful experience by the AEC in connection with a uranium mill that it had operated during the 1950s at Monticello, Utah. The mill was closed in 1959, and two years later the tailings pile was graded and covered with 30 to 60 cm (1 to 2 ft) of rock and soil. Organic and chemical fertilizers were spread over the surface, and the area was seeded with native grasses. In 1965, an examination of the air, water, and vegetation in the vicinity gave no evidence of the transportation of any radioactive material from the stabilized tailings. (Subsequently, however, heavy rains caused some of the tailings to be washed away.) Radiation levels in the air above the pile were much less than the permissible values for the general population. However, because of uncertainty concerning jurisdiction over the tailings remaining after a uranium mill ceased operation, no other tailings piles were stabilized in this manner until the late 1960s.

In December 1966, the Colorado State Board of Health conducted a hearing attended by representatives of the AEC, the PHS, and uranium milling companies and members of the general public. As a result, Colorado promulgated regulations for the stabilization of disused tailings piles to become effective in January 1967. The main requirements were that the piles be covered with vegetation, soil, rock, or stone, or in some other acceptable way to prevent wind and water erosion, and that drainage be provided to prevent access of surface runoff water from reaching the piles.

At a meeting held in December 1966, a federal position concerning the control of mill tailings was proposed by the Federal Water Pollution Control Agency, the PHS, and the AEC, and its terms were unanimously agreed to in July 1967 at a meeting attended by representatives of the seven states bordering on the Colorado River. According to the agreement, uranium mill operators were made responsible for stabilizing and containing inactive tailings piles and for managing active piles to minimize erosion. The primary responsibility for enforcing control programs was to rest with the individual states.

Some efforts were made to stabilize the inactive tailings piles, but it became evident that effective stabilization and reduction of the radon emission would place a severe financial burden on the states. Consequently, Congress passed the Uranium Mill Tailings Radiation Control Act of 1978. One of the purposes of the Act was to establish conditions for government agencies to cooperate with interested states, Indian tribes, and others, "to stabilize and control...[25 inactive mill] tailings [piles] in a safe and environmentally sound manner and to minimize or eliminate radiation health hazards to the public." The Act authorized the payment by the government of 90 percent of the costs to individual states and all the costs for tailings piles on Indian lands.

The major responsibility for fulfilling the requirements of the Act were assigned to the U.S. Department of Energy (DOE), with the states participating in the selection and performance of appropriate remedial actions. Such actions must have the concurrence of the NRC and must be undertaken in accordance with standards prescribed by the EPA. For tailings on Indian lands, the tribe and the U.S. Department of the Interior would be consulted. The stabilization and radon control procedures will vary from one tailings pile to another, but they will generally be along the lines indicated in the next section.

The Act also requires that licensees of presently operating mills make provision for stabilization and control of tailings piles after the mill has closed down. Before issuing a license to operate a uranium mill, the NRC now requires assurance that funds will be available for reclaiming the tailings site when mill activities are terminated. The owner must not permit exposure and release of tailings to surrounding areas. No structure that may be occupied by people or animals may be built on the covered surface, neither may roads, trails, or rights-of-way be established across it. In some circumstances, title of the land may be transferred to the United States or to the state in which the land is located.

Radon from Mill Tailings

The stabilization procedure described in the preceding section decreases wind and water erosion, but the effect on radon emission from the tailings is less marked. In fact, the 15 to 30 cm (6 to 12 in.) of earth cover commonly used reduces the release rate by only 10 to 15 percent. Because of the potential hazard associated with radon and especially with its short-lived decay products, the emissions from uranium mill tailings must be considered.

Assuming that 350 light-water reactors (LWRs) are in operation by the year 2000, a total of about 1×10^9 MT of dry mill tailings will have been accumulated in piles. The estimated rate of emission of radon-222 from these piles with a 30-cm earth cover for stabilization would be about 3.6×10^5 Ci/year. This is only 0.36 percent (or less) of the natural annual release of radon from the soil in the United States. On an annual basis, therefore, radon released to the atmosphere from stabilized tailings piles would not appear to be significant. But since the emission will continue for a very long time, the piles could constitute a problem.

Because of this possibility, further action is required beyond the stabilization by a 30-cm earth cover. The tailings from inactive mills should be placed underground, but above the water table, with enough soil cover to reduce the radon emission rate (per unit area) to not more than twice the background emission from the region. Such a reduction will generally require a depth of 1.8 to 6 m (6 to 20 ft) of soil, depending on local conditions. The radon emission would then be about 1 percent of that from an uncovered pile.

The soil cover just described should be capable of surviving most natural phenomena. An ice age might displace the cover, but it would probably not be catastrophic. The effective lifetime of the protective measure is, of course, uncertain, but it is thought to be several hundred years at least. If, at some later time, the soil cover should be disturbed, it could easily be restored. This places a burden on future generations, but it may not be as great as that associated with the combustion of coal (p. 225).

Potential Health Effects of Radon Emissions

As seen earlier in this chapter, miners exposed to high levels of radon-222, generally more than 100 pCi/litre, for extended periods have a higher than normal incidence of lung cancer. The atmosphere in the United States contains, on the average, about 0.15 pCi/litre of radon; much higher levels, however, exist indoors in brick and concrete buildings and even in the open in several Western states. Whether or not such relatively low levels of radon in the air can cause cancer is uncertain.

According to the linear theory of delayed radiation effects, described in Chapter 7, there will be an increased incidence of lung (and other) cancers from radon exposure in a large population in proportion to the population dose (in person-rems) no matter how low the individual doses may be. If this is the case, then the radon released from uranium mining and milling operations will cause an increased incidence of cancer, although the associated radiation dose to an individual is only a very small proportion of the average natural background dose (Chapter 6).

Most of the uranium mining and milling operations in the United States are conducted in the West, and the radon released is generally carried eastward by the prevailing winds. On the basis of this situation and the linear theory of radiation-induced cancer incidence, it has been calculated that, in a future U.S. population of 300 million, there should be an increase of 2×10^{-5} (0.00002) cancer death of all types for every curie of radon present in the atmosphere.

To estimate the potential effects of radon associated with nuclear power generation, the emissions may be considered in two stages:

1. from active mining and milling operations
2. from dry tailings piles after the mills have ceased operation.

Radon emissions from most inoperative mines are very small and may be neglected.

Mining the uranium ore needed to supply one reactor's "annual fuel requirement" (AFR)[g] releases some 4000 Ci of radon-222; this is approximately the same for both underground and surface mines. In addition, about 1100 Ci per

[g]The AFR is defined as the fuel required by a 1000-MWe LWR operating at an average annual plant factor of 80 percent.

AFR are released from operating mills and from tailings before they dry out and are stabilized. Hence, in actual mining and milling, 5100 Ci of radon are released per AFR. According to the linear theory, assuming 2×10^{-5} cancer death per curie, there should be about 0.1 additional cancer death per AFR in a population of 300 million.

If 350 LWRs are in operation in the year 2000, it appears that, allowing for a latent period of 10 to 30 years, there eventually will be an additional 35 cancer deaths per year due to radon releases from active mines and mills. This number should occur annually for every year the 350 reactors are in operation. On the same basis, however, at least 2000 cancer deaths per year should result from the 10^8 Ci (or more) of radon released from the ground each year in the United States. In view of the assumptions involved in applying linear theory to very small radiation doses spread over a large population, the significance of these calculated values is uncertain.

In any event, mining and milling of uranium will decrease in the 21st century because of the exhaustion of the ores. The release of radon from these sources will then become very small. Emission of radon from inactive mill tailings, however, will continue for a very long time. If the piles are simply stabilized with 30 cm of earth cover, the emission during the year 2000 from accumulated tailings from uranium to supply the needs of 350 LWRs would be 3.6×10^5 Ci, as given earlier. By linear theory, this would imply 7.2 additional cancer deaths per year.

The accumulation of tailings from uranium milling operations may be expected to continue for some years (but not many) into the 21st century, so that the annual radon emission rate will increase. On the other hand, decay of thorium-230 will result in a gradual decrease. If the rate of radon release from thinly covered piles is assumed to remain constant over a period of 100 000 years and *no effective treatment for cancer is developed in the meantime*, there may be a total of 720 000 cancer deaths over this time period (100 000 years) in the United States from LWR operations. To place this number in perspective, it should be compared with the 45 *billion* "natural" cancer deaths that would be expected during the same time period.

Although calculations of increased cancer deaths over many thousands of years may be meaningless, the large numbers prompted the NRC (and the Agreement States) to require the expensive measures that will reduce the radon emissions to about 1 percent of that from uncovered tailings piles. This is about 1.15 percent of the value from the simple stabilized piles. The calculated cancer deaths for a period of 100 000 years would then be reduced to about 8300 or an average of less than 0.1 additional death per year.

For the remainder of this century, coal is the only practical alternative to uranium as the fuel in electric power plants. Coal, however, contains small amounts of uranium, and the ash pits of coal-fired plants continuously emit radon. The emission rate varies with the source of the coal, but it is similar to that, per AFR equivalent, from uranium mill tailings buried in the manner recommended by the NRC. Consequently, the use of coal as a power plant fuel instead of uranium would not reduce the radon emission.

Although maintenance of tailings sites may place a burden on future generations, there are compensations. Combustion of coal to produce steam, when an alternative means is available, deprives these generations of a source of important chemicals. It has been said, with some justification, that coal is too valuable to burn. Furthermore, the introduction of many billions of tons of carbon dioxide into the atmosphere from coal-fired plants could have a serious and long-lasting effect on the climate. Thus, the operation of nuclear power plants in this and the next generation may not be without benefits to future generations.

Use of Mill Tailings in Construction

An unfortunate development occurred in connection with the tailings from a uranium mill located near Grand Junction, Colorado. This mill, which started operating in 1950, was accumulating tailings at an average rate of approximately 100 000 MT/year. Almost from the beginning of mill operation, contractors and others were allowed to remove the sand-like tailings, for which no charge was made. As a result, some 250 000 MT of tailings were taken, mainly for use in the construction of roads, driveways, and sidewalks. However, in addition, a considerable amount, roughly 50 000 MT, was utilized as a base under concrete slabs in homes and other buildings, including a few schools, and as outside backfill around basements.

In June 1966, employees of the Colorado Department of Health observed that uranium mill tailings were being used in Grand Junction as fill under and around buildings being constructed at the time. Measurements made inside some of these buildings indicated elevated radon gas concentrations, and on August 1, 1966, the State of Colorado ordered a halt to the further use of tailings without prior approval of the Department of Health. In subsequent studies, it appeared that about 3000 structures in Grand Junction had tailings under or adjacent to the buildings. Potential hazards existed from the radon (and its daughters) if the gas seeped through the floor, and also from penetrating gamma radiation, emitted by radium-226 and some of its decay products, even if gas seepage did not occur.

Although the federal government disclaimed legal responsibility for the misuse of the uranium mill tailings, in 1972 Congress passed Public Law 92-314 recognizing the need for compassionate action. By the terms of the law, the federal government agreed to provide 75 percent (and the State of Colorado the remaining 25 percent) of the estimated cost of evaluating the need for and of taking appropriate remedial action in the Grand Junction area. The responsibility for conducting the required operations was assigned to the AEC (now the DOE).

The U.S. Surgeon General provided guidelines for average radon daughter concentrations (in WL units) and of gamma-ray exposure rates (in milliroentgens per hour) in buildings for which (a) remedial action is indicated, (b) remedial action may be suggested (depending on the use of the building, location of the tailings, ventilation, etc.), and (c) no remedial action is indicated. The AEC used these guidelines in determining the action to be taken. For dwellings and schools in the first two categories, the tailings were removed and replaced by other materials. In other cases, alternative remedial techniques, such as use of sealants, ventilation, and shielding, as well as tailings removal, were considered.

To prevent the recurrence of situations similar to that at Grand Junction, access to and removal of tailings are now restricted. The piles are fenced in, and warning notices are posted. In addition, coverage of the material with soil and vegetation for stabilization and protection purposes will make removal more difficult. In some sates—Colorado, in particular—written authorization is required for the use of uranium mill tailings.

PRODUCTION AND ENRICHMENT OF URANIUM HEXAFLUORIDE

Uranium Hexafluoride Production

The next step in the production of reactor fuel material is the conversion of the impure uranium oxide (U_3O_8) in the yellowcake into pure uranium hexafluoride (UF_6) required for the uranium-235 isotope enrichment process (p. 15). The procedure for obtaining a license to possess and use source material for the production of uranium hexafluoride is similar to that described earlier for uranium milling.

Two different procedures, which involve the same basic chemistry, are used to convert the yellowcake into uranium hexafluoride. In each case, the essential steps are:

1. reduction of the oxide U_3O_8 to uranium dioxide (UO_2) by means of hydrogen produced by "cracking" (i.e., strongly heating) ammonia gas
2. hydrofluorination of uranium dioxide to uranium tetrafluoride (UF_4) by heating in a stream of hydrogen fluoride gas

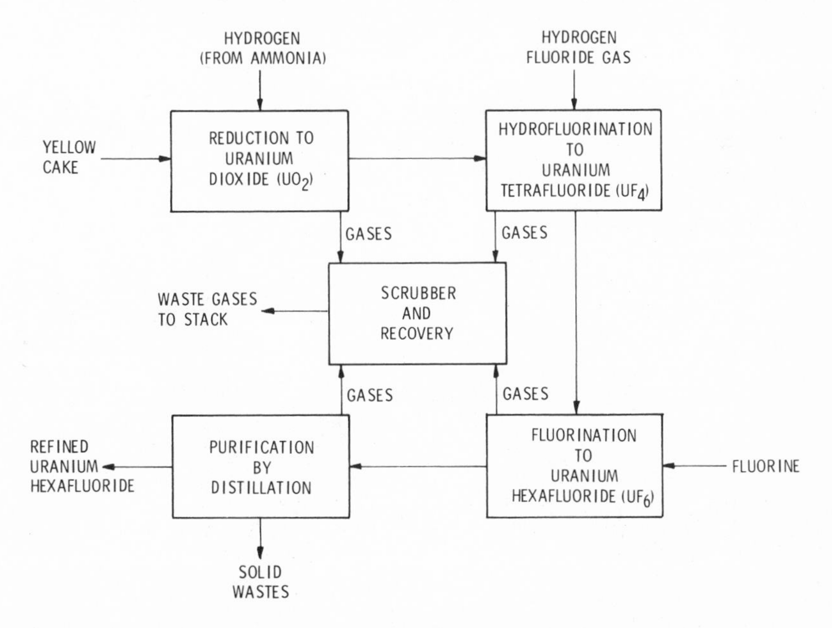

FIG. 8-2. Schematic outline of dry process for the production of uranium hexafluoride.

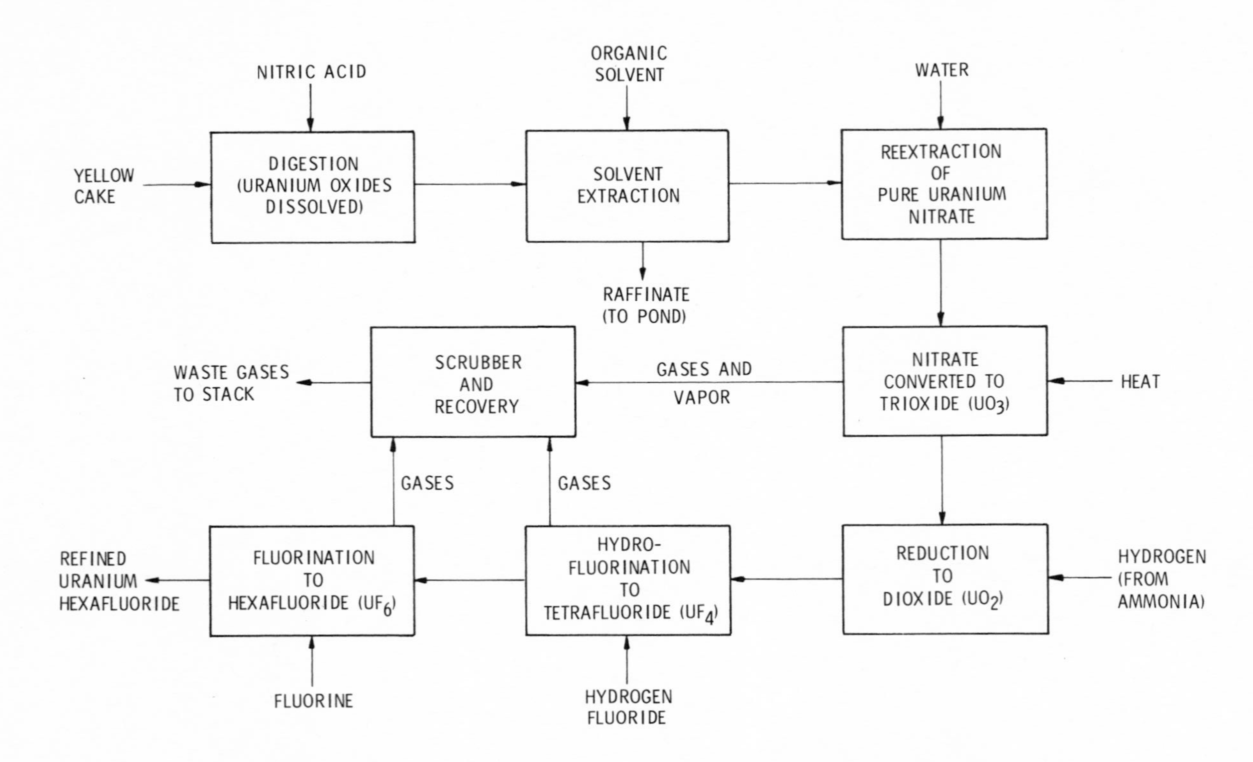

FIG. 8-3. Schematic outline of wet process for the production of uranium hexafluoride.

3. conversion of the tetrafluoride to the hexafluoride (UF_6) by fluorine gas.

In one (*hydrofluor*) process, no purification is carried out until the final stage when the crude uranium hexafluoride is purified. In the other (*solvent extraction*) process, treatments in the early stages lead to the production of pure uranium dioxide and, hence, directly to pure uranium hexafluoride.

The hydrofluor process is characterized by being a "dry" process that involves no liquids. The yellowcake from the uranium mill is heated in a flow of hydrogen gas (and nitrogen) obtained by cracking ammonia. The resulting crude uranium dioxide is first converted into uranium tetrafluoride by means of gaseous hydrogen fluoride and then into the hexafluoride by elemental fluorine gas. The crude uranium hexafluoride is then purified by distillation. The hexafluoride volatilizes readily when heated, and the vapor is condensed to a pure (solid) product; the impurities are not volatile and remain behind (Fig. 8-2).

In the "wet," solvent-extraction process, the yellowcake is digested with hot aqueous nitric acid, which dissolves the uranium (and impurities). Uranyl nitrate is extracted from the solution in water by means of an organic solvent (tributyl phosphate in hexane); the impurities remain in the aqueous raffinate (p. 218). Pure uranyl nitrate is then re-extracted from the organic solution by means of fresh water. The water is evaporated, and the residue is heated to convert the uranyl nitrate into pure uranium trioxide (UO_3), which is then reduced by hydrogen to uranium dioxide. This is heated, first in hydrogen fluoride gas and subsequently in fluorine, as described above, to produce pure uranium hexafluoride (Fig. 8-3).

Environmental Aspects of Uranium Hexafluoride Production

In the hydrofluor process for the production of uranium hexafluoride, aqueous liquid wastes are generated in the preliminary treatment of the yellowcake, in the scrubbers (see below), and in various cleaning operations. These are treated with lime to remove fluorides and are then mixed with large volumes of relatively clean process cooling water before discharge to an adjacent water body. The discharge must meet federal and state requirements for fluorides and other chemicals and NRC standards (10CFR20) for radioactivity concentrations.

Most of the radioactivity in the plant effluent is in gaseous and solid forms. Before discharge, gaseous wastes are passed through filters for control of particulate matter and wet scrubbers for removing fluorides, including uranium hexafluoride vapor, and other noxious gases, such as sulfur dioxide. The discharge to the atmosphere contains traces of radioactive material, but the concentrations must comply with the NRC standards.

The solid waste, consisting mainly of nonvolatile fluoride impurities, is treated with fluorine gas to recover as much as possible of the uranium present. The residue still contains small quantities of radioactive species, especially thorium-232 and uranium-238 and their decay products. These solids are packaged and consigned to a licensed waste burial site (p. 247).

Gaseous wastes from the solvent-extraction process contain noxious chemicals, such as fluorides and nitrogen oxides. These are largely removed by filtration and scrubbing. The aqueous scrubber waste is treated with lime to precipitate fluoride, and the remaining liquid is mixed with process cooling water before discharge. Both gaseous and liquid discharges must comply with federal and state standards for noxious chemicals and radioactivity, as noted above.

The major radioactive waste generated is the aqueous raffinate from the solvent-extraction stage. The raffinate contains significant amounts of thorium-232 and uranium-238 and their decay products and cannot be discharged. The liquid is therefore retained in a pond with a sealed bottom to prevent seepage of radioactivity into the ground. Any acidity remaining in the raffinate is neutralized with lime, and most of the radioactive material present then settles out as a sludge.

Isotopic Enrichment of Uranium Hexafluoride

Isotopic enrichment of the uranium in uranium hexafluoride, to the extent of the 2 to 4 percent of uranium-235 required for LWRs (p. 26), is carried out at the government-owned gaseous diffusion plants described in Chapter 2. The plant sites have a total area of about 600 hectares (1500 acres), of which 170 hectares (425 acres) are occupied by structures, yards, and roadways. These plants also produce highly enriched uranium for weapons and for special reactor fuels, but most of the area can be attributed to commercial nuclear power generation. The plants use large volumes of water for cooling purposes and discharge heat to the surroundings. The major adverse environmental effects of the gaseous diffusion plants, however, are probably those associated with the effluents from coal-fired installations for generating the electricity required to operate the thousands of compressors and pumps. These effluents are regulated by the EPA.

The liquid effluents from gaseous diffusion plants contain small amounts of chemicals arising mainly from the treatment of cooling water and from process cleanup operations. These liquids are diluted to a considerable extent with clean water before discharge; the concentrations in adjacent water bodies are well below permissible NRC limits. Liquids that may contain significant concentrations of uranium from equipment cleanup or from auxiliary production facilities are collected in holding ponds. The radioactivity then deposits as sludge, which can, if necessary, be removed and buried at the plant site. Small quantities of uranium are vented in gaseous form, but the concentration in the surrounding air is not significant.

The centrifuge method for uranium isotope enrichment, which is expected to contribute to the production of nuclear reactor fuels in the late 1980s, will use less electrical power and less cooling water than the gaseous diffusion technique (Chapter 2). The environmental impact of a centrifuge enrichment plant should thus be significantly less than of an equivalent gaseous diffusion plant.

URANIUM DIOXIDE FUEL FABRICATION

Production of Uranium Dioxide Fuel

The final stage in the fabrication of reactor fuel is the production of uranium dioxide pellets from the enriched uranium hexafluoride. Uranium enriched in uranium-235 or other fissile species is called *special nuclear material* (p. 300), and its possession and use are licensed by the NRC (10CFR70, "Special Nuclear Material"). Before constructing a plant for fabricating enriched-uranium reactor fuel, an application for a license must be filed with the NRC. The procedure for obtaining such a license is similar to that described earlier in connection with the licensing of source material.

The first stage in the production of uranium dioxide is the hydrolysis (i.e., decomposition by water) of the hexafluoride to form uranyl fluoride (UO_2F_2); ammonium diuranate ($[NH_4]_2 U_2O_7$) (commonly referred to as ADU) is then precipitated by an aqueous ammonia solution. The solid ADU is separated by a centrifuge or filter and is then dried by heating. The dry solid is calcined and reduced in the presence of hydrogen (from ammonia gas) to yield uranium dioxide (Fig. 8-4). The product, which has the same uranium-235 enrichment as the uranium hexafluoride starting material, is powdered, compacted into small cylindrical pellets, and sintered (i.e., heated to make the particles adhere) until the required density is attained. The pellets are then ground to specified dimension, cleaned, and loaded into zirconium-alloy tubes to form the long, thin fuel rods for LWRs. All operations, especially those involving powders, are conducted in ventilated boxes designed to prevent escape of uranium dioxide to the surroundings.

Because the finished fuel pellets must meet strict density and dimensional specifications, a substantial number are rejected and must be recycled. Some of the material can be ground and reused with minimal processing, whereas the remainder is treated in the scrap recovery operation. Uranium dioxide scrap is dissolved in nitric acid to yield a solution of uranyl nitrate; this is extracted with an organic solvent, leaving the raffinate containing impurities. The uranyl nitrate is re-extracted from the organic medium with water, and uranium tetroxide (UO_4) is precipitated from the aqueous solution by ammonia and hydrogen peroxide. The tetroxide is heated and reduced with hydrogen (from ammonia gas) to regenerate purified uranium dioxide without change in the uranium-235 enrichment.

Environmental Aspects of Fuel Fabrication

All the gaseous effluents (including the ventilation air) from the fuel fabrication plant are drawn through a high-efficiency particulate air (HEPA) filter (p. 83) before discharge to the atmosphere. Gaseous effluents originating in wet processes are scrubbed for the removal of chemicals and often "demisted" to separate water droplets before passing through the filter. The radioactivity level in the discharged gas is then very small.

The waste liquid from the chemical conversion of uranium hexafluoride to the dioxide contains chemical impurities, especially fluorides. The effluent is treated with lime to precipitate calcium fluoride containing traces of uranium and some decay products. The calcium fluoride solids are usually buried at the plant site. In the scrap recovery operation, the major effluent is the raffinate from the organic solvent extraction stage; this liquid is generally held in a retention pond before discharge. Since scrap recovery represents a relatively minor aspect of fuel production, the total amount of radioactivity in the effluent is small.

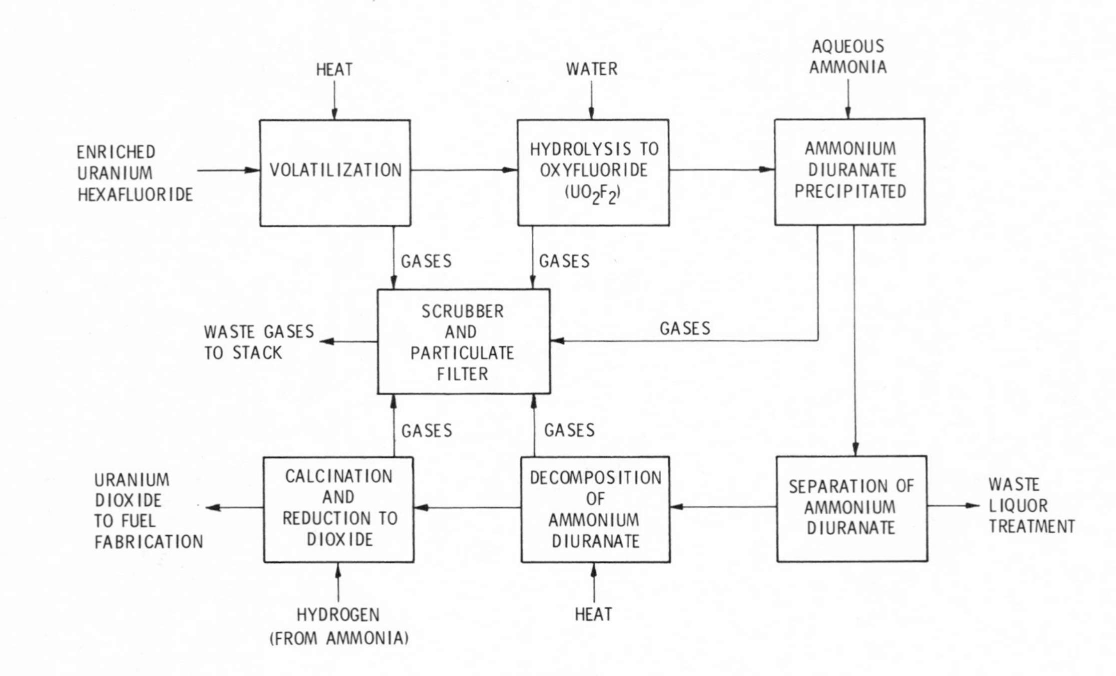

FIG. 8-4. Schematic outline of the production of uranium dioxide from uranium hexafluoride.

Although thorium isotopes are essentially absent from the uranium hexafluoride when it is first produced from yellowcake, thorium-234 (half-life 24 days) is formed by the decay of uranium-238, and the amount increases with time until radioactive equilibrium is approached (p. 213) in about six months.[h] The thorium-234 decays in turn to protactinium-234 (half-life 1.18 min) and then to uranium-234, which has a long half-life. Consequently, if several months have elapsed between the initial production of uranium hexafluoride and the conversion into dioxide, the liquid effluent may be more radioactive than expected from the uranium-238 content.

Another point to be borne in mind is that the product of gaseous diffusion is enriched in the light uranium-234 isotope as well as in uranium-235. The former is not consumed appreciably during reactor operation, as is the latter, and, hence, uranium-234 remains in the spent fuel. In due course, uranium recovered from the fuel when it is reprocessed may be returned to a gaseous diffusion plant for re-enrichment in uranium-235. At the same time, there is further enrichment in uranium-234, and the concentration would greatly exceed the equilibrium value. Since uranium-234 has a much shorter half-life than uranium-238, the activity of a given mass of uranium will increase when (or if) the uranium from spent fuel is recycled. Traces of fission products remaining in the recovered uranium will also contribute to the radioactivity level. Attention is being paid to this situation, although in view of the indefinite postponement of commercial spent-fuel reprocessing, the problem is not an urgent one.

MIXED-OXIDE FUEL FABRICATION

Introduction

Plutonium-239 (mixed with other plutonium isotopes), produced in a nuclear reactor from uranium-238 as a result of neutron capture, would be the fissile material in fast breeder reactors (p. 23). It also might possibly be used as a partial substitute for uranium-235 in LWRs. The fuel material is a mixture of plutonium and uranium dioxides (PuO_2-UO_2), commonly referred to as *mixed oxide*.[i] For fast reactors, the mixture would contain about 15 percent of plutonium dioxide, whereas for LWRs, the content would be 3 to 5 percent. Mixed-oxide fuel elements have been fabricated, either for use in experimental fast reactors or for testing in LWRs.

Because plutonium is potentially a greater biological hazard than uranium, as seen in Chapter 5, the incorporation of plutonium dioxide into the reactor fuel requires techniques and equipment that are different from those used for uranium dioxide fuel. The fabrication technology has been developed by utilizing the extensive experience in handling plutonium and its compounds in the

[h]Roughly seven half-lives are required for 99 percent of the equilibrium amount of any daughter to be attained from a long-lived parent, such as uranium-238. The situation is similar to that discussed on page 167.

[i]This section is based on the assumption that mixed-oxide fuels will be fabricated and used in the United States, but the situation is uncertain. Small amounts of these fuels, however, have been made and tested.

nuclear weapons program. The mixed-oxide fuel fabrication plants are designed to protect the workers and to prevent the release to the environment of quantities of plutonium that could present a hazard to the general public.

Mixed-Oxide Fuel Production

Mixed-oxide fuels are commonly made by mechanically mixing plutonium and uranium dioxide powders. The uranium dioxide is prepared in the manner described earlier, using natural (rather than enriched) uranium hexafluoride. The starting point for plutonium dioxide production is generally plutonium nitrate ($Pu[NO_3]_4$) solution, which is the form in which plutonium would be recovered from spent LWR fuel or from the fuel and blanket of fast breeder reactors.

Nuclear Regulatory Commission regulations prohibit the shipment of plutonium compounds, apart from small quantities, in liquid form (i.e., in solution). Consequently, the plutonium is converted into the solid dioxide prior to shipment. Plutonium oxalate is generally precipitated from the nitrate solution; the oxalate is separated, dried, and calcined to form the dioxide. The product is ground and screened to yield a powder of the desired particle size.

At the fuel fabrication plant, the uranium and plutonium dioxide powders are mixed and pressed into pellets. The pellets are sintered, ground to final dimensions, cleaned and loaded into long, thin tubes of zirconium alloy (for LWRs) or stainless steel (for fast breeders). The procedure is similar to that for fabrication of uranium dioxide fuel elements, except for the precautions taken to minimize release of the plutonium.

The scrap material from mixed-oxide fuel fabrication is described as "clean" or "dirty." Clean scrap, consisting mainly of rejected fuel pellets that do not meet specifications and powders too fine for pressing into pellets, is crushed and calcined in air. The resulting solid is reduced by heating in hydrogen gas (from ammonia) and ground; the powder is then blended in with the fresh mixed-oxide feed.

A small quantity of scrap becomes contaminated with impurities during the fuel element fabrication operations; this is the dirty scrap, which requires chemical treatment. The oxides are dissolved in nitric acid with the addition of a small amount of hydrogen fluoride to facilitate the dissolution. Liquid wastes containing plutonium from other sources in the operation may be added to the solution. The uranium and plutonium are then recovered as nitrates, either separately or together, by extraction with an organic solvent. Addition of ammonia to a solution containing plutonium and uranium nitrates gives a precipitate consisting of plutonium hydroxide and ammonium diuranate. This is separated, dried, and heated in hydrogen gas to regenerate the mixed oxides in pure form.

Environmental Aspects of Mixed-Oxide Fabrication

The process equipment for all stages of mixed-oxide fuel fabrication is designed to contain the potentially hazardous material to the maximum practical extent. The equipment is, in turn, enclosed in *glove boxes*. These are closed chambers with glass (or other transparent) fronts; long-armed gloves, made of rubber or similar material, are sealed into the box. Workers can conduct operations on the equipment in a glovebox by placing their hands and arms in the gloves. Finally, the gloveboxes are enclosed in an outer building that must be capable of withstanding maximum natural phenomena, such as earthquakes, hurricanes, tornadoes, and floods.

Doors, windows, and other openings in the buildings are normally kept closed. The air pressure within the buildings is kept below that of the outside atmosphere so that, if there are any leaks, air is drawn into the building from outside rather than in the opposite direction. Ventilation air flows from the area in which the plutonium contamination is small (i.e., the outer building) into areas of higher contamination (i.e., inside the gloveboxes). This scheme limits the spread of radioactive particles. Alternatively, separate ventilation systems with appropriate pressure differentials may be used. Buildings in which different processes are conducted usually have separate ventilation systems. Emergency electric power is available to operate the ventilation system fans in the event of a failure of off-site power. Air from each confinement area is exhausted through at least two HEPA filters before being discharged.

All exhaust air, as well as the air in the working areas, is continuously monitored for alpha-particle radiation (emitted by plutonium, in particular). Operations are stopped if radiation levels are unexpectedly high, usually long before they become unsafe. Measurements made at mixed-oxide fuel fabrication plants show that the concentration of plutonium dioxide in the air released from the stacks is substantially less than the value permitted by the NRC regulations in 10CFR20.

The most important radioactive liquid wastes from a mixed-oxide fuel production facility are those derived from the treatment of dirty scrap. Other liquids, including laboratory and laundry wastes, which may be contaminated with plutonium, are collected and analyzed. If the plutonium content is sufficiently small, the liquid may be diluted with clean water and discharged. Liquid wastes containing plutonium are evaporated; the vapor may be discharged with ventilation air, or it may be condensed and reused in the plant.

If the evaporator residues contain sufficient plutonium, they may be treated with the dirty scrap, and the plutonium recovered. Otherwise, the residues are solidified in cement (or other material) and buried at an approved site or disposed of in the manner described in the next section. Air containing nonradioactive waste gases, mainly nitrogen oxides, ammonia, and hydrogen fluoride, is scrubbed and filtered before discharge to the atmosphere. The discharge must be in compliance with federal and applicable state air quality standards.

TRU Wastes from Mixed-Oxide Fuel Fabrication

Various solid wastes generated in mixed-oxide fuel fabrication are contaminated with plutonium and other transuranic (TRU) elements. Most of the isotopes of these elements are alpha-particle emitters and could constitute a health hazard if they entered the human body by inhalation or ingestion (Chapter 5). Some of the TRU nuclides have radioactive half-lives of several thousand years; hence, solid wastes containing significant quantities of TRU elements will be separated from other wastes for special handling.

The so-called *TRU wastes*, also sometimes referred to as *alpha wastes*, will be stored temporarily on site and eventually consigned to a federal repository for final disposal (Chapter 10). There are other sources of TRU wastes (p. 259), but if mixed-oxide fuels are used to any substantial extent, the fabrication plants will make the major contribution.

Because of their large initial volume, some form of compaction of the TRU wastes is desirable. In this respect, the solid wastes from fuel fabrication may be divided into combustible and noncombustible categories. The combustible TRU waste includes gloves, plastic bags, and paper that have been exposed in the gloveboxes, as well as protective clothing, plastic sheets and bags, paper, and rags that have become contaminated in other parts of the plant. The framing material used in HEPA filters may also be combustible. The chief noncombustible solid TRU wastes are metal containers, metallic equipment and ductwork, glovebox structural materials, cladding from rejected fuel elements, and HEPA fiber-glass filter material.

The volume of the combustible wastes can be reduced by incineration. Considerable effort is under way to develop methods for safely incinerating combustible materials contaminated with TRU elements. The incinerator off-gases will be filtered to remove entrained particulates, and the residual ash will be incorporated in cement (or other material), either directly or after treatment for plutonium recovery, for safe storage and shipment.

Compaction of noncombustible wastes could be achieved by dismantling or otherwise reducing the dimensions of large pieces of equipment, etc. The smaller pieces could then be compressed mechanically into containers, such as 55-gallon drums.

Bibliography—Chapter 8

"Environmental Survey of Uranium Mill Tailings Piles." *Radiological Health Data Report* 11 (1970): 511. See also, *Radiological Health Data Report* 12 (1971): 17.

Goldsmith, W. A. "Radiological Aspects of Inactive Uranium-Milling Sites: An Overview." *Nuclear Safety* 17 (1976): 722.

Gotchy, R. L. *Health Effects Attributable to Coal and Nuclear Fuel-Cycle Alternatives* (NUREG-0332). Washington, D.C.: U.S. Nuclear Regulatory Commission (1977).

International Atomic Energy Agency. *Proceedings of the Conference on Nuclear Power and Its Fuel Cycle.* Vienna: International Atomic Energy Agency, 1977.

______. *Proceedings of the Panel on Radon in Uranium Mining.* Vienna: International Atomic Energy Agency, 1975.

Organization for Economic Cooperation and Development. "Management, Stabilization, and Environmental Impact of Uranium Mill Tailings." Washington, D.C.: OECD Publications Center (1978).

Pechin, W. H., *et al. Correlation of Radioactive Waste Treatment Costs and the Environmental Impact of Waste Effluents in the Nuclear Fuel Cycle—Fabrication of LWR Fuel* (ORNL-TM-4902). Oak Ridge, Tenn.: Oak Ridge National Laboratory, 1975.

Sears, M. B., *et al. Correlation of Radioactive Waste Treatment Costs and the Environmental Impact of Waste Effluents in the Nuclear Fuel Cycle—Milling of Uranium Ores* (ORNL-TM-4903). Oak Ridge, Tenn.: Oak Ridge National Laboratory, 1975.

Travis, C. C., *et al.* "Natural and Technologically Enhanced Sources of Radon-222." *Nuclear Safety* 20 (1979): 722.

U.S., Atomic Energy Commission. *Environmental Survey of the Nuclear Fuel Cycle* (WASH-1248). Washington, D.C., 1974.

U.S., *Code of Federal Regulations*, Title 10 (Energy).
Part 40. Licensing of Source Material;
Part 70. Special Nuclear Material.

U.S., Congress, Joint Committee on Atomic Energy. *Hearings on the Use of Uranium Mill Tailings for Construction Purposes.* Washington, D.C., 1971.

U.S., Congress, Joint Committee on Atomic Energy, Subcommittee on Research, Development, and Radiation. *Hearings on Radiation Exposure of Uranium Miners.* Washington, D.C., 1967.

______. *Hearings on Radiation Standards for Uranium Mining.* Washington, D.C., 1969.

U.S., Environmental Protection Agency. *Environmental Analysis of the Uranium Fuel Cycle—Part I: Fuel Supply* (EPA-520/9-73-003B). Washington, D.C., 1973.

______. *Environmental Surveys of Uranium Mill Tailings Piles and Surrounding Areas, Salt Lake City, Utah* (EPA-520/6-74-006). Washington, D.C., 1974.

U.S., Environmental Protection Agency and U.S. Energy Research and Development Administration. *Study of Inactive Uranium Mill Sites and Tailings Piles, Phase I* (Summary Report [updated]). Washington, D.C., 1976.

U.S., Federal Radiation Council. *Guidance to the Control of Radiation Hazards in Uranium Mining* (Report No. 8 [revised]). Washington, D.C., 1967.

U.S., Nuclear Regulatory Commission. *Draft Generic Environmental Impact Statement on Uranium Milling* (NUREG-0511). Washington, D.C., 1979.

______. *Final Generic Statement on the Use of Recycled Plutonium in Mixed Oxide Fuel in Light Water-Cooled Reactors* (NUREG-0002), *Washington, D.C., 1975.*

______. *Guide to the Contents of Applications for Uranium Mill Licenses* (Regulatory Guide 3.5). Washington, D.C.

9

Radioactive Effluents from Nuclear Facilities

EFFLUENTS FROM LIGHT-WATER REACTORS

Sources of Radiation

The major reactor types used in nuclear-electric power plants in the United States in the foreseeable future will be pressurized water reactors (PWRs) and boiling water reactors (BWRs)—that is, light-water reactors (LWRs). This chapter is therefore concerned with the release of radioactivity to the environment from LWR installations and from the reprocessing plants where spent LWR fuels are treated for the recovery of material that could be used as reactor fuel.

The sources of radioactivity generated during the operation of an LWR may be considered as falling into three categories; these are: (a) fission products, (b) tritium, and (c) neutron activation products. Tritium is formed to some extent in fission, but since it also arises from other nuclear reactions, it is convenient to treat it separately.

Fission Products. The main fission products are defined here as those formed in the great majority of fissions. As a result of radioactive decay, the fission products after a short time constitute a complex mixture of more than 300 different species (nuclides), most of which are radioactive (Chapter 2). Many of these radionuclides have such short half-lives and are formed in such small proportions that they may be ignored for the present purpose. Radioactive nuclides with very long half-lives decay so slowly that they also can usually (but not always) be neglected as far as reactor effluents are concerned. Some of the more important radioactive fission products (including tritium)—from the standpoint of their half-lives, the amounts produced in reactor operation, and their biological effects—are listed in Table 9-I. Elements that are normally gases are given in the first part, and those that are solid at ordinary temperatures are given in the second part. The latter are generally found as compounds dissolved in the liquid effluent from a nuclear facility.

Although iodine is a solid at ordinary temperatures, it vaporizes readily so that part of the iodine in the fission products often appears together with the gases in the effluent from a reactor. The gases may also carry small quantities of radioisotopes of other normally solid elements, some of which (e.g., rubidium, cesium, strontium, and barium) are formed by the radioactive decay of krypton

TABLE 9-I
SOME IMPORTANT FISSION PRODUCTS

	Half-life	Radiation		Half-life	Radiation
Gases			*Solids*, cont'd.		
Tritium	12.3 yr	beta	Yttrium-91	59.0 d	beta, 0.3% gamma
Krypton-83*m*[a]	1.9 h	gamma	Tellurium-129	70.0 min	beta, 20% gamma
Krypton-85*m*[a]	4.4 h	beta, gamma	Tellurium-131*m*[a]	1.3 d	82% beta, 31% gamma
Krypton-85	10.8 yr	beta, 0.4% gamma	Tellurium-131	25.0 min	beta, 68% gamma
Krypton-87	1.3 h	beta, gamma	Tellurium-133	13.0 min	beta
Krypton-88	2.8 h	beta, 35% gamma	Iodine-131	8.0 d	beta, gamma
Xenon-133*m*[a]	2.3 d	gamma	Iodine-133	20.0 h	beta, gamma
Xenon-133	5.3 d	beta, 37% gamma	Iodine-135	6.7 h	beta, 37% gamma
Xenon-135*m*[a]	16.0 min	gamma	Cesium-134	2.0 d	beta, gamma
Xenon-135	9.1 h	beta, gamma	Cesium-136	14.0 d	beta, gamma
Xenon-138	18.0 min	beta, gamma	Cesium-137	30.0 yr	beta, 89% gamma
Solids			Cesium-138	32.0 min	beta, 73% gamma
Rubidium-88	18.0 min	beta	Barium-140	13.0 d	beta, gamma
Strontium-89	53.0 d	beta, 0.01% gamma	Lanthanum-140	1.7 d	beta, gamma
Strontium-90	28.0 yr	beta	Cerium-144	290.0 d	beta, 13% gamma
Yttrium-90	2.7 d	beta	Praseodymium-144	17.0 min	beta, 2.5% gamma

[a]The letter *m* refers to a metastable (high-energy) state of the indicated nuclide; metastable states usually decay by emitting their extra energy as gamma rays.

and xenon isotopes of short half-life. These solids in finely divided form contribute to the particulates that are associated with the effluent gases.

Tritium. Tritium, the radioactive isotope of hydrogen, is produced to a small extent as a result of nuclear fission; on the average, about one fission of uranium-235 in ten thousand is accompanied by the formation of tritium.[a] In addition, tritium is generated in the coolant water by the capture of neutrons in the deuterium present in all natural waters. An important source of tritium, especially in LWRs, is the interaction of neutrons of high energy with boron. This element may be present as a burnable poison (in BWRs), it may be dissolved (as boric acid) in the water for shim control (in PWRs), or it may be used in the control rods, as described in Chapter 2. In some PWRs, tritium has resulted from the reaction of neutrons with lithium-6; the latter is a stable isotope occurring in the lithium hydroxide sometimes used for water treatment. Currently, if lithium is utilized for this purpose, the lithium-6 content is reduced in order to minimize the formation of tritium.

In PWRs, substantial amounts of tritium are produced by interaction of neutrons with the boron shim control. To avoid adding to this, it is the practice not to use boron in the control rods. Hence, these rods commonly consist of an alloy of cadmium, silver, and indium. BWRs, on the other hand, do not utilize soluble boron shim control; hence, the control rods generally contain boron (in the form of carbide) as the neutron-absorbing material.

Neutron Activation Products. Activation products are formed in the reactor coolant by interaction of neutrons with the oxygen nuclei in the water molecules, with oxygen, nitrogen, and argon in dissolved air, and with various impurity elements present (e.g., as a result of corrosion of metallic components).

[a]About twice as many tritium nuclei are formed in the fission of plutonium-239.

In a sense, tritium formed in the water, from deuterium, boron, or lithium, is an activation product. The majority of radionuclides produced by the interaction of neutrons with oxygen, nitrogen, and argon have relatively short lives and are formed in small quantities. Consequently, only three, nitrogen-13, nitrogen-16, and argon-41, make a significant contribution to the radioactive effluent from a nuclear power plant (BWR). In addition, carbon-14 may present a long-range problem (p.175).

The most important activation products are those arising from neutron reactions with various elements (e.g., iron, chromium, nickel, cobalt, and manganese) that enter the water as a result of corrosion and erosion of steel and other alloys used in the reactor vessel, pumps, piping, valves, and steam generator. In BWRs, corrosion and erosion in the turbine, turbine condenser, and feedwater heaters can also contribute to the activation products because the condensed water is returned to the reactor vessel by way of the feedwater heaters. In a PWR, however, the condenser and feedwater form a separate circuit that includes the steam generator, but not the reactor.

The identity and radioactive characteristics of the activation products of major significance are given in Table 9-II. Zinc-65, with a half-life of 245 days, has been found as an activation product in the water effluent from a BWR. It originated in a zinc-containing alloy used in the feedwater heaters (p.66); this alloy was later replaced by stainless steel. The solid elements mentioned in the table occur mainly as insoluble oxides suspended in the water. They can thus be partially removed by filtration.

Fission Products in Reactor Water

In normal operation of an LWR, the fission products, including tritium generated in fission, are retained within the uranium dioxide fuel and its zirconium-alloy cladding (p.68). There is evidence that tritium gas can penetrate stainless steel, but since zirconium alloys have come into general use as the cladding material, the escape of tritium gas from intact fuel rods has been greatly diminished. However, a few fuel rods may have small holes ("pinholes") in the cladding that arise from welding defects or from localized corrosion during reactor operation. Gaseous and easily vaporized fission products, such as

TABLE 9-II
ACTIVATION PRODUCTS

	Half-Life	Activity
Gases		
Nitrogen-13	10 min	beta
Nitrogen-16	7 s	beta, gamma
Argon-41	1.8 h	beta, gamma
Solids		
Chromium-51	28 d	beta, 9% gamma
Manganese-54	300 d	gamma
Manganese-56	2.6 h	beta, gamma
Cobalt-58	71 d	15% beta, gamma
Cobalt-60	5.3 yr	beta, gamma
Iron-59	45 d	beta, gamma

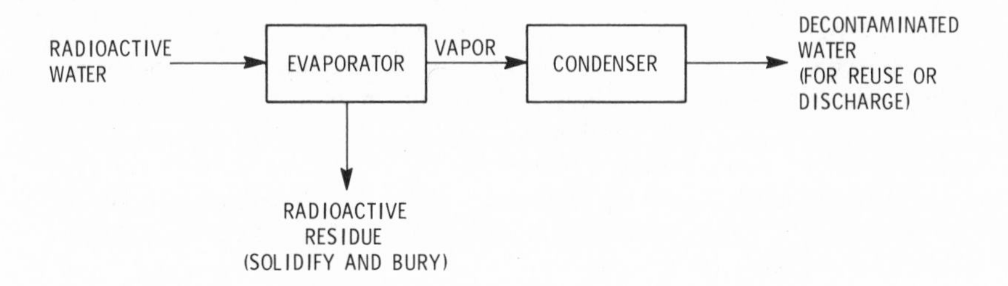

FIG. 9-1. Schematic outline of decontamination by evaporation.

tritium, krypton, xenon, and iodine, and some dissolved solid products can then escape into the reactor water. In designing the water purification system, to which reference will be made shortly, it is assumed that a number of the fuel rods will develop pinholes in the cladding.

After fabrication of the fuel elements, traces of uranium dioxide may remain on the outside surface; this is referred to as *tramp uranium*. In the reactor, the uranium-235 present undergoes fission, and the fission products, both gaseous and solid, are released into the surrounding water. The amounts of tramp uranium are generally quite small, but they constitute a possible source of radioactivity in the liquid effluent from the nuclear power plant.

Radwaste Systems

The system for the treatment of various radioactive liquids and gases prior to discharge to the environment is called the *radwaste system*. Its purpose is to reduce radioactivity levels in the plant effluents to such an extent that they will satisfy the regulations and guidelines of the U.S. Nuclear Regulatory Commission (NRC). As explained in Chapter 5, concentrations of radioactive nuclides in water and air at the nuclear plant boundaries must not exceed the values specified in Title 10 of the *Code of Federal Regulations*, Part 20 (10CFR20). Furthermore, radiation doses from effluents to people in unrestricted areas must be kept as low as is reasonably achievable (Chapter 6). The details of the radwaste systems vary from one reactor installation to another because of differences in local circumstances and preferences of the plant designer. The general descriptions given here, however, are applicable to all plants using reactors of the same type (i.e., PWR or BWR).

A radioactive liquid (i.e., water containing dissolved radioactive material) is commonly *decontaminated*, that is, it has its radioactivity decreased, in two different ways.[b] One is by evaporation, in which the water is boiled off and the steam is condensed. Most of the radioactive material remains behind in the evaporator residue and is disposed of in a controlled manner (Fig. 9-1). The condensed water is essentially (although not completely) free of dissolved solids, but it will contain nearly all of the tritium. The latter is largely present as tritiated water (HTO)—that is, water (H_2O) in which one atom of ordinary hydrogen

[b]In a sense, the decay of radioactivity in the course of time is a kind of decontamination.

(H) has been replaced by a tritium (T) atom. In the evaporator, the tritiated water is vaporized and subsequently condensed with the ordinary water.

The second way of decontaminating a radioactive liquid is by means of a *demineralizer*. The latter contains what is known as an *ion-exchange resin*, similar to the material often used in household (and industrial) water softeners. When a radioactive solution is passed through the demineralizer, the resin removes and retains much of the dissolved matter. The elements cesium, yttrium, and molybdenum are removed relatively slowly by demineralizers, and tritium is essentially unaffected.

As a general rule, demineralization, although less effective than evaporation, is the preferred decontamination procedure because of its simplicity. However, if the total amount of dissolved solids, both radioactive and nonradioactive, is moderately large (or quite large), the ion-exchange resin would soon become saturated and would have to be regenerated or replaced frequently. Evaporation is then the preferred process. In some situations, where considerable decontamination is to be achieved, both procedures may be used, with evaporation generally preceding demineralization.

The gaseous effluent from a nuclear power plant contains the noble gases (i.e., krypton and xenon), iodine, and particulate matter. The particulate matter is decreased by passing the gas through a high-efficiency particulate air (HEPA) filter (p.83) prior to discharge, and part of the iodine is removed by a charcoal (or other) filter. The noble gases are difficult to remove (see, however, p.175), and so they are held up for a time to permit the activity to decrease by natural radioactive decay.

Liquid Wastes from a PWR Plant

Chemical and Volume Control System. The chemical and volume control system (or CVCS) of a PWR power plant is not strictly part of the radwaste

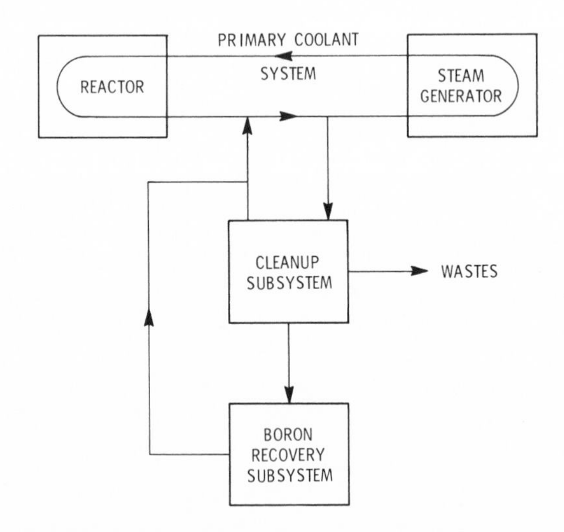

FIG. 9-2. Schematic outline of chemical and volume control system of a PWR.

system, but it plays a role in decreasing the amount of radioactivity in the liquid effluent. The CVCS consists of two subsystems: namely, the reactor coolant water cleanup subsystem and the boron recovery subsystem (Fig. 9-2). A bypass in the main (primary) coolant circuit permits some of the reactor water to be diverted continuously through the cleanup (i.e., purification) subsystem. The water is cooled and passed through a demineralizer to remove dissolved substances, including fission and activation products (but not boric acid), before being filtered and returned to the coolant circuit. Most of the radioactivity is thus retained by the ion-exchange resin in the demineralizer.

During the course of operation of a PWR, the concentration of the boric acid used for shim control must be decreased. In the early stages, some of the purified water from the cleanup subsystem is withdrawn periodically for removal of boric acid in the boron recovery subsystem. The water is evaporated, and the steam is condensed; the condensate, essentially free from boron, is then returned to the primary circuit. The boric acid remains in the evaporator as a concentrated solution, which is drawn off periodically and stored for reuse as required.

When there is an accumulation of radioactive material in the boric acid residues, they are disposed of with solid wastes (p.246). In the later stages of the core lifetime, the concentration of boric acid in the primary coolant circuit is fairly low, and evaporation for boron recovery is not economical. A special deborating demineralizer is then included in the reactor coolant cleanup subsystem to decrease the boron concentration.

Clean and Dirty Wastes. In most PWR facilities, the liquid wastes are divided into three main categories: (a) *clean* wastes, (b) *dirty* wastes, and (c) *laundry* wastes. The terms *clean* and *dirty* as used here refer to the chemical purity of the water and not to the amounts of radioactivity. The clean wastes, also called *primary system wastes*, originate in the primary coolant circuit and have the highest level of radioactivity of all the liquid wastes. Apart from the radioactivity, however, the water is exceptionally pure.

The clean wastes are made up of excess water from the CVCS, pump-seal and valve leakages, and any other liquids that may have leaked or been released from the primary coolant system. The liquids are collected in a tank where they are held for several (up to 30) days to permit radionuclides of short half-life to decay. The contents of the tank are filtered and then decontaminated by evaporation or demineralization or both. Part of the decontaminated water is retained for use as primary coolant. After being tested for radioactivity, the remainder is diluted with clean water (e.g., from the turbine condenser) and discharged at a controlled rate into an adjacent water body (Fig. 9-3). When the condenser water is recirculated through a cooling tower (Chapter 13), only a small volume is available for dilution purposes. Additional decontamination of the effluent might then be required prior to discharge.[c]

In the dirty liquid wastes, the water is not originally of reactor quality, but its radioactivity is relatively low. These wastes include liquids from various floor drains and sumps, laboratory drains, and from cleanup areas. The wastes are

[c]The Rancho Seco (PWR) plant of the Sacramento (California) Municipal Utility District is designed to eliminate radioactive liquid effluents. The liquid wastes are to be evaporated, and the water condensed from the steam is to be reused. The evaporator residues will be solidified and disposed of with the solid wastes.

collected in a holding tank and filtered; if the radioactivity level is low enough, the liquid may be diluted and discharged. Otherwise, the dirty wastes may be decontaminated by evaporation (and possibly by demineralization) before discharge. In some cases, the dirty wastes are mixed with clean (primary) wastes and decontaminated.

Laundry wastes consist mainly of water from the plant laundry, where protective clothing is washed, and from showers. Because of the presence of detergents, which might interfere with the operation of demineralizers and evaporators, the relatively small volume of laundry wastes is kept separate from other wastes. Since the radioactivity level is low, the water is stored for a time in a holdup tank, filtered, mixed with clean water, and discharged after testing.

Steam Generator Blowdown. As the PWR plant operates, the concentration of salts in the secondary system water gradually increases; this can occur as a result of the presence of normal impurities in the water, of leakage from primary to secondary in the steam generator (see Fig. 2-7), and especially of leakage of cooling water through defective tubes in the condenser (see Fig. 1-1). To prevent the buildup of solids on the steam generator tubing, which would decrease the operating efficiency, water is withdrawn at a rate of about 38 litres (10 gal)/min from the secondary (or steam) side of the generator and replaced by fresh water. The withdrawn water (or *blowdown*) enters a flash tank at a lower pressure where about half flashes into steam. If the steam generators are completely free from internal leaks, there is no radioactivity in the blowdown. Both liquid water and steam can be discharged without further treatment.

In practice, however, small holes develop in the miles of tubing in the steam generators; radioactive material in the primary coolant can leak into the secondary system and thus appear in the blowdown. In the earlier PWRs, the blowdown water was discharged after mixing with uncontaminated condenser water, but in later plants the water is either treated along with the clean wastes or demineralized separately before discharge. In some PWR installations

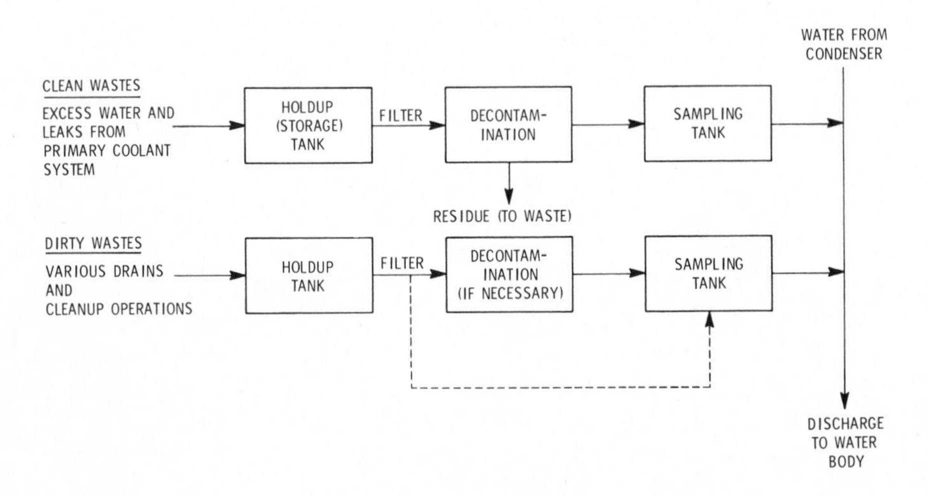

FIG. 9-3. Schematic outline of liquid radwaste system of a PWR.

blowdown is avoided by using a cleanup demineralizer in a bypass of the secondary circuit.

With adequate decontamination, moderate leakage rates (e.g., up to about 76 litres [20 gal]/day) in the steam-generator tubing can be tolerated. If the leakage rate becomes too large, the faulty steam generator is shut down, and the leaks are plugged. High leakage rates not only mean that additional treatment is necessary to reduce the radioactivity level in the blowdown, but the boric acid from the primary coolant deposits as solid on the steam generator tubing and is difficult to remove.

Gaseous Wastes from a PWR Plant

Primary System Gases. Gaseous wastes from a PWR installation are conveniently considered as primary system gases, secondary system gases, and building ventilation gases. The primary system gases include fission product gases (and vapors) vented from the CVCS and from the liquid waste holdup tanks. In addition, hydrogen and nitrogen gases are present for the following reasons. As a result of the intense neutron and gamma radiation within and near the reactor core, water may be decomposed into its component gases, hydrogen and oxygen. In a PWR this decomposition, called *radiolysis*, is suppressed by the deliberate addition of a certain amount of hydrogen gas to the primary coolant water. Furthermore, nitrogen gas is commonly used to purge (i.e., to remove) the air from the reactor vessel in order to avoid the possibility of forming an explosive mixture of hydrogen and atmospheric oxygen.

In many PWRs, the primary system gases are collected in a storage tank and compressed into one of several decay tanks where the gases are held for a period of about 60 days, on the average (Fig. 9-4). The only radioactive species remaining in appreciable amount are then krypton-85, xenon-133, iodine-131, and tritium. These residual gases are passed through filters to remove particulate matter and then mixed with large volumes of filtered ventilation air before discharge through a roof vent or stack. In more recent PWR plants the volume of primary waste gas is decreased by removing the hydrogen component. The gas is mixed with oxygen and passed through a catalytic recombiner; here the hydrogen and oxygen unite to form water.

Secondary System Gases. The secondary system gases are mainly

1. those released during steam generator blowdown
2. air ejector gases, which are removed continuously from the low-pressure (exhaust) side of the turbine by means of a steam jet
3. gland-seal effluent, consisting of steam used to seal the turbine gland and prevent the entry of air.

Radioactivity in the secondary waste gases would arise from leakage in the steam generator tubes. If there is little or no leakage, the radioactivity level will not be significant. The secondary gases may then be discharged without treatment after mixing with ventilation air. In many PWRs, however, only the gland-seal effluent is discharged in this way. After condensing the steam, the other gases are passed through charcoal to decrease the iodine content; they are then filtered to remove particulate matter and discharged with the ventilation air by way of a roof vent or stack.

Ventilation Air. In the third category of gaseous wastes are the large volumes of ventilation air, most of which is from auxiliary buildings where the radioactivity level is very low. The air is often discharged directly after filtration, but in some plants charcoal is used to remove much of the iodine before discharge. Because of the low radioactivity level and large volume, the ventilation air is useful for reducing the concentrations of radionuclides in the other gaseous effluents.

The closed containment structure containing the reactor primary system (Chapter 4) is purged a few times a year with clean air. Before discharge with the other gases through the plant vent (or stack), the air is passed through charcoal and particulate filters. In some PWR installations the accumulation of radioactivity in the containment atmosphere between purges is decreased by circulating the air continuously through charcoal filters, called "kidneys," to remove iodine and through particulate filters to retain suspended solids.

Solid Wastes from a PWR Plant

The solid wastes from a PWR plant arise mainly from spent ion-exchange (demineralizer) resins, discarded filter material (including charcoal), and evaporator residues. The latter are often slurries rather than solids, but they are generally solidified by being mixed with cement (or cement and vermiculite). Solid wastes are enclosed in 55-gal steel drums and held in a shielded area for a few months to permit partial radioactive decay.

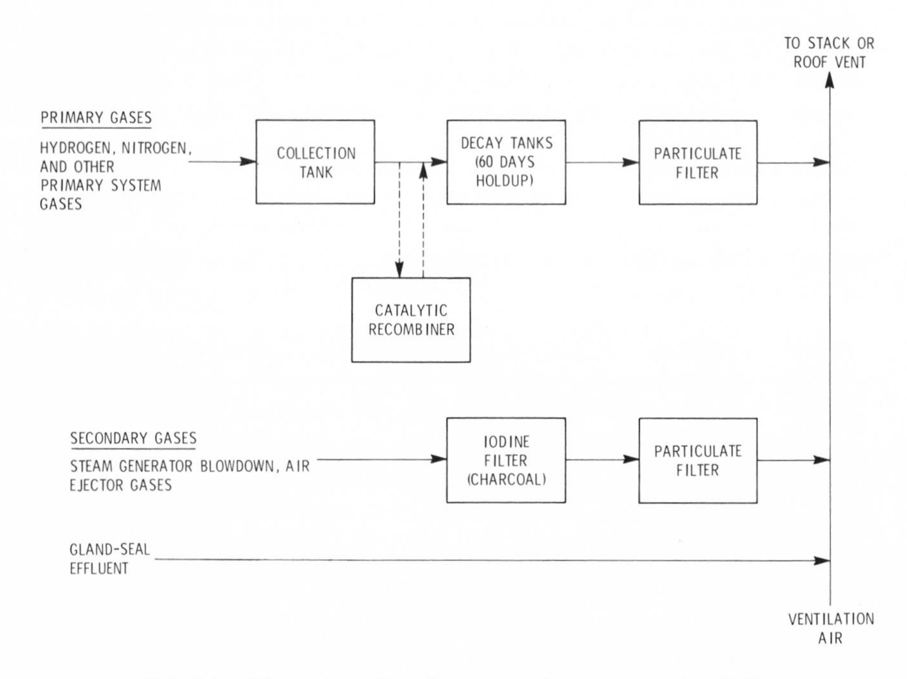

FIG. 9-4. Schematic outline of gaseous radwaste system of a PWR.

The drums are eventually shipped in accordance with U.S. Department of Transportation and NRC regulations to a state-licensed burial site. Such sites, to which access is severely restricted, were located near West Valley, New York; Morehead, Kentucky; Sheffield, Illinois; Richland, Washington; Beatty, Nevada; and Columbia, South Carolina.The sites in New York, Kentucky, and Illinois were closed down in 1979 and the others have restricted their operations. Because of the reluctance of licensed sites to accept radioactive wastes from other states, it is possible that the federal burial sites may have to be developed, especially for wastes from medical facilities. Solid wastes from nuclear power operations may have to be buried within the plant sites.

Tritium Disposal from a PWR Plant

Tritium presents a special problem in PWR radioactive waste disposal because substantial quantities of this radionuclide are formed in the primary coolant from neutron interaction with boron. As already noted, the tritium, as tritiated water, becomes an integral part of the reactor water from which it is not separated by evaporation or demineralization. Some of the tritium is discharged with the gaseous effluents, but most remains in the PWR liquid waste. When ample volumes of water are available (e.g., from an adjacent river) to dilute the liquid waste and so reduce the concentration below the required limit (see Table 5-VII), the tritium can be discharged at a controlled rate with the liquid effluent. Alternatively (or additionally), the water may be evaporated and the tritiated water vapor discharged to the atmosphere with the gaseous effluents.

In many modern PWR plants most of the decontaminated liquid waste is not discharged, but is reused in the primary system. There is consequently a steady buildup of tritium in the reactor water. Eventually, the concentration reaches a level at which it might represent a hazard to operators when the reactor vessel is open for refueling. Part of the water is then withdrawn for disposal. Presently, the water is mixed with cement (or other material), and the resulting solid is buried with other solid wastes.

Alternative methods of tritium disposal, which are being studied but have not yet been approved, are prolonged storage of the tritiated water in large tanks, injection into deep wells considerably below the levels accessible for human consumption, and discharge into the ocean, where extensive dilution would occur. Since the half-life of tritium is just over 12 years, about 90 percent decay will occur in 40 years.

Comparison of PWR and BWR Wastes

The radioactive wastes produced in a BWR power plant differ from those generated in a PWR for three main reasons. In the first place, BWRs do not use boric acid in the reactor water as a shim control; there is consequently much less tritium formed. Second, since steam is generated in the reactor vessel, radioactive gases (and vapors) that have escaped from defects in the fuel cladding or that were formed by neutron activation will be carried by the steam into the

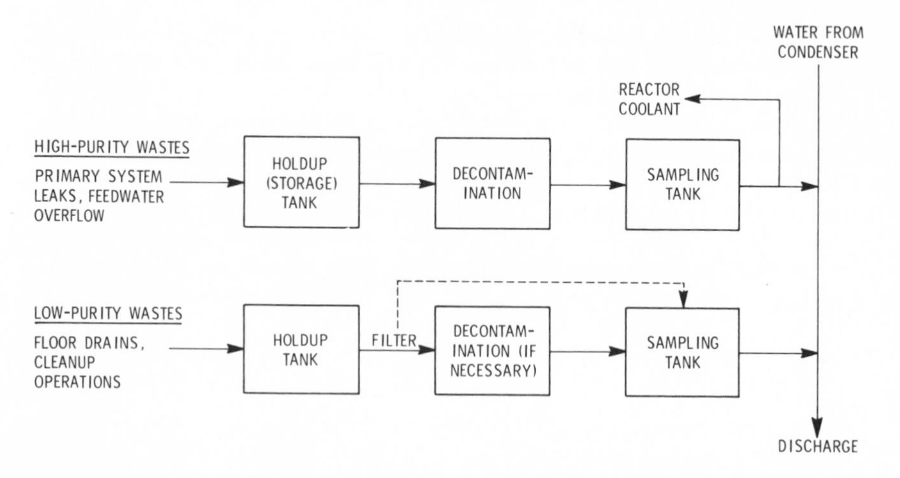

FIG. 9-5. Schematic outline of liquid radwaste system of a BWR.

turbine. Finally, in a BWR the design does not permit the use of hydrogen gas to suppress radiolytic decomposition of the water; hence, the gases removed from the low-pressure side of the turbine contain substantial volumes of hydrogen and oxygen.

As in a PWR, the dissolved radioactive material in the coolant water is maintained at a low level by continuously bypassing some of the water through a cleanup system. The water is first cooled and then filtered and demineralized; the decontaminated water is returned to the reactor coolant system.

Liquid Wastes from a BWR Plant

High-Purity and Low-Purity Wastes. The liquid radwaste system is not the same for all BWR plants, but the following descriptions may be taken to be fairly typical of the procedures required to comply with NRC regulations. The liquid wastes may be regarded as falling into three general classes somewhat similar to those in a PWR. They are called: "high-purity" or "low-solid" wastes, which are equivalent to the clean wastes of a PWR; "low-purity" or "high-solid" wastes, similar to the dirty wastes of a PWR; and laundry wastes. The laundry wastes will not be considered further because they are treated in the same manner as already described for PWR laundry wastes.

The high-purity (or clean) wastes, consisting of the purest water, but with the highest radioactivity concentration, arise from reactor coolant water that has leaked from the primary system equipment (e.g., pumps and valves), the overflow from the feedwater tank, etc. The liquids are collected in a storage (or holdup) tank, where partial decay of the radioactivity occurs. The wastes are then filtered and demineralized. After being sampled for radioactivity, part of the decontaminated liquid may be discharged after dilution with clean turbine condenser water, if available, but most is retained for reuse as reactor coolant (Fig. 9-5).

The low-purity (or dirty) wastes from various floor drains are also collected in a storage tank. In some BWR installations these liquids are allowed to stand for a time and are then filtered and discharged with the condenser water, provided the radioactivity concentration is sufficiently low. The preferred practice, however, is to follow filtration by either demineralization or evaporation. The water may then be discharged after dilution or, if it has been purified by evaporation, used as makeup for the reactor coolant as required.

Chemical Wastes. In many BWR plants the chemical wastes, from laboratory drains, equipment decontamination, and solutions used to regenerate demineralizer resins, are neutralized in a chemical treatment tank and then mixed with the low-purity wastes. Alternatively, after neutralization, the chemical wastes are evaporated; the condensate is discharged with the plant effluent, whereas the residues are disposed of with the solid wastes.

Gaseous Wastes from a BWR Plant

Air-Ejector Gases. Three types of gaseous wastes are associated with BWR operation: steam-jet air-ejector gases, gland-seal effluent, and building ventilation air. Of these, the first, consisting of gases removed by a steam jet from the exhaust side of the turbine, is the most important. It carries nearly all (more than 99 percent) of the radioactive noble gases and iodine escaping from the fuel, as well as large amounts of hydrogen and oxygen produced by radiolysis of the reactor water. The total volume of steam-jet air-ejector gases from a BWR is thus considerably greater than that from a PWR.

In BWRs of more recent design the air-ejector gases are first passed through a catalytic recombiner in which the hydrogen and oxygen gases combine to form water (Fig. 9-6). After condensation of the water vapor, the remaining gases have about one-fifth of the original volume. These gases enter a long, wide, con-

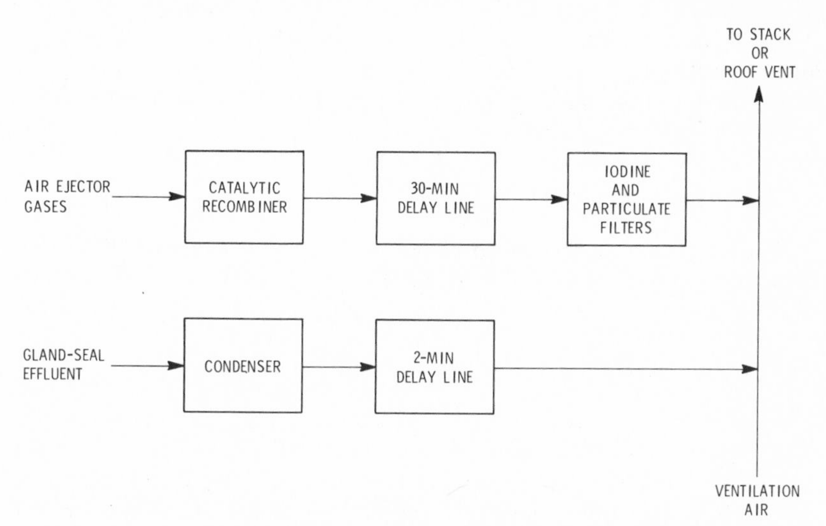

FIG. 9-6. Schematic outline of gaseous radwaste system of a BWR.

voluted pipe called a *delay line*, where they are held up for about 30 min to permit radionuclides with short half-lives to decay. From the delay line the gases pass on to charcoal beds at ambient temperature. The charcoal selectively removes (adsorbs) the noble gases and then gradually releases (desorbs) them. The net effect is to cause these gases to move slowly through the charcoal beds, which thus serve as an effective holdup system. Radioisotopes of iodine are largely retained by the charcoal. The gases emerging from the charcoal beds are filtered to remove particulate matter, are mixed with large volumes of ventilation air, and are tested for radioactivity before discharge to the atmosphere.

After the treatment just described, the gaseous effluent from a BWR plant can satisfy the NRC's regulations and guidelines. Nevertheless, the radioactivity of the noble gases released from a BWR installation is usually greater than from a PWR plant of the same electrical capacity (see Table 9-III). Hence, consideration is being given to methods for decreasing the noble gas activity in BWR gaseous effluents. Among the possibilities being studied are low-temperature absorption by charcoal to achieve longer holdup times and separation of the noble gases by liquefaction with subsequent storage (see also p. 175).

Gland-Seal Effluent. The turbine gland-seal effluent gases from a BWR plant are more highly radioactive than those from a PWR because of the greater radioactivity of the steam used to provide the seal. However, since the total radioactivity content is relatively small, the gases remaining after condensation of the steam are held up for about 2 minutes. They are then mixed with ventilation air before discharge, provided the radioactivity level is sufficiently low. A proposal has been made to use clean (nonradioactive) steam, generated especially for the purpose, to provide the gland seal; this would eliminate a minor, but not insignificant, radioactive effluent.

Ventilation Air. Building ventilation air has a low level of radioactivity and large volumes are discharged; this air serves to reduce the concentration of noble gases and iodine in the gaseous effluent entering the atmosphere. On occasion, especially when the reactor is being refueled, the radioactivity level in the reactor building (or secondary containment) becomes fairly high. The ventilation air is then routed through a standby gas treatment system, consisting of iodine and particulate filters, before discharge.

Solid Wastes from a BWR Plant

The solid wastes from a BWR power plant are quite similar to those produced in a PWR installation. The treatment of such wastes is thus the same as described earlier.

Radioactive Releases from Nuclear Power Plants

The NRC requires that each nuclear power plant be equipped to measure continously the amount of radioactivity released into the air and water. These measurements must be reported to the NRC every six months, and the reports are available for public inspection. Once a year the NRC publishes a report summarizing the data. In this way, the public is assured that radioactive releases from nuclear power plants are kept below the limits in the Technical

TABLE 9-III
RADIOACTIVITY IN EFFLUENTS FROM NUCLEAR POWER PLANTS IN 1977
(Ci)

	AIRBORNE		LIQUID	
	Noble Gases	Iodine and Particulates	Fission and Activation Products	Tritium
BWRs				
Browns Ferry 1	$<17 \times 10^4$	0.10	1.2	24
Cooper Station	0.13×10^4	<0.02	0.75	9.0
Dresden 1, 2, 3	83×10^4	12	1.0	5.1
Millstone Point 1	62×10^4	4.9	0.53	4.4
Monticello	0.7×10^4	0.09	0	0
Nine Mile Point 1	0.4×10^4	0.2	0.3	2.5
Oyster Creek	18×10^4	9.1	0.1	19
Peachbottom 2, 3	7×10^4	0.29	2.2	71
Pilgrim 1	41×10^4	0.69	3.4	33
Quad Cities 1, 2	2.6×10^4	1.7	1.3	26
Vermont Yankee	0.4×10^4	<0.02	0.16	0.84
PWRs				
Fort Calhoun	3 800	<0.02	0.36	160
R.E. Ginna	3 200	<0.03	0.07	120
Indian Point	16 000	<0.06	3.0	370
Kewaunee	2 430	<0.03	1.3	295
Maine Yankee	286	<0.01	0.44	150
Oconee 1, 2, 3	35 600	0.54	36	190
Point Beach 1, 2	1 130	<0.01	1.5	1000
H.B. Robinson	476	<0.01	0.33	685
San Onofre 1	154	<0.01	9.8	1790
Surry 1, 2	19 000	0.12	66	410
Three Mile Island 1	16 600	<0.04	0.2	190
Turkey Point 3, 4	23 300	1.0	3.9	920
Yankee Rowe	125	<0.01	0.02	140
Zion 1, 2	32 200	0.05	1.0	720

Specifications (see Chapter 3), which are based on the design guidelines for radiation levels that are as low as is reasonably achievable, defined in 10CFR50, Appendix I (see Chapter 6). In any event, radiation doses to the public must comply with U.S. Environmental Protection Agency (EPA) standards for the uranium fuel cycle (p. 142).

The amounts of radioactivity (in curies) in airborne and liquid effluents released during 1977 from nuclear power plants that had generated substantial amounts of electricity for at least three years are given in Table 9-III.[d] (The curie is a unit of radioactivity defined on page 132.) In almost all cases, the total releases were below the limits in the applicable regulations (e.g., 10CFR20) and in the Technical Specifications for the particular installation.

As expected, relatively large amounts of noble-gas activity are in the effluents from BWRs. However, most of this activity arises from the short-lived species krypton-87, krypton-88, and xenon-133 (see Table 9-I), which decay in a

[d]*Radioactive Materials Released from Nuclear Power Plants: Annual Report 1977*, NUREG-0521, U.S. Nuclear Regulatory Commission, Washington, D.C. (1979).

short time. Less than 1 percent of the activity is contributed by the long-lived krypton-85. Hence, only a small fraction of the total noble-gas activity will remain in the environment to expose the public at large.

The NRC, however, is also concerned with the possible radiation doses received by people living in the neighborhood of nuclear power plants who would be exposed to the radionuclides of both short and long half-lives. For example, if a person had remained continuously for a whole year at one point on the boundary of the Oyster Creek plant, he would have received a dose of approximately 50 mrems. Although such a situation is not a realistic possibility, this and some other BWR plants will add equipment that will reduce releases of radioactive materials to unrestricted areas to levels that are as low as is reasonably achievable.

The activity of iodine-131 released to the atmosphere is very small compared with that of the noble gases. Nevertheless, the release of radioiodines is carefully controlled and monitored because of its involvement in the grass → cow → milk → thyroid chain (p. 155). Only a few power plants discharged more than 1 curie (Ci) of radioiodines during the year. In some cases, where cows graze near the plant site, corrective action may be required to assure that the thyroid dose to any child is less than 15 millirems per year (mrems/year).

The activity of the liquid effluents was, on the whole, greater for PWRs than for BWRs. Nearly all of the activity in the PWR releases is due to tritium. This hydrogen isotope does not concentrate in the food chain, nor does it do so in the body (p. 149, *et seq.*); hence, the low concentrations in the liquid effluent are not a significant health hazard. Some of the fission and activation products may enter (and accumulate in) the body by way of drinking water or eating fish that live near the plant discharge. Although the activities of fission and activation products released in the liquid effluents are well below those permitted by the Technical Specifications, they may have to be reduced in some cases to satisfy the guide that no nearby resident will receive a whole-body dose exceeding 3 mrems/year from these effluents.

Environmental Surveys of Nuclear Plants

Prior to the operation of a nuclear power plant, the licensee must conduct a radiological (and biological) survey of the environment to establish a basis for comparison when the plant is operating. The survey may last for two or more years while the plant is under construction. It involves taking periodic samples of air, water, soil, river or lake bottom material, vegetation, milk, small animals, fish and other forms of aquatic life, and such other items as may be appropriate to the location. These samples are analyzed in a laboratory for their content of radioactive species. Ambient radiation levels, representing the doses that could be received by individuals, are measured at various places around the plant.

Essentially the same radiological surveillance program is conducted when the facility becomes operational. In this way, any significant changes associated with the plant effluents would be detected. By arrangement with the NRC, several states conduct independent radiological surveys in the environments of nuclear plants. Among other things, these surveys provide assurance that the radiation doses received by people living within an 80-kilometer (km) (50-mile) radius of the plant are indeed as low as is reasonably achievable.

During 1967 and 1968, scientists from the Bureau of Radiological Health (U.S. Public Health Service) carried out a detailed radiological monitoring program of the effluents and environment of Commonwealth Edison's Dresden Nuclear Power (BWR) Station, near Morris, Illinois, approximately 80 km southwest of Chicago. The station at that time consisted of one unit, a 200-MWe BWR, in operation and two large BWRs (800 MWe each) under construction.

Liquid wastes from the station are diluted with turbine condenser water and discharged into the Illinois River. Because the river receives sewage from Chicago, it is unfit near Morris for fishing or for use as a public water supply.

The critical external radiation exposure pathway appeared to be radiation from the short-lived, gamma-emitting radioisotopes of krypton and xenon gases in the plume from the stack. Measurements with radiation monitoring instruments indicated that a person who remained continuously on the ground about 1.6 km (1 mile) from the stack, and was immersed in the plume and without any protection, would receive an external radiation dose of 5 to 15 mrems/year from the plume. The actual value is somewhat uncertain since there are natural variations in the background radiation as measured from aircraft in the general vicinity. Because of dispersal of the gases in the atmosphere and the decay of the radioactive species of short half-life, the dose on the ground would decrease markedly with distance from the plant. Furthermore, it is only downwind that radiation doses appreciably higher than the background would be received.

The critical pathway for internal radiation exposure from fission products is by way of radioiodine in milk. Large samples of milk were collected from a herd of Holstein cows belonging to a dairy 3.2 km (2 miles) west of the Dresden plant. If there were any radioiodine in these samples, it was too small to be detected with the sensitive instruments used. The thyroid glands of three heifers that had grazed for several weeks about 2.4 km (1.5 miles) east of the nuclear installation were found to contain small amounts of radioactive iodine isotopes. The level of the radioactivity was too small to have any detectable effect on the heifers, and people do not eat thyroids. It should be recalled that in milk cows about half the absorbed iodine tends to concentrate in the thyroid gland with the other half in the milk.

The first public use of water from the Illinois River downstream from the Dresden plant is made at Peoria, Illinois, about 160 km (100 miles) distant. No radioactivity attributable to the reactor facility was detected in either surface or drinking water at the Peoria treatment plant. Furthermore, there was no significant radioactivity above the background level in rain water, soil, cabbage, grass, corn husks, deer, rabbit, and fish in the vicinity of the power plant. Snow and corn kernels collected about 0.8 km (0.5 mile) south (downwind) from the stack appeared to have higher than expected concentrations of strontium-89 and cesium-137, respectively. However, only a single sample was taken in each case, and the significance of the results was considered uncertain. Snow and corn kernels from other locations around the Dresden installation had lower concentrations of the indicated radioisotopes.

A comprehensive radiological study, similar to that just described, was made by the Bureau of Radiological Health of the environment of the PWR power plant of the Yankee Atomic Electric Company located on the Deerfield River, near Rowe, Massachusetts. The radioactive liquid effluent, diluted with turbine condenser water, is discharged into the Sherman Reservoir.

Since PWRs emit less gaseous radioactivity than BWRs, the external radiation dose received on the ground from the stack effluent was very small. Tritium was found in the water discharged from the plant, but the concentration was a fraction of 1 percent of the limit permitted by the operating license. No significant concentrations of fission and activation products were detected after dilution in the Deerfield River, but traces of such radioisotopes as manganese-54, cobalt-84, cobalt-60, and cesium-137 were found in water moss and sediment near the plant. Radiation doses that would have resulted from the ingestion of fish from the Sherman Reservoir were too small to be measured, but it was estimated that they would be less than 1 mrem/year.

In 1970, when the EPA was formed, the radiation monitoring (and several other) activities of the Bureau of Radiological Health were transferred to the EPA's Office of Radiation Programs. Environmental radiation studies have been made under the auspices of the EPA of several nuclear power plants. The EPA also maintains a continuing radiation surveillance program over the entire United States (p. 176).

RADIOACTIVE WASTES FROM SPENT-FUEL REPROCESSING [e]

Introduction

Roughly once each year a nuclear power reactor is shut down for refueling. About a third (in a PWR) or a fourth (in a BWR) of the fuel is removed and, after some rearrangement, is replaced by fresh fuel. The spent fuel rods contain most of the uranium-238 and roughly one-third of the uranium-235 originally present, together with all the fission products formed during operations apart from the small proportion that may have escaped through cladding defects. In addition, the spent fuel contains plutonium-239, resulting from the capture of neutrons by uranium-238 (p. 16), and smaller amounts of other isotopes of plutonium and of the transuranium elements neptunium, americium, and curium.

The main purposes of a fuel reprocessing operation are to remove the highly radioactive fission products and unwanted transuranium elements, and to separate the plutonium from the uranium. After conversion to uranium hexafluoride, the uranium can be reenriched in uranium-235 in a gaseous-diffusion plant (p. 15) to make it suitable for use as fresh reactor fuel. The plutonium could be used as a partial replacement for uranium-235 in an LWR, or as the fuel in a fast breeder reactor (Chapter 2).

[e]Because of the 1977 Presidential decision to defer indefinitely the reprocessing of spent reactor fuel, there are no radioactive wastes for the present from commercial reprocessing plants. However, this section is included in the event that a form of spent-fuel reprocessing is found to be desirable (or necessary) at some future time.

Fuel Reprocessing Plants

Prior to 1966, there were three locations in the United States where the chemical reprocessing of nuclear reactor fuels was carried out on a substantial scale. All were operated under direct contract with the U.S. Atomic Energy Commission (now the U.S. Department of Energy [DOE]). Reprocessing facilities at the Hanford site, near Richland, Washington, and at the Savannah River Plant, near Aiken, South Carolina, have been used mainly for the recovery of plutonium for nuclear weapons. In the Idaho Chemical Processing Plant, at the Idaho National Engineering Laboratory, highly enriched uranium fuels, from naval propulsion and experimental reactors, are reprocessed.

The first commercial reprocessing plant for spent fuels from nuclear reactor power stations started operation in 1966 at West Valley in western New York on a tract of land owned by the state. The plant suspended reprocessing operations early in 1972 to permit expansion of capacity and to make process improvements; later, it was decided to close it down permanently. A second commercial installation (Midwest Fuel Recovery Plant) was constructed near the Dresden BWR plant in Illinois, but because of unforeseen engineering problems, it will not become operational.[f] Another plant (Barnwell Nuclear Fuel Plant) near Barnwell, South Carolina, adjacent to the government-owned Savannah River Plant exclusion area, has been almost completed. However, the future of this plant is uncertain, especially in view of the decision to defer fuel reprocessing indefinitely in the United States (p. 165).

An LWR power plant with a design capacity of 1000 megawatts (electric) (MWe) discharges, on the average, about 30 metric tons (MT) (33 short tons) of spent fuel per year. A single reprocessing plant capable of handling 1500 MT of spent fuel annually, such as the one at Barnwell, would thus be required for roughly 50 of these power plants. Hence, there will never be a large number of spent-fuel recovery installations.

Licensing of Reprocessing Plants

Construction and operation of fuel reprocessing plants are subject to many of the same licensing and regulation requirements as are nuclear reactor facilities. The applicant for a permit to construct the plant and subsequently for a license to operate it must submit to the NRC descriptions of the plant site, the design basis of the plant, and its operating procedures.[g] Detailed Safety Analysis Reports must consider a wide variety of abnormal situations that might arise during normal operations and under potential accident conditions. An Environmental Report must also be submitted describing the expected impact of the installation on all aspects of the environment. The applications for a construction permit and for an operating license are subjected to the reviews and hearings described in Chapter 3.

[f]The installation, now called the Morris Operation, is used to store spent reactor fuel elements.

[g]The requirements are stated in 10CFR50, the proposed Appendix P, entitled *General Design Criteria for Fuel Reprocessing Plants.*

During operation, the reprocessing plant must comply with the NRC standards in 10CFR20 for the maximum permissible concentrations of radionuclides in air and water at the site boundaries and for the radiation dose that might result from the daily intake of radioactive material in air, water, or food (Chapter 5). In any event, the radioactivity levels in effluents are required to be "as low as is reasonably achievable," although the design levels have not yet been defined quantitatively, as they have for LWR facilities. The EPA uranium fuel cycle standards are, however, applicable.

The licensee of a fuel reprocessing plant is required to establish a radiation monitoring system, both inside and outside the plant, in order to detect the discharge of an unacceptable quantity of radioactive effluent, either gaseous or liquid. The NRC license stipulates that appropriate action, which may include complete plant shutdown, be taken if effluent specifications cannot be met.

Just as for a reactor installation, the reprocessing plant licensee must conduct an extensive environmental surveillance program both before and after the plant starts operation. From time to time, the NRC and EPA, either directly or through a contract with a state organization, make independent checks of external radiation doses and of radioactivity levels in the air, surface and drinking water, river sediment, soil, vegetation, milk, and fish in the general area of the plant. When the commercial reprocessing plant at West Valley was operating, the New York State Department of Health regularly carried out a program of environmental surveillance. This included daily samples of raw milk, which were tested, in particular, for radioiodine.

Treatment of Spent Fuel

After removal from the reactor, the spent fuel [h] is stored under water at the power plant for a period of at least six months to permit decay of the radioactive species of shorter half-life. The assemblies are then shipped a few at a time to the reprocessing plant in strong, shielded containers (Chapter 11). The spent-fuel assemblies are removed from the containers at the plant and are stored in a pool of water, which serves as a radiation shield and also takes up the radioactive decay heat from the fission products in the spent fuel. The details of the subsequent treatment of the spent fuel vary to some extent, but the principles outlined in Fig. 9-7 are generally the same.

The fuel assembly is lifted from the container by remote operation and is taken to an enclosed, shielded cell. Here some of the hardware may be removed, and the fuel rods are sheared mechanically into short pieces. The pieces of solid fuel, including the zirconium-alloy (Zircaloy) cladding, are transferred to a dissolver, where they are leached with (and subsequently soaked in) hot nitric acid to dissolve out the spent-fuel material, leaving the cladding. The latter is removed, washed, and checked for radioactivity to make sure that essentially all the spent fuel has been removed. The resulting *hulls* (and hardware), which contain traces of uranium, plutonium, and other transuranium elements, are

[h] The description of spent-fuel treatment given here is based on experience with plants operated for the DOE and with the West Valley, New York, commercial plant, and also on the design of the Barnwell plant.

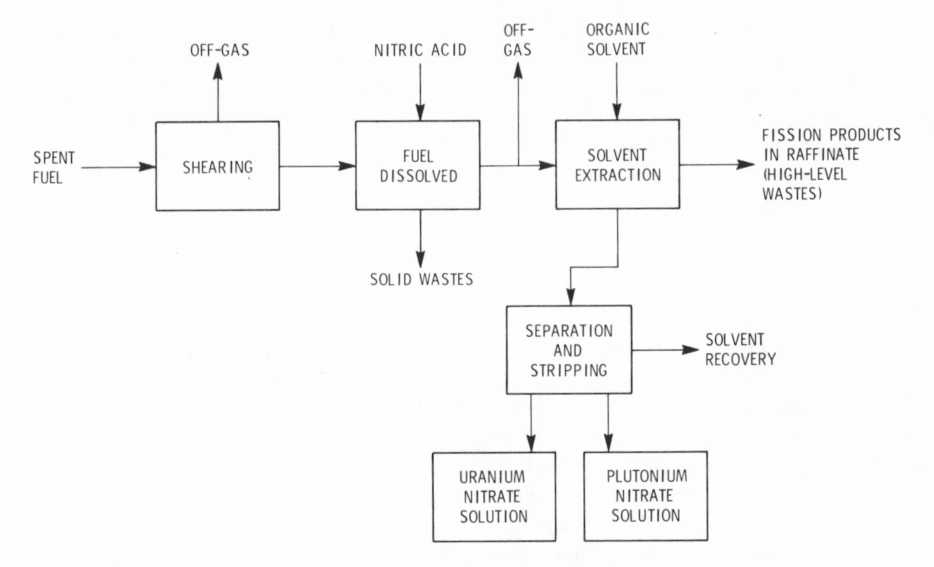

FIG. 9-7. Schematic outline of spent-fuel reprocessing plant.

part of the plant's solid waste (p. 246). The off-gases from the rod shearing and dissolving operations pass to a treatment system, described below.

The (aqueous) nitric acid solution of the spent fuel formed in the dissolver contains uranium, plutonium, small amounts of other transuranium elements, and nearly all the fission products as chemical compounds (i.e., nitrates). This constitutes the feed solution for the next stage of the process—namely, solvent extraction (p. 229). The uranium and plutonium nitrates, in oxidized states, are extracted from the solution into an organic liquid, usually tributyl phosphate (TBP) diluted with a kerosene-like hydrocarbon. The residual aqueous medium (raffinate) contains almost all (99.9 percent) of the fission products as well as part of the tritium as tritiated water. The treatment of the raffinate as a liquid waste is described below. The plutonium nitrate in solution is first converted into the reduced state, when it is no longer soluble in the organic liquid, and is extracted into an aqueous acid solution. The uranium nitrate remaining in the TBP-hydrocarbon solution is then stripped with dilute nitric acid. Each of the resulting product streams, one containing the plutonium nitrate and the other the uranium nitrate, is purified by further solvent extraction (or in other ways) and is concentrated by evaporation. The plutonium nitrate later is converted into solid plutonium dioxide prior to shipment to a mixed-oxide fuel fabrication plant (p. 233). The uranium nitrate is converted either into dioxide, for blending with plutonium dioxide in a mixed-oxide fuel, or into hexafluoride, for subsequent reenrichment in uranium-235.

Treatment of Fuel Reprocessing Wastes

Gaseous Wastes. The off-gases from shearing and dissolving the spent fuel contain oxides of nitrogen (from the nitric acid) together with almost all of the krypton and xenon (i.e., noble gases), a large fraction of the radioiodines, and a small part of the tritium formed in the fuel. A substantial proportion of the tritium may be retained in the cladding hulls as a combination of zirconium and tritium. The residual tritium, mainly as tritiated water, remains in the nitric acid solution. Small amounts of semivolatile ruthenium may also be present. In addition, particulate matter may be carried by the off-gases from the dissolver.

The gases are first passed through a water (or sodium hydroxide) scrubber to remove oxides of nitrogen (and nitric acid vapor). A large part of the iodine, in both elemental and organic forms, is absorbed by scrubbing with a solution of mercuric nitrate in fairly concentrated nitric acid; almost all of the remaining iodine is taken up by a silver zeolite, that is, a zeolite (ion-exchange mineral) impregnated with silver nitrate. At least 99.5 percent and possibly 99.9 percent of the iodine present in the off-gases is removed in this manner. The residual (mostly noble) gases then pass through a HEPA filter for removal of particulate matter. Finally, the gas is mixed with large volumes of ventilating air, tested for radioactivity, and discharged through a tall stack (Fig. 9-8). If necessary, ruthenium can be removed prior to discharge (p. 269).

Liquid Wastes. The treatment of liquid wastes in a fuel reprocessing plant depends on whether they have a low level or a high level of radioactivity. The low-level wastes include fuel cask decontamination water, plant laundry, floor drains, and fuel storage pool water. These wastes contain small amounts of fission products and some tritium as tritiated water. The liquid is evaporated, and the water vapor is usually discharged to the atmosphere through the plant stack. In some cases, the vapor may be condensed and the resulting water reused in the plant. There is thus no discharge of liquid effluent to the environment.

The evaporator residues may be sent to the high-level waste system, or they may be solidified and buried at an approved site.

The aqueous raffinate remaining after removal of uranium and plutonium from the fuel solution contains essentially all of the fission products formed dur-

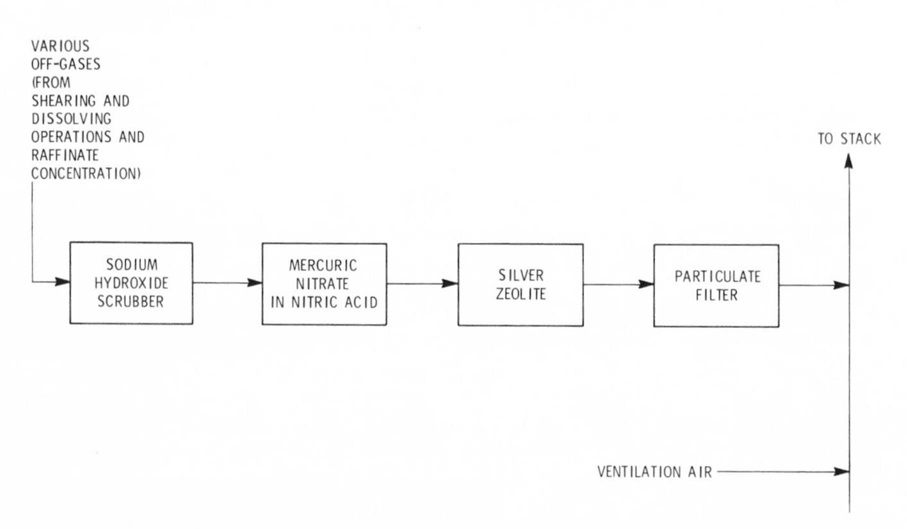

FIG. 9-8. Schematic outline of off-gas treatment in a spent-fuel reprocessing plant.

TABLE 9-IV

ESTIMATED RELEASES FROM FUEL REPROCESSING PER 1000 MW-YR IN AN LWR

	Estimated Releases (Ci)
Gases and Vapors	
Tritium	23 000
Carbon-14	30
Krypton-85	500 000
Iodine-129	0.04
Particulates	
Fission products	0.23
Uranium	5×10^{-5}
Transuranics	0.03

ing reactor operation, as well as small amounts of plutonium and other transuranium elements. This constitutes the major portion of the high-level liquid waste. The liquid is evaporated to decrease the volume, and the waste is ultimately solidified, as noted below. Nitric acid recovered from the vapor is concentrated and is reused in the fuel dissolver. Condensed water forms part of the low-level wastes, and uncondensed gases (and particulates) are treated with the other off-gases before discharge to the atmosphere.

In the reprocessing plants for weapons material at Hanford and Savannah River, the high-level liquid wastes, which are initially acidic, have been made alkaline and stored in carbon-steel tanks. These tanks have leaked occasionally, but the leakage has not constituted a hazard. In the Idaho plant, however, the acidic wastes, prior to solidification, are held in stainless-steel tanks; no leaks have occurred in any of the tanks since they were first used in 1961.

The NRC regulations require that the high-level liquid wastes (if any) be converted into solid form within five years of processing the spent fuel (p. 264). The acidic solutions appear to be easier to solidify than the alkaline (or neutral) forms; hence, the acidic high-level wastes from the commercial fuel reprocessing plant at Barnwell were to be stored temporarily in a 1135-cubic metre (m^3) (300 000-gal) stainless-steel tank. The tank, equipped with water-cooling coils for removal of radioactive decay heat, is contained in a concrete vault with a stainless-steel liner. A plant has been designed that does not include a large storage tank for high-level wastes; these wastes would be solidified soon after generation.

Solid Wastes. Apart from the solidified high-level wastes referred to above, which are treated more fully in the next chapter, there are several other kinds of solid waste from a spent-fuel reprocessing facility. If the wastes are not contaminated by plutonium or other transuranium (TRU) elements, they may be sent to an approved site for burial. Solid waste containing TRU elements include fuel-cladding hulls, particulate filters, discarded equipment and tools, and contaminated trash. These solids may be stored temporarily in containers enclosed in impervious clay or concrete at the plant site, pending eventual shipment to a federal repository. Management of the solid TRU wastes is also considered in Chapter 10.

Radioactive Releases from Fuel Reprocessing Plants

Because no spent-fuel reprocessing plants of recent design have been operated, the amounts of radioactive releases are not known. However, the estimates in Table 9-IV are based on Table 5-3 of the NRC regulations in 10CFR51 as modified in NUREG-0116 (October 1976), except that complete decay of iodine-131 is assumed. The values are for airborne releases that would be associated with the reprocessing of fuel from a 1000-MWe LWR plant operating at full capacity for a year (i.e., 1000 MWe-years); there are no liquid effluents. Presumably, after January 1, 1983, steps will have to be taken to reduce the discharge of krypton-85, iodine-129, and transuranics in order to comply with EPA standards (p. 143). The expected radiation doses to people living near a spent-fuel reprocessing plant are given in Chapter 6.

Prevention of Accidental Releases of Radioactivity

Apart from the deliberate, controlled release of radioactive effluents from a fuel reprocessing plant during normal operation, there is a possibility that an accident might lead to the uncontrolled dispersion of fission products. The most likely causes of such an accident are nuclear criticality (p. 14), fire, and explosion. Precautions are taken in the design and operation of the plant to prevent,or, at least, to minimize the probability of accidents and to confine to the plant site the consequences of such an unlikely accident. As is the case with reactor power plant licensing (Chapter 3), the Safety Analysis and Environmental Reports submitted to the NRC by the applicant to construct and operate a fuel processing plant must include discussions of a whole range of conceivable accidents.

Since the spent fuel still contains a substantial amount of uranium-235, plutonium-239, and other fissile species, it is conceivable that at some stage in the recovery process, a condition of criticality could be attained. Several different procedures are used to prevent the attainment of criticality. For example, the quantity and concentration of fissile material that is present in any vessel (or in adjacent vessels) is limited. In some cases, the dimensions of the process equipment are such that a critical condition is impossible for the particular solutions being processed. Another means for preventing criticality is to include neutron absorbers, such as boron or cadmium, either in solution or in solid form in the vessels used in the plant. In any event, in the application for a license, a serious criticality situation must be postulated and analyzed; it must be shown that it would cause no unacceptable off-site radiation doses.

The hydrocarbon that serves as diluent for the organic solvent (TBP) in the extraction stage is especially chosen for its high flash point. The probability of fire is thus reduced. Additional protection against fire is provided by the use of heat detection devices, automatic sprinklers, and other conventional methods. Appropriate design of the plant prevents the occurrence of potentially explosive conditions. If a fire should cause an explosion, the destructive effects would be minimized by ductwork that attenuates the accompanying pressure wave.

If, despite all precautions, criticality, a fire, or an explosion should occur within the plant, the escape of radioactive airborne material would probably be the major hazard. Fuel reprocessing plants have a number of barriers to minimize (or prevent) the escape of radioactivity. A safety analysis is performed to show that the multiple barriers will indeed provide the necessary protection for the general public.

The first set of barriers includes the process vessels, interconnecting piping, and the off-gas system. Should these be breached, a second barrier would be provided by the thick concrete structure of the cells containing the vessels. The main purpose of the thick concrete is to serve as a radiation shield, but it can also limit the effect of an explosion in a process vessel. Another barrier is the industrial-type building that houses the process equipment.

The development of a leak in a storage tank for high-level liquid wastes is an accident that could lead to the release of radioactivity to the environment. In modern plants designed for the reprocessing of spent fuel from nuclear power facilities, the chances of any significant escape of radioactivity from a stainless-steel storage tank is very small. If a leak occurred, the escaping liquid would be retained in the steel-lined concrete vault, where its presence would be quickly detected.

Bibliography—Chapter 9

Bebbington, W. P. "The Reprocessing of Nuclear Fuels." *Scientific American* 235, no. 6 (1976): 30.

Finney, B. C., *et al. Correlation of Radioactive Waste Treatment Costs and the Environmental Impact of Waste Effluents in the Nuclear Fuel Cycle—Nuclear Fuel Reprocessing* (ORNL-TM-4901). Oak Ridge, Tenn.: Oak Ridge National Laboratory, 1975.

International Atomic Energy Agency. *Proceedings of the Conference on Nuclear Power and Its Fuel Cycle.* Vienna: International Atomic Energy Agency, 1977.

______. *Proceedings of the Symposium on the Management of Radioactive Wastes from the Nuclear Fuel Cycle.* Vienna: International Atomic Energy Agency, 1976.

Kahn, B.; Schleien, B.; and Weaver, C. "Environmental Experience with Radioactive Effluents from Operating Nuclear Power Plants." *Proceedings of the Fourth United Nations Conference on the Peaceful Uses of Atomic Energy* 11 (1971): 559.

Keilholtz, G. W. "Krypton-Xenon Removal Systems." *Nuclear Safety* 12 (1971): 591.

National Academy of Engineering. "Environmental Protection: Radiological Engineering Aspects of Power Plants and Their Fuel Cycles." *Engineering for Resolution of the Energy-Environment Dilemma*, Chap. 9, Washington, D.C., 1972.

National Academy of Sciences-National Research Council, Committee on Radioactive Waste Management. *The Shallow Land Burial of Low Level Radioactively Contaminated Solid Waste.* Washington, D.C., 1977.

"Noble Gas Symposium at Las Vegas." *Nuclear Safety* 15 (1974): 302.

Oak Ridge National Laboratory. *Siting of Fuel Reprocessing Plants and Waste Management Facilities* (ORNL-4451). Oak Ridge, Tenn.: Oak Ridge National Laboratory, 1970.

Reinig, W. C., ed. *Proceedings of the Symposium on Environmental Surveillance in the Vicinity of Nuclear Facilities.* Springfield, Ill.: Charles C. Thomas, Publisher, 1971.

Rust, J. H., and Weaver, L. E., eds. *Nuclear Power Safety.* Elmsford, N.Y.: Pergamon Press, 1977.

U.S., Atomic Energy Commission. *Environmental Survey of the Nuclear Fuel Cycle* (WASH-1248). Washington, D.C., 1974.

———. *The Safety of Nuclear Power Reactors and Related Facilities* (WASH-1250). Washington, D.C., 1973.

U.S., Energy Research and Development Administration. *Alternatives for Managing Wastes from Reactors and Post-Fission Operations in the LWR Fuel Cycle* (ERDA-76-43), Vol. I. Washington, D.C., 1976.

———. *Proceedings of the International Symposium on the Management of Wastes from the LWR Fuel Cycle* (CONF-760701). Washington, D.C., 1976.

U.S., Environmental Protection Agency. *Environmental Analysis of the Uranium Fuel Cycle*, Part II, Nuclear Power Reactors; Part III, Nuclear Fuel Reprocessing (EPA-520/9-73-003 C and D). Washington, D.C., 1973.

———. *Environmental Radiation Protection Requirements for Normal Operations of Activities in the Uranium Fuel Cycle*, Environmental Statement. Washington, D.C., 1975.

U.S., Nuclear Regulatory Commission. *Environmental Survey of the Reprocessing and Waste Management Portions of the LWR Fuel Cycle* (NUREG-0116). Washington, D.C., 1976.

———. *Measuring and Reporting Effluents from Nuclear Power Plants* (Regulatory Guide 1.21). Washington, D.C.

———. *Radioactive Materials Released from Nuclear Power Plants* (NUREG-0367). Washington, D.C., 1978. See also, *Nuclear Safety* 19 (1978): 628.

10

Management of High-Level and TRU Wastes

INTRODUCTION

Historical Background

From the time the commercial nuclear power program was initiated in the United States in the late 1950s, it had been accepted that the spent fuel would be reprocessed in the manner described in the preceding chapter. The major objective was to recover plutonium and unused uranium, both of which could be reused in the production of reactor fuels. During the reprocessing operation, the fission products would be separated and would remain in the high-level liquid wastes. Because of the persistent radioactivity of these wastes, which would continue for thousands of years, special care would have to be taken to isolate them from the human environment. Considerable effort has thus been devoted to the development of procedures for managing the high-level wastes from nuclear fuel reprocessing plants.

In April 1977, however, President Carter decided that, in order to discourage the separation of plutonium, which could be used as the explosive in nuclear weapons, spent commercial fuel reprocessing in the United States would be postponed indefinitely. The problem then is, what is to be done with the highly radioactive spent fuel elements? In many respects, this problem is similar to that concerned with the isolation from the environment of solidified high-level reprocessing wastes.

Although there will be no reprocessing of spent fuels from commercial nuclear power plants in the immediate future,[a] it is conceivable that the situation may change in due course. The amount of spent fuel accumulated may become so great that reprocessing may be necessary to reduce the quantity of radioactive material to be managed. This reduction is possible because the fission products (as oxides) constitute roughly 3 percent by weight of the spent fuel. Furthermore, a number of techniques have been suggested for reprocessing fuel in a manner that would recover a material that could be made into fresh reactor fuel but not be suitable for nuclear weapons. In all these circumstances, high-level (fission product) wastes would be generated.

[a]The Barnwell Nuclear Fuel Plant referred to on page 255 was not expected to start operation before 1980 even if reprocessing had not been postponed.

This chapter is concerned with the management of high-level reprocessing wastes as well as of intact (or untreated) spent reactor fuel elements. As described below, it had been planned to convert the high-level liquid wastes into solid form. These solid reprocessing wastes may be categorized with spent fuel elements as high-level solid wastes. The mass (or volume) of spent fuel to be handled would be greater than that of the equivalent solid reprocessing wastes, but the radioactivity (and heat generation) per unit mass (or volume) would be correspondingly smaller.

The limited quantities of high-level wastes from the only commercial nuclear fuel reprocessing plant were stored in liquid form in tanks.[b] With the increasing development of nuclear power, however, the volume of these wastes would be so large that extensive tank storage would be necessary. Consequently, in the early 1960s, the U.S. Atomic Energy Commission (AEC) decided that the highly radioactive liquid wastes from fuel reprocessing plants should be converted into solid form. The volume of the solid waste would be considerably less than that of the original liquid, although the radioactivity of a given mass (or volume) would be greater. Storage of the solid would be much safer than the liquid, since there is less danger of the escape of radioactivity to the environment, especially if the solid is in a form that is not easily leached by water. Furthermore, control of a facility for storing solid wastes should be easier than for a large tank farm.

According to the requirements of Title 10 of the *Code of Federal Regulations*, Part 50 (10CFR50), Appendix F, originally promulgated by the AEC in 1970, high-level radioactive wastes at commercial reprocessing plants were to be solidified within five years of processing the spent fuel. Hence, the volume of liquid wastes held in storage tanks at the plant site would not exceed the quantity generated during a period of five years. The solidified product was to be shipped to a federal repository within ten years of spent-fuel reprocessing.

TRU and Iodine Wastes

If spent fuel is reprocessed, there are other wastes, apart from the bulk of the fission products, that may have to be isolated from the environment because of their persistent radioactivity. One of these is the transuranium (TRU) waste, part of which would be generated in fuel reprocessing plants, as seen in Chapter 9. The remainder of the TRU waste would be produced only if mixed-oxide fuels were to be fabricated (Chapter 8); this would require the separation of plutonium from the spent fuel. The other important waste is iodine-129; although it is a fission product, it is volatile and tends to separate from the other fission products.

The TRU wastes have been defined provisionally as solid materials with an activity of more than 10 microcuries per kilogram (μCi/kg) of alpha-emitting transuranium elements. The most important of these elements are plutonium, americium, and curium. Nearly all of their isotopes are potentially hazardous alpha-particle emitters with long half-lives (Chapter 5). Consequently, they must be kept segregated for thousands of years.

[b]The high-level wastes generated in weapons plutonium production plants have also been stored initially in liquid form since 1945 (p. 255).

Several iodine isotopes are formed in fission, but all except iodine-129 decay almost completely within a year or so. Iodine-129, however, has a half-life of 17 million years. As much as possible of the iodine is removed before discharge of airborne effluent, especially from fuel reprocessing plants. Because of the limited lifetime of such plants, compared with the long half-life of iodine-129, it is evident that a more permanent storage would be required than could be provided at the reprocessing plant.

Federal Waste Repositories

High-level and other long-lived wastes, regardless of whether they consist of untreated spent fuel elements or are produced in reprocessing, would require long-term surveillance. It appears to be mandatory, therefore, that such wastes be placed in a federal repository under permanent government control, rather than in a commercial repository. The U.S. Department of Energy (DOE), as successor to the AEC, is responsible for the actual development of environmentally safe waste management technology and for establishing and managing federal repositories for high-level wastes.

The licensing of such repositories, like that of other nuclear power-related facilities, is the responsibility of the U.S. Nuclear Regulatory Commission (NRC). Appropriate regulations, standards, and guides for the safe management of high-level wastes are being developed by the NRC. Furthermore, the NRC is required by the National Environmental Policy Act of 1969 to assess the environmental impact associated with waste management activities at the repositories.

PROPERTIES OF HIGH-LEVEL WASTES

Liquid Reprocessing Wastes

Although no commercial spent reactor fuel reprocessing plants are currently in operation, a considerable amount is known about reprocessing. This is based on experience with one commercial plant and with other reprocessing plants operated by DOE contractors (p. 255). The information given here is based on that experience.

About 5000 litres (1300 gal) of primary high-level liquid waste solution are produced in the reprocessing of 1 metric ton (MT) (1000 kg, or 2205 lb) of spent uranium fuel. The volume is reduced to roughly 1100 litres (260 gal) by evaporation before storage in a stainless-steel tank (p. 259). Prior to the actual solidification process, to be described shortly, the volume is further reduced by evaporation to some 380 litres (100 gal). Typically, the liquid waste from reprocessing 1 MT of fuel contains about 70 kg (154 lb) of dissolved solids in the form of nitrates of fission products, TRU elements, and impurities.

The high-level wastes include 99.9 percent (or more) of the nonvolatile fission products in the spent fuel, as well as roughly 0.5 percent of the uranium and plutonium that is not extracted and remains in the waste. In addition to the familiar plutonium-239 isotope, formed from uranium-238 by neutron capture and beta decay, as described in Chapter 2, several other isotopes are produced by neutron capture and in other ways.

The spent fuel also contains appreciable amounts of a number of isotopes of the heavier transuranium elements, americium and curium. These are formed mostly by neutron captures and beta decays. For example, starting with plutonium-239 (^{239}Pu), various possible reactions, leading to isotopes of americium (Am) and curium (Cm), are shown below; the symbols n, β, and α indicate neutron capture, beta decay, and alpha decay, respectively.[c] Essentially all of the americium and curium originally present in spent fuel remains in the reprocessing waste solution.

Atomic Number										
94	^{239}Pu	$\xrightarrow{n}$	^{240}Pu	$\xrightarrow{n}$	^{241}Pu	$\xrightarrow{n}$	^{242}Pu	$\xrightarrow{n}$	^{243}Pu	$\xrightarrow{n}$
	α		α		$\downarrow\beta$		α		$\downarrow\beta$	
95					^{241}Am	$\xrightarrow{n}$	^{242}Am	$\xrightarrow{n}$	^{243}Am	$\xrightarrow{n}$
					α		$\downarrow\beta$		α	
96							^{242}Cm	$\xrightarrow{n}$	^{243}Cm	$\xrightarrow{n}$
							α		α	

Solid High-Level Reprocessing Wastes

Several different forms of high-level solid reprocessing wastes have been produced by treating the liquid wastes in various ways. The two most common general types are obtained by the following procedures, which have been demonstrated on a fairly large scale: (a) drying and calcination, and (b) vitrification.

In drying and calcination, the water is first removed, and the residual solid is heated to drive off volatile matter; the product is called a *calcine.* If the high-level solid wastes are to be stored in the calcine form, they must be heated to a temperature of about 900 °C (1650 °F) to cause complete decomposition of the nitrates. The high-temperature calcine, consisting mainly of oxides, can then be stored in sealed cylinders without the risk of developing a high internal gas pressure.

In the vitrification process, the current practice is to form a calcine at a temperature of 500 to 600 °C (930 to 1110 °F). A glass-forming *frit,* such as a mixture of borax and silica, is added, and the blend is heated to 1000 to 1100 °C (1830 to 2010 °F); upon cooling, a glass-like (or vitrified) borosilicate product is obtained. Methods are being developed for one-stage vitrification by adding the frit to the concentrated liquid waste and heating the resulting slurry in a furnace of special design.

Depending on the process details, the calcine is either a powder or a mixture of powder and granules, as distinct from the solid glassy product of vitrification. The calcine has the drawbacks of a lower heat conductivity and a greater leachability by water than the glass. A high thermal conductivity of the solid is

[c]The product of alpha-particle decay is a nuclide two units less in atomic number and four units less in mass number; this is difficult to depict in the representation given here.

desirable in order to prevent the development of excessively high internal temperatures as a result of heat generation from radioactive decay. Furthermore, if the solid is not easily leachable, chances are reduced that radioactive material will be dissolved out by water and escape to the environment

On the other hand, the calcine contains a larger proportion of fission and transuranium products and has a smaller total volume than does the vitrified borosilicate form. For each metric ton of uranium fuel reprocessed, for example, the calcine occupies 0.03 to 0.05 cubic metres (m^3) (1.0 to 1.7 ft^3) compared with 0.06 to 0.08 m^3 (2.1 to 2.8 ft^3) for a borosilicate glass.[d]

The solid wastes should be chemically stable even when subjected to continuous irradiation by alpha and beta particles and gamma rays. Such stability means there will be no generation of gases, which might affect the integrity of the solid or the container during storage. In addition, there should be some assurance that the basic structure of the solidified waste will not undergo change.

Borosilicate glass appears to be the preferred form for the solidification and storage of high-level radioactive wastes in France, Germany, and the United Kingdom. It is also of considerable interest in the United States, particularly because of the good thermal conductivity and low leachability of the glass. At high temperatures, however, such as might develop from radioactive decay heat during storage, the glass may become devitrified—i.e., changed to a nonglassy (microcrystalline) state. It is then more readily leachable than the crystalline calcination product. This drawback may be overcome by preventing an excessive temperature rise in the glass (e.g., by reducing the fission product content or by increasing the distance between the containers in storage).

Because of the potential instability of glasses to heat and radiation, attention is being paid to the conversion of high-level wastes into ceramics, somewhat similar to minerals. The ceramics, formed at high temperatures, are crystalline in nature and are expected to be stable. One class of ceramic waste product is called *supercalcine.* It is made by adding several different elements, including calcium, aluminum, cesium, strontium, molybdenum, zirconium, and rare earths, as oxides, as well as phosphate and silicate, to the liquid wastes prior to calcination. The ceramic product may contain as many as nine different mineral-like substances. In the somewhat related *synroc* (for synthetic rock) process, the objective is to produce a ceramic resembling minerals that have proved stable under natural conditions over geologic time. Efforts are also being made to increase the resistance of ceramic particles to attack by water (or salt solution). One way is to coat ceramic pellets with protective layers, such as carbon and aluminum oxide. Another method involves the production of cermets in which the calcine particles are incorporated in a metal matrix.

Processes for Waste Solidification

Several procedures have been demonstrated, both in the United States and in other countries, for drying and calcining the concentrated high-level liquid reprocessing wastes. Of these, two in particular have been favored in the U.S. They are: (a) spray calcination, and (b) fluidized-bed calcination.

[d]One cubic meter (m^3) is equivalent to 1000 litres.

In *spray calcination*, the liquid waste is sprayed through an atomizing nozzle into the top of a cylindrical tower that is heated in a furnace (Fig. 10-1). The walls of the cylinder are maintained at a temperature of about 700 °C (1290 °F). As the spray descends, water is driven off the liquid droplets, and the resulting solid particles are calcined. The product is collected in a stainless-steel canister at the bottom of the tower. Some of the powder is carried off with the vapors and gases generated during the drying and calcination phases. It is removed from the off-gases by stainless-steel filters, and the deposits are periodically blown off the filters by a blast of air or steam.

If the calcined powder is to be stored, the canister is heated to 900 °C (1650 °F) in a bake-out furnace to decompose all the remaining nitrates, leaving a solid consisting mainly of oxides of fission products and transuranium elements. On the other hand, if the calcine is to be vitrified for storage, a silica-borax frit is fed to the stainless-steel canister in which the calcine is being collected, as shown in Fig. 10-1. The mixture is heated to about 1000 to 1100 °C (1830 to 2010 °F) and upon cooling, it becomes a monolithic mass of glass. The canister is then sealed for subsequent storage or permanent disposal.

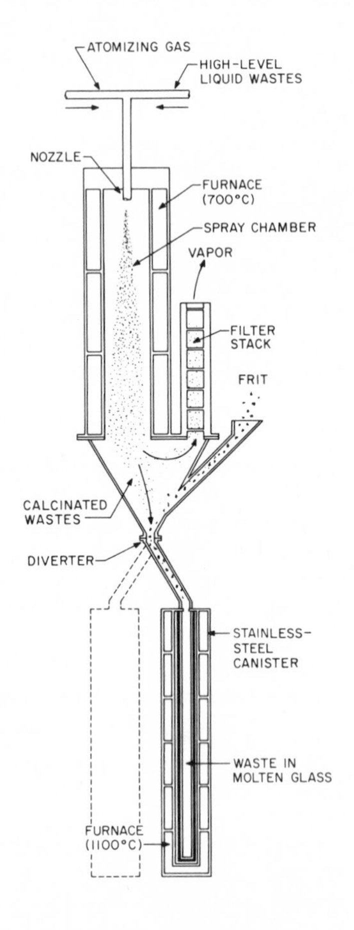

FIG. 10-1. Spray calcination and vitrification of high-level liquid radioactive wastes.

The *fluidized-bed calcination* process was developed at the Idaho Chemical Processing Plant, where it has been used since 1963 for the solidification of liquid wastes from the reprocessing of spent fuels from naval propulsion and experimental reactors (p. 255). In general, the radioactivity levels in these wastes have been lower, and the compositions have been different from those to be expected from the treatment of spent fuels from nuclear power plants. Tests indicate, however, that the fluidized-bed calcination procedure can be adapted to liquid wastes of the latter type.

The liquid waste is fed continuously into a calciner containing a bed of small nucleation particles, which may consist of the solid obtained by drying the waste; the bed is heated internally to 500 to 600 °C (930 to 1110 °F) by the combustion of kerosene in oxygen. A stream of air passing upward through the bed causes the particles to be "fluidized" so that they flow like a liquid. Intimate contact between the hot particles and the incoming liquid waste causes drying and calcination to occur. Part of the calcine is deposited on the fluidized-bed particles, and part is in smaller particulate (or powder) form. Hence, the product collected from the calcining vessel and the off-gas consists of a mixture of granules and powder in the size range of roughly 0.05 to 0.5 millimetre (mm) (0.002 to 0.02 in.).

In the Idaho Chemical Processing Plant, calcined wastes are retained on-site in vented steel bins, and it is not necessary to decompose the nitrates completely by further heating. However, if the calcine is to be stored in a repository, it would be collected in a stainless-steel canister and subjected to bake-out at 900 °C (1650 °F). Alternatively, the calcine could be vitrified by adding a glass-forming frit and heating in the manner described above.

Treatment of Off-Gases

Regardless of the procedures used, the off-gases from the concentration, drying, and solidification of high-level liquid wastes must be treated before discharge to the atmosphere. The main constituents to be considered are oxides of nitrogen (from the thermal decomposition of nitric acid and nitrates) and radioactive isotopes of iodine and ruthenium. These two elements differ in the respect that elemental iodine and its compounds present in the waste are very volatile, whereas ruthenium is described as being "semivolatile." The extent of volatilization is highly dependent on the temperature and other conditions of liquid waste solidification.

The off-gases are first scrubbed with water to remove as much as possible of the nitrogen oxides. Nitric acid may be recovered from the scrub liquid. If necessary, more efficient methods for removing nitrogen oxides, which have been developed for other industries, could be used.

If the calcination process results in the vaporization of more ruthenium than can be safely discharged to the atmosphere, the off-gas is treated next for removal of this element. The ruthenium in volatile form can be greatly reduced by passing the off-gases through silica gel, and most of the particulate form is retained by high-efficiency particulate air (HEPA) filters. Since the main isotope, ruthenium-106, has a half-life of only one year, the amount in storage (or in the atmosphere) soon approaches a limit.

Iodine is absorbed from the off-gas before passage through the HEPA filter and subsequent discharge by way of a stack. A solution of mercuric nitrate in nitric acid and a silver-zeolite bed may be used for this purpose, as in spent-fuel reprocessing (p. 258). The iodine removed is mainly the long-lived isotope iodine-129, which would probably be consigned to a federal repository, as indicated earlier.

Quantity of Solid Reprocessing Wastes

The reprocessing of spent fuel from a nuclear power facility with an electrical capacity of 1000 megawatts (MWe) and a plant factor of 70 percent (p. 4) would generate some 5 to 7 MT (5.5 to 7.7 short tons) of solidified high-level wastes occupying a volume of roughly 1.8 to 2.5 m^3 (63 to 88 ft^3) per year. The

TABLE 10-I

ESTIMATED ACCUMULATED SOLID HIGH-LEVEL WASTES

YEAR	VOLUME	
	m^3	ft^3
1985	1,000	35,000
1990	2,500	88,000
1995	4,900	170,000
2000	8,400	300,000

actual mass and volume would depend on the nature of the solid (i.e., calcine or glass).

Because of their higher thermal conductivity, vitrified wastes can be stored in larger canisters than calcines. A canister for vitrified solid may be expected to have an internal diameter of about 30 centimetres (cm) (1 ft) and a length of 3 m (10 ft). Its volume would then be 0.2 m^3 (7 ft^3). Hence, an estimated 10 to 12 canisters would be required to contain the solidified waste from one year's operation of an average 1000-MWe nuclear power plant.

In view of the uncertainties associated with spent-fuel reprocessing, no estimates can be made of the amounts of solidified high-level waste that will have to be sent to a federal repository in the foreseeable future. The estimates given in Table 10-I were based on the assumption that fuel reprocessing on a substantial scale would start in 1980 and that by 1985 a quantity of solid waste would have accumulated, although not necessarily shipped to a repository.

Radioactivity of Solid Reprocessing Wastes

The solid high-level reprocessing wastes contain many radioactive species, but only a few are significant because of their activity, persistence (i.e., long half-life), and potential hazard if released to the environment. These are listed in Table 10-II together with their half-lives and type of radiation emitted. Radionuclides of short or moderate half-lives are not included because within 10 years their activities will have become small enough to be negligible in comparison with those of the other radioactive species present in the waste. The three nuclides in the upper part of the table are fission products, whereas the others are isotopes of transuranium elements. Although plutonium-241 has a substantial activity, it is not included because it has a moderate half-life (13 years) and is a beta-particle emitter; thus, it is less of a potential hazard than the other plutonium isotopes (Chapter 5.)

In selecting a location and developing a procedure for the final disposition of high-level solid wastes, two essential requirements must be satisfied: first, dispersal of the heat generated during radioactive decay and, second, segregation from the environment for the safety of the public—in particular, to prevent entry of radioactive materials into the air, water supplies, and food chain. Both the rate of heat generation and the potential health hazard decrease with time, but as a result of the differences in activities and half-lives of the radionuclides present in the wastes, the decrease is not uniform.

TABLE 10-II
SIGNIFICANT RADIOACTIVE SPECIES IN SOLIDIFIED HIGH-LEVEL REPROCESSING WASTES

	Half-life (Years)	*Radiation*
Fission Products		
Strontium-90	28	beta
Technetium-99	2.1×10^5	beta
Cesium-137	30	beta and gamma
Transuranics		
Plutonium-238	89	alpha
Plutonium-239	2.4×10^4	alpha
Plutonium-240	6.6×10^3	alpha (and neutrons[a])
Americium-241	460	alpha
Americium-243	8.0×10^3	alpha
Curium-244	18	alpha (and neutrons[a])

[a]The neutrons arise from spontaneous fissions.

Activities in the solid high-level wastes of the species listed in Table 10-II, at various times after removal of the spent fuel from a light-water reactor (LWR) (Chapter 2), are given in Table 10-III. The values are calculated activities in curies (p. 132) of the waste generated in the reprocessing of 1 MT of spent fuel; they are based on the supposition that the waste contains 0.5 percent of the uranium and plutonium in the spent fuel and essentially all the nonvolatile fission products and americium and curium. The increase in activity of plutonium-239 between 1000 and 10 000 years is due to the alpha decay of americium-243 to neptunium-239, followed by beta decay to plutonium-239. In a somewhat similar manner, there is an increase in the plutonium-240 activity at earlier times, arising from the alpha decay of curium-244.

Strontium-90 and cesium-137 are the main contributors to the total activity of the solid waste in the early stages; they are consequently responsible for almost all the heat generated in the solid waste. However, since these nuclides

TABLE 10-III
ACTIVITIES IN CURIES OF SOLIDIFIED HIGH-LEVEL REPROCESSING WASTES FROM 1 MT OF SPENT LWR FUEL

	TIME AFTER REMOVAL FROM REACTOR (YEARS)				
Nuclide	10	100	1000	10,000	100,000
Strontium-90[a]	1.2×10^5	1.3×10^4	—	—	—
Technetium-99	14	14	14	13	10
Cesium-137	8.3×10^4	1.0×10^4	—	—	—
Plutonium-238	14	13	6	—	—
Plutonium-239	1.7	1.7	2.0	3.8	0.6
Plutonium-240	4.5	8.3	8	2	—
Americium-241	180	160	40	4	—
Americium-243	17	17	16	7	—
Curium-244	1.7×10^3	53	—	—	—

[a]Includes the short-lived decay product yttrium-90.

have relatively short half-lives, they will have decayed almost completely by about 800 years. This is indicated by the sharp drop in the total activity, which is shown as a function of time in Fig. 10-2. (Note that both activity and time scales are logarithmic and that successive divisions represent factors of 10.) Hence, after 800 years or so, the heat is generated in the wastes at a very low rate, and heat dissipation ceases to be a problem. A potential hazard still exists, however, from technetium-99 and from the transuranic species of long half-lives. After a few hundred thousand years, the main source of radioactivity is not expected to be nuclides arising from reactor operation, but rather the normal decay products (including radium-226) of the uranium-238 present in the fuel residues.

Characteristics of Spent Fuel

The significant radioactive species in untreated spent fuel include those given in Table 10-II for the solid reprocessing wastes. The clad fuel elements will, however, still contain the radionuclides tritium, carbon-14, and krypton-85, and iodine isotopes, which are volatile and would be released in the reprocessing plant. The activities of strontium-90, technetium-99, and cesium-137 would be the same as in Table 10-III. In addition, about 300 Ci of tritium and 5000 Ci of krypton-85 per metric ton of spent fuel would be present after 10 years. The half-life of tritium is 12.3 years and that of krypton-85 is 10.8 years; hence, the activities after 100 years would be substantially less. The activities of carbon-14 and iodine-129 are small—a fraction of a curie—but the long radioactive and biological half-lives of these radionuclides makes them important.

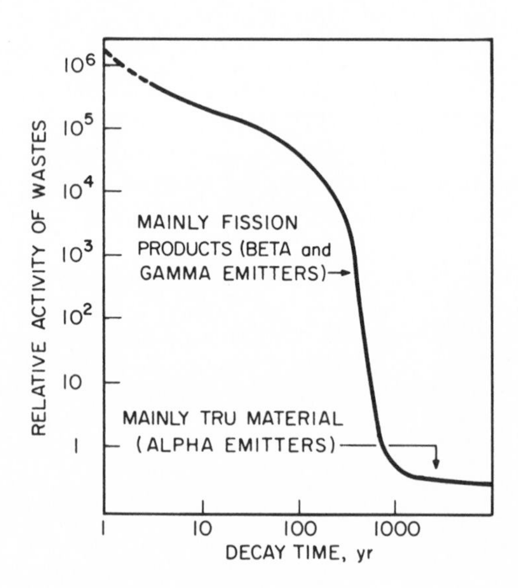

FIG. 10-2. Approximate decay of activity with time of high-level radioactive wastes.

The activities of the americium and curium isotopes in spent fuel are essentially the same as for solid reprocessing wastes given in Table 10-III. Since all the plutonium isotopes are retained in the spent fuel, however, their activities will initially be greater by a factor of 200. There will also be some activity from uranium isotopes, but the total is originally only about 0.25 Ci/MT of spent fuel. Because of their very long half-lives, the activity will decrease very slowly at first, but it will increase later as a result of alpha-decay of plutonium isotopes.

The change in radioactivity and heat release rate of spent fuel is qualitatively the same as for solid, high-level reprocessing waste shown in Figure 10-2. Both will decrease rapidly after about 800 years of storage, but subsequently they will be higher for the spent fuel, largely because considerably more plutonium is present.

WASTE MANAGEMENT TECHNIQUES

Introduction

The techniques for managing (or isolating) solid high-level and TRU radioactive wastes can be considered in two categories: (a) storage and (b) disposal. As used in the present context, the term *storage* implies retrievability, whereas *disposal* means permanent isolation from man's environment in such a way that retrieval is virtually impossible. Storage may be interim, in which event the waste is held temporarily pending removal to a permanent location. An alternative is provisional storage, which, if it proves satisfactory, could be converted into a more permanent (or terminal) storage while retaining the capabilities for retrieval should it prove to be necessary. These alternatives are applicable to both untreated spent fuel or solid reprocessing wastes.

In the early 1960s, the AEC initiated a study, described later, to explore the possibility of storing solid high-level wastes in a moderately deep (300 m[1000ft] or more) geologic formation, specifically a salt bed. The results of this study were promising, but pending the identification of a suitable site for a demonstration program on a larger scale, a project was started in 1972 for the design of an interim surface (or near-surface) storage facility. Although such a facility could be used for either untreated spent fuel or solidified reprocessing wastes, it was assumed at the time that the wastes would be in the latter form. (Because of technical and political uncertainties, this plan was abandoned by the DOE.)

In view of the postponement of spent-fuel reprocessing and the consequent need for isolating untreated spent fuel, there is an increased interest in interim storage. This would allow time for the detailed studies that are necessary for the development of more permanent storage facilities. Furthermore, if reprocessing of spent fuel is deemed to be desirable at a later date, a retrieval from interim storage would be a simple matter.

At the request of the AEC, a panel of the National Academy of Sciences-National Research Council examined the alternatives for a Retrievable Surface Storage Facility (RSSF). The original purpose of the RSSF was to provide storage of the high-level wastes from fuel reprocessing operations for possibly as long as 100 years, but it could have been used equally well for untreated spent

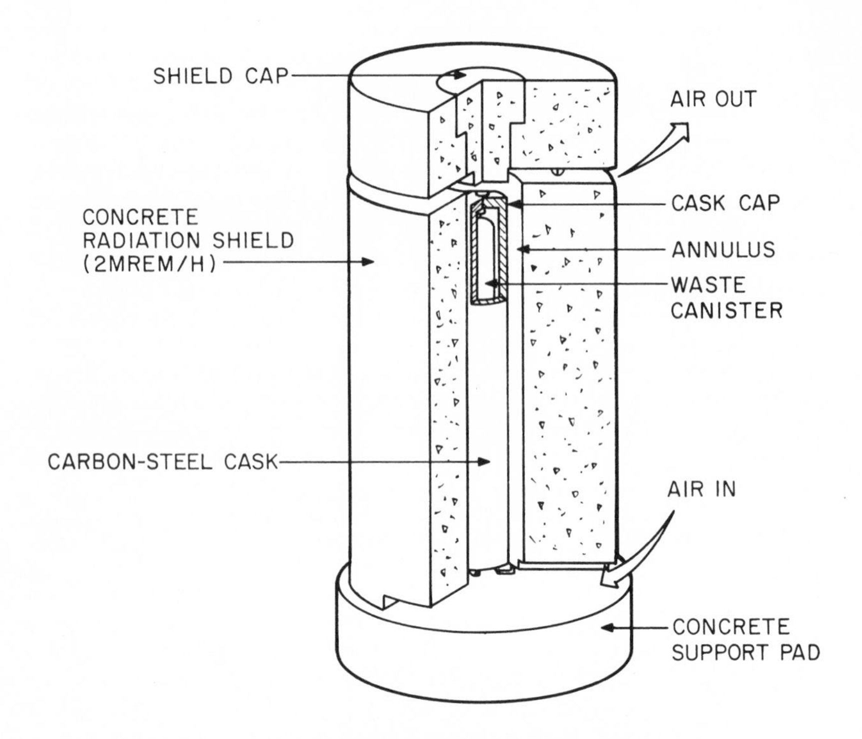

FIG. 10-3. Sealed storage cask concept for solid wastes.

fuel elements. The solid waste (or spent fuel) was to be placed in stainless-steel canisters, and three methods were considered for storing these canisters:

1. in sealed, carbon-steel storage casks each enclosed in a concrete cylinder (Fig. 10-3)
2. in a steel-lined concrete basin filled with water
3. in carbon-steel overpacks placed in a subsurface concrete vault with cooling by natural convection or forced air.

In its report, issued in 1975, the panel concluded that high-level radioactive wastes could be stored safely for many years in any of the ways outlined above, but they recommended an optimized version of the storage cask and concrete cylinder concept. However, the RSSF plan has since been shelved, although a modified form of water-basin storage for spent fuel elements is receiving attention, as seen in the next section.

In addition to the more immediate plans for the isolation of spent fuel or solidified reprocessing wastes, some consideration has been given to long-range (or advanced) concepts for the disposal of high-level wastes. These concepts include permanent geologic isolation on land, beneath the ocean floor, or in ice sheets, and elimination from existence on earth. Such advanced concepts, which are outlined at the end of the chapter, all require extensive research to determine if they are practical and environmentally acceptable.

Interim Storage of Spent Fuel

When spent fuel is removed from a reactor, the fuel rod bundles (or assemblies) are transferred to a stainless-steel-lined concrete water tank adjacent to the reactor. The water, which is circulated through a heat exchanger, removes the heat generated by radioactive decay of the fission (and other) products and also serves as a radiation shield. The spent fuel still contains substantial amounts of fissile species and could conceivably attain nuclear criticality in water (p. 14). To ensure that the arrangement will remain subcritical, the fuel assemblies are kept well separated from each other.

It was expected that, after allowing some time for radioactive decay, the spent fuel would be shipped to a reprocessing plant. Since this is not possible now, spent fuel assemblies are accumulating at reactor installations. Consequently, existing spent-fuel storage water tanks are being modified to increase their capacity. Each assembly is inserted in a vertical channel of stainless steel containing a neutron absorber (boron). A subcritical condition can then be maintained with closer spacing between assemblies.

In due course, many spent-fuel storage pools at reactor sites will be filled, even to the increased capacity. Consequently, there is interest in the construction of away-from-reactor (AFR) storage facilities. These facilities are expected to be water-filled concrete basins with boron (dissolved in the water and/or included in the steel structure) to prevent criticality. Existing reactor sites with ample space and inoperative fuel reprocessing plants are possible locations for AFR storage.

No serious consideration has yet been given to the eventual fate of the stored spent fuel elements. If the fuel is not reprocessed, presumably a more permanent storage than AFR facilities would be required. Because spent reactor fuel contains a large proportion (about 96 weight percent) of uranium, and some plutonium, which would be recovered in the reprocessing operation, the mass (and volume) of spent fuel is considerably greater than that of the equivalent solid, high-level reprocessing waste. The storage volume occupied by the spent fuel will depend on the manner in which the assemblies are packaged in containers. It is estimated that the total volume would be increased at least ten-fold (and probably more) compared with reprocessing wastes. However, the radioactivity and heat-generation rate per unit volume would be correspondingly less. On the other hand, the proportion of plutonium isotopes is much greater in the spent fuel.

Terminal Storage in a Geologic Formation

Geologic formations exist that have been physically and chemically stable for millions of years. It seems reasonable to suppose, therefore, that they will remain stable for a long time in the future. Formations of this kind thus offer the potential for providing a terminal (provisional) storage repository (with retrievability) for high-level radioactive solid wastes, either untreated spent fuel or reprocessing wastes. One difference in storing these two types of waste is that, as seen above, the volume to be stored would be much greater for the untreated spent fuel.

In a geologic medium that meets the requirements summarized in the next section, a terminal storage facility would require the excavation of a number of rooms or tunnels. Holes would be made in the floor with metal liners to facilitate insertion, and removal if necessary, of canisters containing the high-level solid wastes (Fig. 10-4). After the canisters were in place, a test and evaluation program would be initiated to determine if there were any adverse effects on the wastes themselves and on the geologic environment.

If the results of the tests were satisfactory, the provisional storage could be converted to the permanent phase by backfilling and sealing the cavities and entrance shaft with mined-out material. Retrievability would still be possible by using conventional mining techniques, which would remove the stored wastes, possibly together with associated geologic material.

As already noted, the terminal storage facility to be operated by DOE would be licensed by the NRC. Similarly, a decision to convert a geologic high-level waste repository from the test (or pilot) phase to the permanent phase would also require NRC approval.

Requirements of Geologic Formations

In assessing the suitability of a particular site for a high-level waste repository in a geologic formation, the first consideration is complete isolation from circulating groundwater. Furthermore, the rock medium should have a very low penetrability for water. The area should not have a history of significant seismic activity; geologic faults and rock fractures should be absent. Faults and

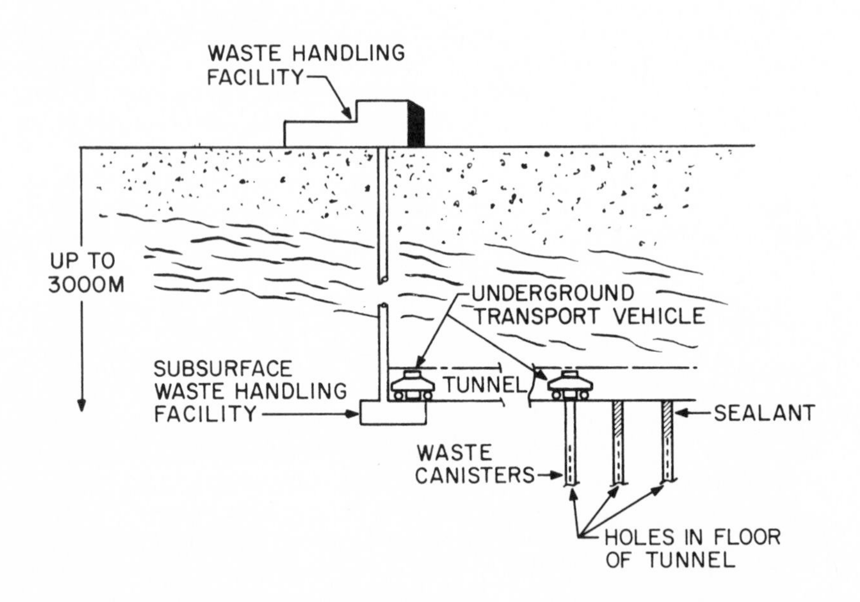

FIG. 10-4. Concept for solid waste emplacement in a mine. (Adapted from WASH-1297, U.S. Atomic Energy Commission, Washington D.C. [1974].)

fractures could provide flow paths for water in an otherwise impermeable medium.

The site should be in an area of gentle relief to minimize accelerated erosion or denudation that might result from climatic effects or changes brought about by preparation for storage or disposal operations. In addition, the site should be far removed from major drainages and from surface water bodies to avoid the possibility of flooding. Areas that have some potential for future exploitation of oil, gas, or other minerals or geothermal energy should be avoided.

The most suitable geologic media for radioactive waste storage (or disposal) are considered to be the following:

1. salt, either in thick beds or stable domes. Salt beds have been the subject of study in the United States (as seen below) and in Germany.
2. hard crystalline rocks, such as granite, whose present form has resulted from the action of high temperature and pressure
3. limestone (and dolomite), a sedimentary rock formed by the deposition and compression of shells of dead aquatic organisms
4. shale, a fine-grained sedimentary rock produced by the consolidation of beds of clay and mud
5. tuff, which consists largely of compacted particles of volcanic origin.

Of the foregoing, salt beds have been studied most completely, and the results of these studies are given first; later, reference is made to other geologic media.

Storage in Salt Beds

Salt beds seem to offer good prospects for the storage (possibly terminal) of high-level radioactive solid wastes. In the first place, these formations are almost unique in the respect that they are always dry. Since salt dissolves readily in water, their very existence shows that the deposits have not been associated with circulating groundwater for millions of years. Thus, it is highly probable that, as a general rule, they will retain their integrity for millions of years in the future. There is essentially no danger that leaching of radioactive species from the waste would cause contamination of groundwater. Furthermore, extensive salt beds often occur in areas of low seismic activity, so that the danger of disruption of the storage facility by earthquakes is negligible.[e]

The physical and mechanical properties are such as to make salt a good medium for a waste repository. Under compression, its strength is similar to that of concrete. Because of its plastic nature, any fractures that might form would tend to heal naturally. Moreover, salt has favorable thermal characteristics (heat conductivity and heat capacity), which would alleviate the problem of localized regions of very high temperature. Salt can be mined readily, and large spaces can be excavated in salt beds, even at depths of 300 m (1000 ft) or more, with only slight compression of pillars left to support the roof.

Between 1963 and 1968, an experimental program, named Project Salt Vault, was conducted by Oak Ridge National Laboratory to study the feasibility and safety of the storage of solid radioactive wastes in a bedded salt deposit. A

[e]A possible drawback to salt beds is that, if water should unexpectedly enter, the salt would readily dissolve to form a solution that is corrosive to steel.

disused salt mine, at a depth of about 300 m near Lyons, Kansas, was chosen for the purpose. A number of rooms were excavated with pillars left between them to support the roof. Test samples were made by enclosing highly radioactive spent fuel elements in steel canisters 15 cm (6 in.) in diameter and about 2 m (6 ft) long. Several such canisters, in two separate batches, were inserted by remote operation in vertical holes in the salt some 4 m (13 ft) deep. After six months, the canisters were removed successfully.

The test program revealed a number of problems, which required further investigation. It was found that there are small, brine-filled cavities in the salt. These cavities tend to migrate toward a heat source—i.e., the waste canisters. If the brine should come into contact with the steel canister, it could cause cracking (of stainless steel) due to corrosion or generalized rusting (of mild steel), but with little loss of integrity. The corrosion could be retarded, if necessary, by cladding the canisters with mild steel or aluminum.

In a salt-bed repository, part of the gamma radiation emitted by the fission products in the solid waste would be absorbed in the waste itself, and the remainder would be absorbed in the salt. It is known that radiation energy, instead of being released continuously as heat, can sometimes accumulate in a solid until its release is triggered, often by an increase of temperature. The sudden release of heat, either in the solid waste or in the salt, could cause an increase of temperature accompanied by a decrease in the strength of the salt. The available evidence suggests that the radiation energy storage would be small; nevertheless, the matter is receiving further study.

A minor, but not negligible, problem would be disposal of the salt removed from the mine in excavating storage chambers and passageways for operations. Part could perhaps be stored in adjacent disused mines, if available, and some could be sold as crushed salt or brine for industrial uses. The remainder would have to be transported elsewhere. If the repository were to be used for terminal storage, about half the excavated salt could be used for backfilling and sealing the filled chambers.

After completion of the Salt Vault test, plans were made for a pilot demonstration program on a larger scale in the same mine in order to provide further information on the suitability of salt beds for the isolation of high-level radioactive solid wastes. The project was abandoned in 1972, however, largely because of an unexplained loss of water in hydraulic operations conducted a few years earlier in a nearby salt mine. This loss raised the possibility that there might be unidentified voids through which water could enter the repository. Furthermore, there were several abandoned old exploratory oil and gas wells in the area that might not have been identified and plugged.

Alternative salt deposits are being studied as potential sites for storage of radioactive solid wastes; one such site is located in southeastern New Mexico, where there are extensive salt beds. Consideration is also being given to salt anticlines [f] in Utah and to salt domes (i.e., intrusions in hard rock) in Louisiana and Mississippi.

[f] An anticline is a layered (or bedded) rock in which the layers slope downward in opposite directions from a crest.

Other Geologic Formations

Other stable geologic formations, such as limestone, granite, shale, and tuff, are also of interest for the location of repositories for high-level radioactive wastes. The construction of many large caverns in limestone, granite, and shale for the storage of liquid and gaseous petroleum products has provided useful related experience. This experience indicates that these geologic media merit further study for radioactive waste storage.

Crystalline rocks (e.g., granite) generally have a low permeability for water and, unlike salt, they are insoluble in water that might penetrate through cracks. Many crystalline rock formations are not associated with oil or gas, so that there are no penetrations; neither are there likely to be any in the future. These rocks normally have considerable strength, and large openings can be made with little roof support.

Carbonate (limestone and dolomite) rocks are strong, like the crystalline rocks, and large openings can be readily constructed. In some situations, carbonate rocks are subject to water erosion. However, storage caverns have been constructed in media in which there are no signs that any weathering or erosion has occurred since the rocks were deposited.

Shales have the great merit of being highly impermeable to water and, in fact, more than half the existing storage caverns in the United States are in shale deposits (e.g., in Ohio, Illinois, Iowa, and Oklahoma). On the other hand, the shale deposits do not have the great strength of other rocks; with adequate roof support, however, mined caverns can easily be constructed in shale. If a site for a high-level waste repository were to be selected in a shale formation, it would probably be in a region where the organic content of the shale was small so that there would be no interest in oil extraction in the future.

Thick-bedded, fine-textured tuff deposits, generally of the welded variety, offer some prospects for a waste repository. The strength of this material is similar to that of shale.

Transuranium Wastes

Nearly all of the common isotopes of the transuranium elements are alpha-particle emitters with long half-lives; if ingested or inhaled, they tend to deposit in certain body organs. Consequently, the transuranium-contaminated wastes, commonly known as TRU wastes (p. 264), could be a significant potential hazard for many years if they should enter the air or water supplies. Hitherto, these wastes have been stored in steel drums and buried at a shallow depth, so that they can be retrieved. Eventually, however, the TRU wastes, like the high-level wastes, will have to be transferred to a federal repository.

An important difference between the high-level wastes and the TRU wastes is that the rate of heat generation from radioactive decay per unit mass is very much less from the TRU wastes. Consequently, the latter can be stored in larger packages, and they can be closer together, since heat dissipation is a relatively minor problem.

The cladding hulls and hardware from spent-fuel reprocessing, one of the major potential sources of TRU wastes, contain beta-gamma, as well as alpha-particle, emitters. The beta-gamma activity arises from various radioactive species produced by neutron capture while in the reactor, especially by alloying and impurity elements in the Inconel and stainless-steel hardware. Hence, when being handled, cladding hulls and hardware packages require gamma-ray shielding, and possibly neutron shielding if appreciable amounts of curium-244 are present.

Most of the beta-gamma activity, but little of the alpha activity, is in the hardware, which constitutes less than one tenth of the total volume. It might be advantageous, therefore, to separate the hardware from the hulls; then the packages containing the hulls would require less gamma-ray shielding.

The storage of TRU wastes could be simplified if the volume can be reduced. As stated in Chapter 8, procedures are being developed for the controlled incineration of combustible material, thereby decreasing the volume to be stored. The cladding hulls and hardware from the reprocessing of 1 MT of spent uranium fuel occupy roughly 0.35 m^3 (about 12 ft^3). Hence, a plant reprocessing 1500 MT of spent fuel annually would accumulate a considerable volume of TRU wastes. The volume could be decreased by a factor of 4 to 6 by separation of the hardware and mechanical compression or melting of the remaining zirconium-alloy cladding hulls. A number of chemical, metallurgical, and other techniques are being studied that will simplify the handling and storage of TRU wastes.

Gaseous and Airborne Wastes

Limitations placed by the U.S. Environmental Protection Agency on the amounts of krypton-85 and iodine-129 (Chapter 5), and possibly later on carbon-14, that may be released to the environment may make it necessary to remove these substances from the gaseous effluents of nuclear power and fuel reprocessing plants. The radioactive materials so removed will be retained as wastes, preferably in solid form. At present, there is no requirement that these wastes be transferred to a federal repository. If necessary, however, long-term management could be provided either by surface storage, terminal storage in a geologic formation, or shallow land burial, as may be appropriate for the activity, half-life, and potential hazard.

Multiple-Barrier Concept

The most likely way for radioactivity from a waste repository in a geologic formation to reach the biosphere (i.e., where living organisms exist) would be by water entering the repository. The waste might then be extracted by water and the resulting contaminated liquid might conceivably reach the surface. The geologic formation selected for the repository would be such that the probability of water entering would be extremely small. However, if it should enter, there would be a number of barriers to inhibit the transfer of radioactivity to the biosphere.

First, the waste would be in a form that is not easily leached by water even at high temperature and it would be enclosed in a corrosion-resistant container.

The container would be inserted in a hole with a sleeve (or overpack) of a material that retards the movement of substances dissolved in water. Laboratory studies indicate that a 0.3-m (1-ft)-thick overpack of a mixture of clay and zeolite might delay escape of fission products for more than 1000 years and transuranium elements by 10 000 to 100 000 years. (Zeolites, which are usually sodium and/or calcium aluminum silicates, are commonly used to purify water.)

If contaminated water should be released from the repository, it would still have to reach the biosphere at the ground surface. The depth of the repository at more than 600 m (2000 ft) would generally ensure a long transit time. Moreover, as a result of various chemical and physical processes (e.g., ion exchange and sorption) the minerals through which the water must travel would retard the movement of the dissolved radioactive species. This retardation effect has been observed in connection with the residues of underground nuclear explosions at the Nevada Test Site and in the uranium deposits in Gabon, West Africa, where a naturally occuring fission chain reaction almost 2 billion years ago left many tons of fission products and transuranic residues in the ground.

LONG-RANGE DISPOSAL CONCEPTS

Introduction

Long-range (i.e., advanced) concepts for the irretrievable disposal of high-level radioactive wastes will require considerable study to determine if they can be implemented. One or the other of the proposed methods might eventually prove to be simpler and possibly provide better isolation than the terminal storage procedure already described.

The more promising advanced concepts fall into two broad classes:

1. disposal in geologic formations
2. removal from existence on earth.

Geologic disposal may be (a) in deep continental formations, (b) in the ocean floor (or seabed), or (c) in polar ice sheets. There are two categories of removal from earth, namely: (a) extraterrestrial disposal (i.e., in space) and (b) transmutation of nuclides having a long-term risk potential into nuclides having a short-term risk potential.

Deep Geologic Land Disposal

Among the concepts for disposal in deep geologic formations on land are the following:

1. matrix of drilled holes

2. rock melting
3. hydrofracturing.

These concepts are advantageous in the respect that all operations can be conducted from the surface, and mining would not be required.

1. An array (or matrix) of holes, to a depth up to about 6 km (3.7 miles), would be drilled in a stable geologic formation with no cracks or fractures that would permit water to circulate. A number of canisters containing the solidified waste would then be lowered into each hole. When a predetermined height is reached, the hole would be sealed completely to the surface (Fig. 10-5). The spacings of both holes and canisters would be such as to permit dissipation of the radioactive decay heat without melting the rock. In this event, retrievability might be possible.

2. One form of the rock melting concept is similar to that described above, except that the nature of the rock and the spacing of the holes is such that melting of both rock and solid waste would occur. The mixture of molten rock and waste would eventually solidify. Alternatively, the waste may be enclosed in a container of refractory material so that the surrounding rock melts, but the container does not. The container and its contents would then descend slowly through the molten rock. Another possibility is to discharge the concentrated liquid waste from reprocessing directly into a deep hole; the decay heat would cause the water and other volatile matter to be driven off, leaving a solid. This would eventually melt and mix with molten rock in the vicinity. In each case, the hole used to introduce the waste—solid or liquid—would ultimately be sealed to the surface.

3. Hydrofracturing is widely used to stimulate oil and gas production and for the disposal of industrial wastes. It involves the fracturing at depth in a geologic medium around a bore hole by pumping into it a fluid (e.g., water) with a solid (e.g., fine sand) in suspension. Liquid high-level reprocessing wastes might then be injected through the bore hole into the fractures. Cement or other grouting material could be added to the liquid so that it would solidify and thus fix the waste in the geologic medium.

Seabed Geologic Disposal

Large areas of ocean are available with depths that would provide isolation and safety from natural phenomena, such as storms, as well as from accidental disturbance. Furthermore, most of the seabed is inaccessible and is of little or no value to man. There may be geologic formations beneath the ocean that are more suitable for the disposal of high-level solid wastes than are continental formations. Any plan for seabed disposal would, of course, have to be approved internationally.

The major oceanic provinces considered as disposal sites are: (a) continental margins, (b) the midoceanic ridge, and (c) the ocean floor. Of these, the ocean floor, at depths of 5 to 11 km (about 3 to 7 miles), contains some of the most promising areas for disposal. Such areas include the abyssal hills and rises that occur in the middle of subocean plates. They are seismically stable, are below the regions of wind-driven and surface-current motion, and are relatively unproductive biologically.

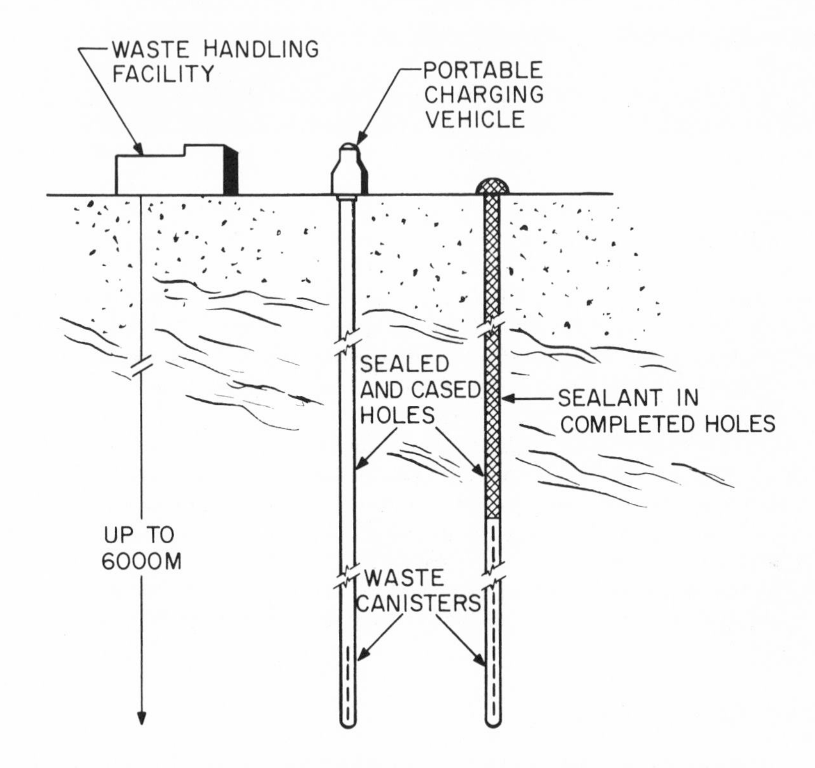

FIG. 10-5. Concept for solid waste emplacement in a matrix of drilled holes. (Adapted from WASH-1297, U.S. Atomic Energy Commission, Washington, D.C. [1974].)

It should be emphasized that the proposed seabed disposal would involve careful emplacement of fully contained solid waste in the ocean basement rock. It must not be confused with "dumping" radioactive wastes in the ocean. Such dumping is prohibited by the United States and several other countries.

Polar Ice Sheet Disposal

Ice sheets are large, thick permanent layers of ice overlying polar land masses, especially in Greenland and Antarctica. They are located in regions far from man's normal activities and where there is little likelihood of future development. The temperature is always below the freezing point of water, so that ice sheets have a natural capability to dissipate radioactive decay heat. In principle, therefore, the continental ice sheets could provide, subject to international agreement, an approach to the disposal of high-level solid radioactive wastes. The technical feasibility, however, depends on the long-term stability of the ice sheets.

Some concepts for disposal of high-level solid wastes in ice sheets are (a) meltdown, (b) anchored emplacement, and (c) surface disposal. The last two procedures have the potential for retrievability of the waste canisters for many years.

In the meltdown concept, a hole, 50 to 100 m (160 to 320 ft) deep would be drilled in the ice, and a canister of waste would be inserted. The radioactive decay heat will cause the ice to melt, and the canister will sink gradually under the influence of gravity until it reaches bedrock, at a depth of 3 to 4 km (2 to 2½ miles), within 5 to 10 years.

For anchored emplacement, the canister would be inserted in a hole in the ice, just as for meltdown, but it is prevented from sinking below 200 to 500 m (650 to 1600 ft) by cables connected to surface anchors. The canisters would be potentially retrievable for about 200 to 400 years. Eventually, new snow and ice would accumulate on the surface, so that the canisters would effectively sink and become inaccessible. It is expected that they would reach bedrock in about 30 000 years.

Surface disposal provides the option of ready retrievability for many years. The waste canisters would be placed in a structure supported above the surface of the ice on pilings or piers. Decay heat would be dissipated by air cooling provided by natural draft. It is estimated that access to the storage structure for retrieving the canisters would be possible for up to 400 years. Subsequently, as snow and ice accumulated and encroached on the structure, the decay heat would melt the ice, and the whole facility would melt down through the ice sheets.

Extraterrestrial Disposal

Disposal of high-level wastes in space, away from the earth, could provide complete isolation from man's environment. Among the space trajectories considered are (a) very high earth orbit, (b) solar orbit other than that of the earth and the planets, (c) solar impact, and (d) escape from the solar system. The last two of these possibilities seem most likely to assure true final disposal. However, a highly reliable rocket vehicle capable of transporting wastes to the sun would have to be developed.

Because of the cooling and shielding requirements and the necessity for a strong container that could survive an aborted mission, extraterrestrial disposal of all solid wastes, even the smaller volume of reprocessing wastes, would be impractical. The payload could be greatly decreased, however, if extraterrestrial disposal were restricted to the transuranium elements, which constitute less than 1 percent of the total mass, leaving the fission products for storage (or disposal) on earth. Since the fission products decay almost completely in about 800 years, whereas several of the transuranium nuclides have long half-lives, extraterrestrial disposal of the latter would decrease the time period over which isolation of the wastes must be assured on earth.

Realization of disposal in space thus depends on the ability to achieve the chemical separation (or *partitioning*) of the high-level reprocessing wastes into appropriate fractions. The fraction to be disposed of extraterrestrially would have to contain essentially all the plutonium, americium, and curium present in the high-level liquid reprocessing wastes, but only a very small proportion of fission products. Laboratory studies for achieving the partition of high-level wastes are in progress, but many difficulties remain to be overcome before a satisfactory procedure can be demonstrated.

The partitioning of reprocessing wastes and extraterrestrial disposal of the transuranium elements would still leave substantial quantities of long-lived TRU wastes for disposal on earth. Moreover, the partitioning operation would itself generate additional TRU wastes. A remotely possible solution to the problem would be to recover nearly all the transuranium elements from the TRU wastes and then dispose of them in space.

Transmutation of Transuranium Elements

The transmutation concept, like extraterrestrial disposal, depends on the ability to partition the high-level reprocessing wastes and therefore has the limitation mentioned above. The separated transuranium elements would be exposed to neutrons in a reactor and thereby converted (or transmuted) into fission products. Fissile nuclides would be fissioned directly by the neutrons, whereas nonfissile species would first be converted into fissile nuclides by neutron capture (sometimes followed by beta decay) and would then undergo fission. The net result would be the transmutation of the long-lived transuranium elements into fission products with mostly shorter half-lives. The latter would, of course, still need to be isolated from the environment for several hundred years.

INTERAGENCY REVIEW

In March 1978, President Carter appointed an Interagency Review Group (IRG) on Nuclear Waste Management to make recommendations relating to an administrative policy for the long-term management of nuclear wastes. The IRG, composed of representatives of 14 government agencies under the chairmanship of the Secretary of Energy, presented its final report in March 1979. Although this report covers many aspects of radioactive waste management, it is appropriate to end this chapter with some of the conclusions and recommendations applicable to high-level wastes.

After comparing the status of the various alternative means for disposing of these wastes, as already described, the IRG concluded that the first disposal facilities should be in a mined repository in a geologic formation. "Present scientific and technological knowledge is adequate to identify potential repository sites for further investigation. No scientific or technical reason is known that would prevent identifying a site that is suitable for a repository provided that a system view is utilized rigorously to evaluate the suitability of sites and designs, and in minimizing the influence of future human activities . . . Reliance on conservative engineering practices and multiple independent barriers (see page 280) can reduce some risks and compensate for some uncertainties."

Near-term research and development programs should be designed to provide a choice of potential repository sites in different geologic environments. Instead of concentrating on a single repository, the program should create the option of having two (or possibly three) repositories in different parts of the country. No specific dates were mentioned in the IRG report, but the possibility of site selection by the mid 1980s and completion of the first repository some 10 years later has been mentioned. In the meantime, federal AFR storage facilities for spent fuel might be required (p. 275).

In the view of the IRG, dealing with institutional issues in a nuclear waste management program is as important and difficult — probably more so — than the technical problems. It was recommended, therefore, that state and local agencies and the general public be closely involved in all stages of the development of national high-level waste disposal facilities.

Bibliography—Chapter 10

American Nuclear Society. "Policy Statement on Away-from-Reactor Storage of Spent Nuclear Fuel." *Nuclear News* 23, no. 2 (1980): 129.

______. "Policy Statement on High-Level Radioactive Waste Disposal." *Nuclear News* 22, no. 14 (1979): 190.

Angino, E. E. "High-Level and Long-Lived Radioactive Waste Disposal." *Science* 198 (1977): 885.

Battelle Pacific Northwest Laboratories. *Incentives for Partitioning High-Level Waste* (BNWL-1927). Richland, Wash.: Battelle Pacific Northwest Laboratories, 1975.

Blomeke, J. O.; Nichols, J. P.; and McClain, W. C. "Managing Radioactive Wastes." *Physics Today* 26, no. 8 (1973): 36.

Bradshaw, R. L., and McClain, W. C., eds. *Project Salt Vault: Demonstration of the Disposal of High-Level Solidified Wastes in an Underground Salt Mine* (ORNL-4555). Oak Ridge, Tenn.: Oak Ridge National Laboratory, 1971. See also *Nuclear Safety* 11 (1970): 130.

Campbell, M. H., ed. *Proceedings of the Symposium on High-Level Radioactive Waste Management.* Washington, D.C.: American Chemical Society, 1976.

Chekalla, T. D., and J. E. Mendel, eds. "Ceramics in Nuclear Waste Management." *Proceedings of the Joint American Ceramic Society and U.S. Department of Energy Symposium* (CONF 79-04-20). Washington, D.C., 1979.

Claiborne, H. C. *Effect of Actinide Removal on the Long-Term Hazard of High-Level Waste* (ORNL-TM-4724). Oak Ridge, Tenn.: Oak Ridge National Laboratory, 1975.

Cohen, B. L. "The Disposal of Radioactive Wastes from Fission Reactors." *Scientific American* 226, no. 6 (1977): 21.

______. "High-Level Radioactive Waste Management from Light Water Reactors." *Reviews of Modern Physics* 49 (1977): 1.

"Completing the Nuclear Fuel Cycle." *EPRI Journal* 2 (1977): 6.

Cowan, G.A. "A Natural Fission Reactor." *Scientific American* 235, no. 1 (1974): 36.

de Marsily, G. "Nuclear Waste Disposal: Can the Geologist Guarantee Isolation?" *Science* 197 (1977): 519.

Fried, S., ed. "Radioactive Waste in Geologic Storage." American Chemical Society Symposium Series, no. 100. Washington, D.C., 1979.

International Atomic Energy Agency. *Proceedings of the Symposium on the Management of Radioactive Wastes from the Nuclear Fuel Cycle.* Vienna: International Atomic Energy Agency, 1976.

Kubo, A. S., and Rose, D. J. "Disposal of Nuclear Wastes." *Science* 182 (1973): 1205.

Lennemann, W. L. "Management of Radioactive Wastes from AEC Fuel-Reprocessing Operations." *Nuclear Safety* 14 (1973): 482.

McCarthy, G. J., ed. "Scientific Basis for Nuclear Waste Management." *Proceedings of the Materials Research Society Conference.* New York: Plenum Publishing Corporation, 1979.

National Academy of Sciences-National Research Council, Committee on Radioactive Waste Management. *Interim Storage of High-Level Radioactive Wastes.* Washington, D.C., 1975.

Oak Ridge National Laboratory. *Siting of Fuel Reprocessing Plants and Waste Management Facilities* (ORNL-4451). Oak Ridge, Tenn.: Oak Ridge National Laboratory, 1970.

Rochlin, G. I. "Nuclear Waste Disposal: Two Social Criteria." *Science* 195 (1977): 23.

Schneider, K. J., and Platt, A. M., eds. *Advanced Waste Management Studies, High-Level Radioactive Waste Management Alternatives* (BNWL-1900). Richland, Wash.: Battelle Pacific Northwest Laboratories, 1974.

U.S., Atomic Energy Commission. *High-Level Radioactive Waste Management Alternatives* (WASH-1297). Washington, D.C., 1974.

U.S., Department of Energy. *Report of the Interagency Review Group on Nuclear Waste Management* (TID-29442). Springfield, Va.: National Technical Information Service, 1979. See also, *Nuclear Safety* 20 (1979): 706.

______. *Storage of U.S. Spent Reactor Fuel* (DOE/EIS-0015D). Washington, D.C., 1978.

U.S., Energy Research and Development Administration. *Alternatives for Managing Wastes from Reactors and Post-Fission Operations in the LWR Fuel Cycle* (ERDA-76-43), Vols. 1 through 5. (Vol. 1. Summary). Washington, D.C., 1976.

______. *Proceedings of the Internatonal Symposium on the Management of Wastes from the LWR Fuel Cycle* (CONF-760701). Washington, D.C., 1976.

U. S., Geological Survey. *Geologic Disposal of High-Level Radioactive Wastes: Earth Science Perspectives.* Washington, D. C., 1978.

U.S., Nuclear Regulatory Commission. *Environmental Survey of the Reprocessing and Waste Management Portions of the LWR Fuel Cycle.* (NUREG-0116). Washington, D.C., 1976.

______. *Generic Environmental Impact Statement in Handling and Storage of Spent Fuel from Light-Water Power Reactors* (NUREG-0404). Washington, D.C., 1978. See also, *Nuclear Safety* 20 (1979): 54.

______. *The Management of Radioactive Waste: Waste Partition As an Alternative* (NF-CONF-001). Washington, D.C., 1976.

11

Transportation and Safeguarding of Nuclear Materials

REGULATIONS OF PACKAGING AND TRANSPORTATION

DOT and NRC Regulations

Several of the stages associated with the generation of nuclear power require the packaging and transportation of radioactive materials. The latter range from the essentially hazard-free compounds of natural uranium, with a very low level of radioactivity, to the highly radioactive spent-fuel assemblies and, eventually, plutonium and solidified wastes from reprocessing plants, if any are operated in the future. These materials have to be shipped from one location to another by road, rail, or water. No matter which mode of transportation is used, the possibility must be taken into account that an accident might cause radioactive material to be dispersed in the environment. Stringent safety standards and regulations have been developed to assure that containers for materials having an appreciable level of radioactivity will retain their integrity even if exposed to the forces of a severe accident. Furthermore, the shipping of these containers is subject to strict regulation.

In the United States, regulations concerning the safe transportation of radioactive materials were first promulgated in 1948 by the Interstate Commerce Commission (ICC). When the U.S. Department of Transportation (DOT) was formed in 1967, the safety functions of the ICC were transferred to it. Before this event, however, the U.S. Atomic Energy Commission (AEC) had published in 1966 its own regulations, which established safety standards for the packaging of both fissile and nonfissile radioactive substances. These regulations were based largely on the standards that had been developed by the International Atomic Energy Agency, of which the United States is a member nation. Then, late in 1968, the DOT issued a new set of regulations, prepared in cooperation with the AEC, that represented a substantial revision of the preexisting ICC regulations. The regulations are amended from time to time as appears desirable in the interest of greater safety.

By the Atomic Energy Act of 1954, the Transportation of Explosives Act of 1960, and subsequent acts of the U.S. Congress, both the AEC (now the U.S.

Nuclear Regulatory Commission [NRC]) and the DOT have regulatory jurisdiction over the safe packaging and shipping of radioactive materials. However, the overlap has not resulted in any problems because of the close working relationship between the two agencies. This relationship is expressed as a joint Memorandum of Understanding, which is modified as may be required by the circumstances.

The NRC regulations are published as Title 10 of the *Code of Federal Regulations*, Part 71 (10CFR71). The DOT regulations for packaging and transportation by rail and highway are recorded in Title 49 of the *Code of Federal Regulations*, Parts 170-189; the rules for packaging are in Parts 173.389-173.399. These regulations duplicated each other to some extent, but where they overlap, they are essentially in complete agreement. The NRC role is mainly concerned with assuring the integrity of the containers to be used for the shipment of substantial amounts of radioactive (including fissile) materials, whereas the DOT is responsible for regulating the packaging and transportation of all radioactive materials. The DOT established safety standards for labeling, loading, unloading, and storing containers, and for their management in transit. In addition, the DOT regulates the mechanical condition of the shippers' equipment, the qualifications of personnel, and the operating instructions they receive in the handling of radioactive shipments.

The design of a container to be used for the transportation of radioactive materials must be approved by the NRC. The applicant for such approval must demonstrate that the container design is in conformity with the requirements of 10CFR71, about which more will be said shortly. The application must contain a detailed description of the package and its proposed contents. There must also be a complete evaluation of the safety aspects, especially under conditions that could cause damage, and a statement of the planned procedures for regular and periodic inspection of the container during its use. Only after the application has been thoroughly reviewed and there is assurance that the design meets the stringent safety standards, would the NRC issue a Certificate of Compliance with 10CFR71 for the particular container to be used for a specified purpose.

The DOT regulations contain a detailed classification of various radioactive and fissile materials, according to their potential hazard, with limitations on the contents based on the type of packaging. The objective of these regulations is to ensure that radioactive materials are transported safely. Furthermore, there is an essential requirement that packages containing fissile material will not attain nuclear criticality, in the sense described in Chapter 2, even if immersed in water. In other words, there must be assurance that fissile material would never be in such a configuration, either during normal transportation or as the result of an accident, that a nuclear fission chain reaction can be sustained.

Types of Packages

According to DOT regulations, uranium ores and concentrates of these ores, such as yellowcake (p. 218), and chemical compounds of natural (i.e., not enriched) uranium, including uranium hexafluoride, may be packaged as "low specific activity materials." The specific activity of a radioactive material is the activity, usually expressed in curies (p. 132), per unit mass (weight) of the material. The essential requirements are that such materials be enclosed in strong, tight packages from which there would be no leaking under conditions normally incident to transportation. There must be no significant removable contamination on the exterior surface of the package and the external radiation levels must be low, which they invariably are for materials of low specific activity.

Provided such materials are transported in vehicles, other than aircraft, assigned for the sole use of a single shipper, as is usually the case, all that is required is that the shipment be marked as "Radioactive—LSA," where the letters LSA stand for "low specific activity." If the LSA material is shipped in a vehicle containing other materials, special packaging and labeling are necessary.

When the radioactivity level does not permit shipment under LSA conditions, the DOT requires that packages meet either *Type A* or *Type B* specifications. In general, Type A packages are limited to relatively small amounts of radioactive (and fissile) material that would not have serious consequences if released in an accident. Type B packages are used for shipment of larger amounts of radioactive (and fissile) material and are designed to contain such material under conditions of severe shipping accidents.

Other than natural (unenriched) uranium and its compounds, the great majority of radioactive materials that have to be packaged and shipped in connection with the operation of nuclear power plants require Type B packages. The NRC 10CFR71 standards provide for performance criteria for these Type B packages that parallel the DOT requirements in 49CFR173.

Packaging Requirements for Radioactive Materials

The DOT regulations for the shipping of radioactive (other than LSA) materials in Type A and Type B packages include numerous safety requirements, but only a few of immediate interest are mentioned here. The package must be designed so that heat generated within it by the decay of radioactive materials present will not affect the integrity of the package under normal transportation conditions. The temperature of accessible surfaces must not exceed 50 °C (122 °F) in the shade when fully loaded, assuming still air at ambient temperature. If the package is transported in a vehicle used solely by a single shipper, the maximum accessible temperature must not exceed 82 °C (180 °F).

The package must include adequate radiation shielding to limit the external dose rates. If the vehicle, other than an aircraft, is assigned for the sole use of the consignor and is unloaded by the consignee from the vehicle in which it was originally loaded, the maximum permissible radiation dose rates are as follows[a]:

1. 1000 (mrems/h) at 0.915 metres (m) (3 ft) from the external surface of the package (closed transport vehicle only)
2. 200 mrems/h at any point on the external surface of the vehicle (closed transport vehicle only)
3. 10 mrems/h at 1.83 m (6 ft) from the external surface of the vehicle

[a]The DOT and NRC regulations give masses in pounds and distances in feet and inches; they have been converted here into the nearest values in metric units.

4. 2 mrems/h in any normally occupied position in the vehicle; this provision does not apply to private motor carriers.

The NRC and DOT standards for packaging include requirements that must be met both during normal transportation and as the result of a transportation accident. The package must be constructed so that, if it were subjected to the various qualifying conditions given below, which are considered to be representative of (or equivalent to) severe environmental and other situations that might be encountered in normal transportation, the following requirements would be met:

1. no radioactive material would be released
2. the effectiveness of the packaging would not be substantially reduced
3. no mixture of gases or vapors would form in the package that could, through any credible increase of pressure or an explosion, significantly reduce the effectiveness of the package.

The qualifying (environmental and test) conditions, which are to be applied separately, are as follows:

1. direct sunlight at an ambient temperature of 55 °C (130 °F) in still air
2. an ambient temperature of – 40 °C (– 40 °F) in the shade and still air
3. atmospheric pressure of half the normal pressure of the atmosphere
4. vibration normally incident to transport
5. a water spray sufficiently heavy to keep the entire exposed surface of the package (except the bottom) continuously wet during a period of 30 minutes
6. between 1½ and 2½ hours after the conclusion of the water spray test, a free drop through the distance specified below onto a flat, essentially unyielding horizontal surface in a position in which maximum damage is expected

Package Weight		Drop Distance	
kg	lb	m	ft
less than 4545	less than 10 000	1.22	4
4545 to 9090	10 000 to 20 000	0.915	3
9090 to 13 636	20 000 to 30 000	0.61	2
more than 13 636	more than 30 000	0.305	1

7. a free drop onto each corner of the package in succession (or, in the case of a cylindrical package, onto each quarter of each rim) from a height of 0.305 m (1 ft) onto a flat, essentially unyielding horizontal surface. (This test applies only to relatively light packages containing small amounts of fissile materials.)
8. impact of the hemispherical end of a solid steel cylinder 3.18 cm (1¼ in.) in diameter and weighing 5.9 kg (13 lb), dropped vertically from a height of 1.02 m (40 in.) onto an exposed surface of the package that is expected to be the most vulnerable to puncture when horizontal

9. for packages not exceeding 4545 kg (10 000 lb) in weight, a compressive load, applied uniformly against the top and bottom of the package in the position in which it is normally transported, for a period of 24 hours.

Accident Damage Test Conditions

In addition to the foregoing requirements for normal transportation, which must be met by both Type A and Type B packages, Type B packages must be able to withstand severe transportation accidents, as simulated by a series of damage tests. These tests must be applied in sequence to determine their cumulative effect on a package (or array of packages). As a result of applying the Type B test conditions, the reduction in shielding should not be sufficient to increase the external radiation dose rate to more than 1000 mrems/h at a distance of 0.915 m (3 ft) from the surface of the container. There should be no release of radioactivity, except specified limited quantities of gases or contaminated coolant.

Furthermore, during and after the hypothetical severe transportation accident, as represented by the test conditions, the package should remain subcritical under the following circumstances:

1. the fissile material is in the most (nuclear) reactive credible configuration
2. moderation occurs by water to the most reactive credible extent
3. there is reflection by water on all sides, in all cases consistent with the damaged condition of the package and its contents.

The accident damage test conditions to be applied to the package in sequence are given below:

1. a free drop through a distance of 9.15 m (30 ft) onto a flat, essentially unyielding horizontal surface, striking the surface in a position for which maximum damage is expected
2. a free drop through a distance of 1.02 m (40 in.) striking, in a position for which maximum damage is expected, the top end of a vertical cylindrical mild steel bar, 15.2 cm (6 in.) in diameter and at least 20.3 cm (8 in.) long, mounted on an essentially unyielding horizontal surface
3. exposure to a thermal (simulated fire) test in which the heat input to the package is not less than would result from exposure to a (heat) radiation environment of 802 °C (1475 °F) for 30 minutes with an emissivity coefficient of 0.9, assuming the surfaces of the package to have an absorption coefficient of 0.8.[b] The package shall not be cooled artificially until 3 hours after the test period, unless it can be shown that the temperature in the interior has begun to fall earlier

[b]A perfect (ideal) radiation emitter and a perfect (ideal) absorber would have respective coefficients of 1.0. In practice, however, radiation emitters and absorbers are far from perfect; the conditions specified are thus quite stringent.

4. for a fissile material package, immersion in water to the extent that all portions of the package are under at least 0.915 m (3 ft) of water for a period of not less than 8 hours.

The "normal transport" and "hypothetical accident" conditions are not intended to imply actual conditions that would be encountered. Rather, their purpose is to provide a means for reproducing in the laboratory and in the field the same type and degree of damage that might reasonably be expected to occur under either normal transport or severe accident conditions, as the case might be. In other words, the tests are not actual conditions, but they represent an attempt to simulate the damage that would occur under such conditions.

In applying for a license for a particular package, it is not necessary that the package be submitted to all the tests described above. As a result of experience with different types of containers and materials and from various laboratory experiments and theoretical treatments, the required information in several areas can be obtained in other ways. These include the use of scale models and acceptable calculational methods, supplemented, if necessary, by field observations on components or packages of similar design. In any event, a license is not issued unless the requirements for the package can be established in one way or another.

TRANSPORTATION OF NUCLEAR REACTOR MATERIALS

Transportation of Fuel Materials and Assemblies

Although the general performance requirements are the same for all Type B packages, the designs of individual packages vary widely according to the material being shipped. For example, enriched uranium hexafluoride (with less than 5 percent of uranium-235) is packaged and shipped in strong steel cylinders enclosed in a protective shipping case designed to minimize damage from impact and fire. These cylinders range up to 0.76 m (30 in.) inside diameter (i.d.) and 2.06 m (81 in.) in length and contain up to 2270 kg (5000 lb) of uranium hexafluoride. They are normally carried by truck, with up to five cylinders per load.

Enriched uranium dioxide powder is shipped to fuel fabrication plants in cylindrical containers about 13 cm (5 in.) i.d., each enclosed in a 208-litre (55-gal) steel drum. The intervening space is filled with shock-absorbing and thermal-insulating material, such as foamed polyurethane, foamed glass, or vermiculite. Each drum contains about 110 kg (240 lb) of uranium dioxide.

Because the level of radioactivity of uranium hexafluoride or oxide is low and because very little internal heat is generated by decay, no special precautions are needed to maintain the radiation dose rate and temperature levels within the DOT specifications. An essential requirement, however, is proof that the shipment, which may consist of more than one package, would not become critical in an accident.

The conditions for packaging unirradiated (i.e., fresh) fuel assemblies for the common water-cooled reactors (Chapter 2) are generally similar to those for slightly (less than 5 percent) enriched uranium compounds. Apart from the structural strength of the package, the essential requirement is that there should be no danger of the attainment of criticality. This can be readily achieved by limiting the number of fuel assemblies in a load. However, a neutron absorber, such as boron-steel or other alloy or compound of boron, is often included in the package to ensure that the arrangement remains subcritical even in water. The containers are designed to prevent damage to the fuel assemblies by vibration or shock during shipment. Because of their length—5.2 m (17 ft) or more—the containers are transported in a horizontal position. A typical loaded container may weigh from 454 to 4082 kg (1000 to 9000 lb).

Transportation of Spent Fuel

The irradiated (i.e., spent) fuel assemblies, which must be transported from the reactor installation to a federal repository (Chapter 10) or to a fuel reprocessing plant, are highly radioactive. Even after allowing them to decay for 150 days or more before shipment, the fission products present are still extremely radioactive. They emit intense gamma radiation and also attain a high temperature as a result of the internal absorption of beta particles and some gamma rays from the decay of the radioactive materials.

The essential requirements of a spent-fuel shipping cask are that it should provide radiation shielding sufficient to reduce the exterior dose rates to specifications given earlier, that the surface temperature not exceed the DOT

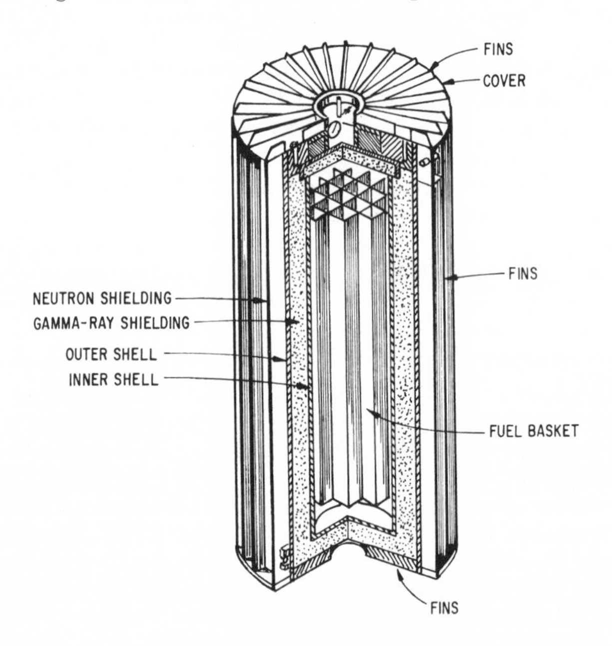

FIG. 11-1. Cutaway diagram of a spent-fuel shipping cask showing the principal components.

permissible value, and that the physical and mechanical properties are adequate to satisfy the conditions specified for Type B packages.

A schematic representation of a cylindrical spent-fuel shipping cask is shown in section in Fig. 11-1; the actual design might vary in some respects from that depicted, but the fundamental principles would be the same. Basically, the shipping cask has a central framework, or "basket," in which the spent fuel assemblies are held firmly. The basket is surrounded by a steel inner shell, about 2.5 cm (1 in.) thick and an outer shell about 3.8 cm (1½ in.) thick; the space between the shells is filled with a gamma-radiation shielding material. The shells are made of special steels that provide adequate strength down to the required temperature of −40 °C (−40 °F). The distance between the shells ranges from roughly 13 to 30 cm (5 to 12 in.), depending on the nature of the shielding material. The latter may be steel, but lead is preferred because it is easily fabricated, has a high density, and is inexpensive; the thickness of a lead shield would be about 23 cm (9 in.).

There is a growing interest in the possible use of depleted uranium metal, made from the residual uranium hexafluoride (tails) of the gaseous diffusion enrichment process (p. 15), to provide gamma-ray shielding in spent-fuel shipping casks. The fact that the metal contains less uranium-235 than normal uranium metal has no detectable effect on its physical and mechanical properties. The thermal conductivity and heat capacity (specific heat) are similar to those of lead; hence, it is equally effective for transferring radioactive decay heat from the fuel assemblies to the outside surface. The mechanical properties of uranium metal also make it suitable for use in a shipping cask.

The main disadvantage of uranium as compared with lead is the higher cost of the material and of its fabrication. On the other hand, for a cask of given capacity, uranium can provide the same gamma-ray shielding as lead with a reduction of 25 percent in weight. Another advantage of uranium is that its melting point (1130 °C [2066 °F]) is considerably higher than that of lead (327 °C [621 °F]); hence, uranium is much less likely to melt if the shipping vehicle should be involved in a fire.

With improvements in the fabrication of reactor fuel materials, it is expected that the fuel assemblies will remain in the reactor for substantially longer times than is now possible; that is to say, the "burnup" potential of the fuel will be increased. This would have the beneficial effect of reducing the cost of electric power. But spent fuel with a high burnup will have accumulated substantial amounts of isotopes of transuranium elements (p. 266). Some of these isotopes, especially curium-244, undergo spontaneous fission at a significant rate, accompanied by the emission of neutrons (see Table 10-II). The radiation shield of the shipping cask must therefore prevent the escape of neutrons, as well as of gamma rays, from the spent fuel. This can be achieved by adding a layer of water or a solid hydrogenous material to serve as a neutron shield outside the normal gamma-ray shield. The total radiation shielding must be such that the sum of the dose rates from gamma rays and neutrons does not exceed the limits imposed by the DOT.

Spent-fuel shipping casks have been designed with diameters from 1.5 to 2.4 m (about 5 to 8 ft) and a length of approximately 6 m (20 ft). The total weight ranges from 25 to almost 100 metric tons (MT) (1 MT = 1000 kg = 2205 lb). The largest casks approved for use in the United States can contain 10 fuel assemblies from pressurized water reactors or 24 of the smaller assemblies from boiling water reactors. The lighter shipping casks are transported by truck; the heavier ones, usually by rail. If the nuclear power plant does not have rail facilities, the heavier casks are carried to the nearest rail siding by trucks, which require special overweight permits. Barge transportation, where available, might have advantages.

Spent-fuel shipping casks generally have fins, which may be straight, running along the length of the cask, or circular, placed around the circumference. In addition, there are radial fins, from the center to the outer edge, at the ends. The fins serve several purposes. In the first place, they help to radiate the decay heat generated by the spent fuel and so prevent the temperature at the outer surface of the cask from becoming too high. Furthermore, by providing what is called a "blocking effect," fins can contribute to the fire resistance of the cask. Another important function of the fins is to act as shock absorbers by their ability to take up considerable amounts of energy if the cask were to be dropped accidentally.

A typical cask of modern design (General Electric Company's IF-300) is shown in Fig. 11-2 mounted on a 100-MT-capacity flatbed for rail transportation. A 1000-megawatt electric (MWe) nuclear power plant would make, on the average, roughly 30 spent-fuel shipments in such casks annually. Radiation shielding is provided by depleted uranium metal for gamma rays and water for neutrons. The outer water layer is retained by a corrugated steel jacket, which is cooled by air from diesel-driven blowers. The corrugations serve the same purpose as fins by helping to radiate heat. There are, however, radial and impact fins at the ends of the cask. The total weight of the cask loaded with fuel elements together with its cooling equipment is about 75 MT.

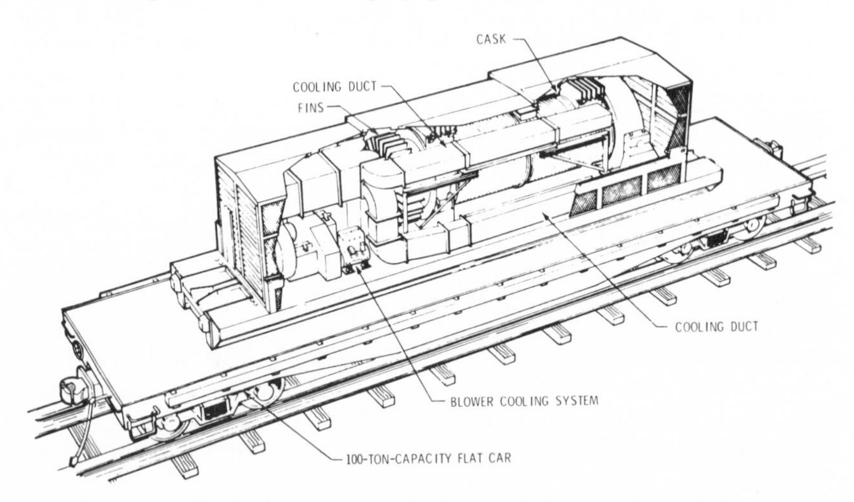

FIG. 11-2. The IF-300 shipping cask for spent fuel mounted on a 100-ton railroad car. (Adapted from a General Electric Company drawing.)

To prevent sabotage or diversion of spent reactor fuel in transit by rail or road, the NRC must be notified in advance of each shipment and the proposed route must be approved. Law enforcement agencies along the route must be informed prior to shipment. The route must avoid heavily populated areas, as far as is possible; special protective measures are required if such areas cannot be avoided. Stops should be kept to a minimum and at least one person must maintain surveillance during stops. The vehicle driver must be accompanied by at least one escort in the transport vehicle or two escorts in a separate vehicle. The transport or separate vehicle must be equipped with a radiotelephone and a CB radio, and calls must be made at least every two hours to advise of the status of the shipment.

Transportation of Solid Wastes

High-Level Wastes. If spent-fuel reprocessing plants should be operated, solid high-level radioactive wastes would be shipped from these plants to a federal repository (Chapter 10). Such wastes would probably be contained in cylindrical steel canisters, roughly 30 cm (12 in.) in diameter and 3 m (10 ft) in length. Several canisters would be loaded into shipping casks, similar in design to the casks used for spent fuel.

Before shipment, a period of several years would have elapsed since the fission products were formed in nuclear reactor fuel; hence, the radioisotopes of shorter half-life would have largely disappeared. Nevertheless, the decay heat would be considerable because of the large proportion of fission products in the waste, relative to that in the spent fuel.[c] It may be required, therefore, to incorporate some form of auxiliary internal or external cooling in the shipping cask. An alternative may be to use an air blower to keep the exterior cool with a continuous flow of air, as in the IF-100 spent-fuel cask. Furthermore, it may be necessary to limit the contents of a single cask in order to meet the DOT's exterior temperature restrictions.

One proposed design is for a cask weighing about 100 MT that can hold from 10 to 12 canisters of solidified wastes. Gamma-ray shielding is provided by making the cylindrical cask of iron, lead, or depleted uranium. This is surrounded by a neutron shield, possibly borated water, in a steel containment vessel, to reduce the neutron dose to an acceptable level. Metal fins are used to dissipate the radioactive decay heat generated in the waste and to keep the accessible surface temperature of the cask below 180 °C (356 °F).

Medium-Level Solid Wastes. Both reactor and fuel reprocesing plants produce a number of radioactive wastes of moderate activity that are occasionally buried within plant site boundaries. As a general rule, however, these wastes are shipped to a licensed burial ground, referred to on page 247. For this purpose, semisolid wastes, such as filter sludges, spent ion-exchange resins, and evaporator residues, are solidified by mixing with cement, with a combination of cement and vermiculite, or with other suitable material. The solidification is

[c]The spent fuel contains about 3 percent by weight of fission products (as oxides), whereas the solidified high-level waste will probably contain 30 to 50 percent or so.

performed in 208-litre (55-gal) drums of DOT-approved design. The drums are usually enclosed in overpacks that qualify as Type B packages for shipment.

Transportation of Liquids

The DOT does not actually require that the wastes be in solid form, but it does stipulate that when liquid radioactive material is to be shipped, it must be contained in or packaged within a vessel with a leak- and corrosion-resistant liner. Furthermore, enough absorbent material must be provided to absorb at least twice the volume of the liquid. The packaging must be adequate to prevent loss or dispersal of the radioactive contents from the inner containment vessel if the package were to be subjected to a drop of 9.15 m (30 ft.)

Transportation of Plutonium Dioxide

The useful products from a spent-fuel reprocessing plant are uranium of relatively low enrichment (less than 1 percent of uranium-235) and plutonium. The transportation of the weakly radioactive uranium does not present a special problem, but the plutonium does. In accordance with the requirements of 10CFR71, Sec. 71.42, plutonium in amounts with activity exceeding 20 curies (Ci) about 60 grams [g] contained plutonium) shall be shipped in solid form. Hence, essentially all the plutonium recovered at fuel reprocessing plants will have to be converted to solid form, probably plutonium dioxide, prior to shipment (p. 234). This regulation became effective on June 17, 1978; however, commercial reprocessing plants, if any are in operation, will not ship any plutonium until the late 1980s.

A further requirement of the regulation referred to above is that plutonium in excess of 20 Ci shall be packaged in a separate container placed within an outer container that meets the requirements described earlier for radioactive (other than LSA) materials. The inner container shall not release plutonium when the entire package is subjected to the accident test conditions applicable to Type B packages, as well as to the conditions applicable to all radioactive material packaging. Reactor fuel elements, plutonium metal or its alloys, and certain other plutonium-bearing solids are exempt from this regulation. The reason for the exemption is not stated, but it is presumably because the plutonium is not in a dispersible form and therefore does not constitute a potential inhalation hazard (Chapter 5).

Risk of Transportation Accidents

There have been extremely few accidents during the transportation of materials associated with nuclear power plants (or with weapons production). As the generation of power from nuclear fission increases, so also will the number of shipments of potentially hazardous materials, such as spent fuel assemblies and perhaps eventually solidified high-level wastes. It is probable that over the years some of these shipments will be involved in accidents, and an estimate of the expected number of such accidents is of interest.

According to DOT accident statistics for the period 1968-70, the frequency of accidents to railroad cars is 0.5 per million car-km (0.8 per million car-miles). For motor carriers of hazardous materials, the accident frequency is about 1.06 per million truck-km (1.7 per million truck-miles). A reasonable (and conservative) assumption, therefore, is that the accident frequency in the shipment of spent fuel and solid high-level radioactive wastes will, on the average, be not more than 1¼ per million vehicle-km (2 per million vehicle-miles). The great majority of these accidents would, however, be less severe that those simulated in the package qualification tests.

Fires occur once in about 50 rail and road transportation accidents. But at least 85 percent are extinguished within 30 minutes, because either the fuel is exhausted or the fire is extinguished by fire-fighting crews. It is expected, therefore, that a package designed to meet the heat test requirement of a Type B package will withstand practically all fire conditions arising from a transportation accident.

It is important to note that more than 99.5 percent of transportation accidents are classified as minor or moderate. Less than 0.5 percent (i.e., less than 1 in 200) is regarded as severe, extra-severe, or extremely severe. Since shipping casks for spent fuel and high-level wastes are designed to withstand severe accidents, involving a substantial impact followed by a prolonged fire, the chances of a transportation accident leading to the release of a significant amount of radioactivity is extremely small.

The Radiological Assistance Program

In the event of a collision, derailment, turnover, fire, package leakage, or other accident occurring during the transportation of radioactive material, the shipper is required to notify the DOT immediately. Steps must be taken to segregate persons from the material and to conduct such radioactive decontamination of people, the vehicle, and the area as may be necessary. Trained personnel are available upon request through the Radiological Assistance Program, operated by the U.S. Department of Energy (formerly the AEC), to provide assistance or advice in handling radiological incidents. This program coordinates the activities in this connection of about a dozen U.S. government agencies in accordance with the Interagency Radiological Assistance Plan of 1961.

To implement the Radiological Assistance Program, eight regional offices, covering the entire United States, have been established. Each office maintains a continuously manned emergency telephone. If the incident is a minor one, advice can be given by telephone. If not, a radiological assistance team is dispatched to the scene of the accident. The mission of the team is to advise the authorities at the scene of the radiological hazards and of the best means for reducing those hazards by appropriate decontamination procedures. Advisory assistance is continued at the scene until the immediate hazard to public health and safety has been evaluated, and the appropriate local authority can assume responsibility. Several individual states and some larger cities have developed their own capabilities for responding to radiological emergencies. However, advisory help from the Interagency Radiological Assistance Program is always available.

SAFEGUARDS FOR SPECIAL NUCLEAR MATERIALS

Introduction

The term *safeguards* refers to "the security measures undertaken by a licensee [of a nuclear plant] to protect strategic special nuclear material [defined below] from loss, theft, or diversion and to prevent sabotage of a licensee's facilities or activities." Special nuclear material (SNM) is defined by the Atomic Energy Act of 1954 as, principally, "plutonium, uranium enriched in the isotope 233 or in the isotope 235." If the SNM includes uranium-235 (contained in uranium enriched to 20 percent or more of this isotope), uranium-233, or plutonium, or in any combination in a quantity such that (kilograms contained ^{235}U) + 2.5 (kilograms ^{233}U + kilograms Pu) is equal to or more than 5 kg, it is frequently referred to as strategic special nuclear material (SSNM).[d]

The NRC regulations in 10CFR73, entitled "Physical Protection of Plants and Materials," and the Regulatory Guides, Div. 5, entitled "Materials and Plant Protection," deal largely (but not exclusively) with the safeguarding of SSNM at fixed sites, where the materials are produced, used, or stored, and in transit from one site to another. The aspect of safeguards concerned with the prevention of sabotage was discussed at the end of Chapter 4 and treatment here deals with the protection of SSNM from loss, theft, or diversion. It appears that, because public transportation routes must be used for shipment of SSNM, these shipments are inherently more vulnerable to theft than at fixed sites. Hence, the regulations for the security of SSNM in transit, which are given in 10CFR73, Sec. 73.30, will be reviewed first. Since most reactor fuels do not qualify as SSNM, they are exempt from the regulations.

Protection in Transit

When shipped by road or rail, SSNM shall be placed in a container sealed with a tamper-indicating seal. Furthermore, the container shall be locked unless it is enclosed in another container or vehicle that is locked. The outermost container or vehicle shall also be sealed by a tamper-indicating seal.

Shipment by road shall be made in a truck or trailer specially designed to reduce vulnerability to theft. Two acceptable approaches to the design of such vehicles are described in NRC Regulatory Guide 5.31, entitled "Specially Designed Vehicles with Armed Guards for Road Shipment of Special Nuclear Materials." One method is to use a cargo vehicle constructed to resist entry for a substantial period of time by unauthorized persons. The other is to utilize an armored car. Resistance to unauthorized entry is provided by armed guards within the vehicle; the guards are protected by the construction of the armored vehicle.

The road shipment shall be accompanied by at least two armed guards in the cargo vehicle, and shall be escorted by one or two vehicles, each with at least two guards. If communication with the shipper (or his agent) by radiotelephone or radio is possible at all times during transit, only one escort vehicle is required;

[d]The reason for calling this *strategic* material is that in sufficient amount it could conceivably be used to make a nuclear explosive.

otherwise, two such vehicles shall be used. Radiocommunication capability shall be provided between cargo and escort vehicles; in addition, all vehicles shall be equipped with radiotelephones. At rest stops, the cargo vehicle shall be under continuous surveillance by at least two armed guards.

During the course of the journey, calls shall be made to the shipper (or his agent) at least every 2 hours, when radiotelephone or conventional telephone coverage is available, to relay the location of the vehicle and the projected route. The call frequency may extend to 5 hours when this coverage is not available along the preplanned route. If no call is received, the shipper must immediately notify an appropriate law enforcement authority and the NRC local Inspection and Enforcement Regional Office.

A shipment by rail shall be escorted by five guards in the shipment car or an escort car of the train. The guards shall keep the shipment car (or cars) under observation, and they shall detrain at stops, when time permits, to guard the shipment car and check locks and seals. Radiotelephone communication shall be maintained with the shipper (or his agent) at least every two hours, and at scheduled stops if a means of communication has not been available in the preceding five hours. In the event that no call is received, the same notification steps must be taken as for shipment by road.

A regional NRC office must be informed at least seven days in advance of a planned SSNM shipment and also immediately upon arrival of the shipment at its destination.

Protection at Fixed Sites

At fixed sites where SSNM are stored or processed, the physical security requirements, as stated in 10CFR73, Sec. 73.50, are essentially the same as those for protection against possible sabotage (p. 108). In addition, the regulations (Sec. 73.60) require that the material be stored or processed only in a "material access area" devoted exclusively to these purposes. Material access areas shall be located within a protected area to which access is controlled. Material not in process shall be stored in a vault or vault-type room equipped with an intrusion alarm. Each such vault shall be controlled as a separate material access area.

Admittance to a material access area shall be limited to authorized persons and vehicles. Intrusion alarms would indicate unauthorized entry. Packages taken into the area shall be searched for firearms, explosives, etc., and also for substitute items that could be used as counterfeit to permit theft or diversion of SNM. Upon exit from a material access area, packages and vehicles shall be searched for concealed SSNM, either by direct observation or by means of special detectors. Methods for observing individuals within material access areas must be used on a continuing basis to assure that SSNM is not being diverted.

Material Control and Accounting

Additional protection against diversion of SNM of all types is provided by a comprehensive program of internal material control and accounting.[e] This pro-

[e]The details are given in 10CFR70, entitled "Special Nuclear Material," Secs. 70.51-70.58.

gram is designed to detect any loss and to take timely recovery action should such a loss occur. The basic elements of the accounting system are similar to those used in financial accounting; they include double-entry accounting methods, recording of material transfers, periodic physical inventories, and audits. Physical inventories based on actual measurements are required to agree with material balances as given by the accounting records. Plutonium, uranium-233, and uranium highly enriched in uranium-235 are inventoried every two months. For less sensitive SNM, a semiannual physical inventory is required.

The NRC Office of Nuclear Reactor Regulation operates an inspection and enforcement program that includes among its activities the supervision of material control and accounting. Inspection of facilities where SNM is used or stored provides assurance that control and accounting procedures are satisfactory. Actual measurements are made to check the validity of these procedures. The inspectors confirm that physical protection, material control, and records are in compliance with NRC regulations.

It should be mentioned, in conclusion, that the NRC's program for safeguarding nuclear material is under continuous study with the objective of developing improvements. Among the areas being studied are (a) alternative approaches to the provision of security against diversion of SNM, (b) techniques and devices for improving physical protection, and (c) improvement in materials measurement and accounting.

Bibliography—Chapter 11

Brobst, W. A. "Transportation Accidents—How Probable?" *Nuclear News* 16, no. 5 (1973):48.

______. "Transportation of Nuclear Fuel and Wastes." *Nuclear Technology* 24 (1974): 343.

International Atomic Energy Agency. *Proceedings of the Symposium on Safeguarding of Nuclear Materials.* Vienna: International Atomic Energy Agency, 1976.

Langhaar, J. W. "Casks for Irradiated Fuel: A Look at the Cask Designers' Guide." *Nuclear Safety* 12 (1971): 553.

Lovett, James E. *Nuclear Materials—Accountability Management Safeguards.* La Grange Park, Ill.: American Nuclear Society, 1974.

Meyer, W., *et al.* "The Homemade Nuclear Bomb Syndrome." *Nuclear Safety* 18 (1977): 427.

Rhoads, R. E., and Johnson, J. F. "Risks in Transporting Materials for Various Energy Industries." *Nuclear Safety* 19 (1978): 135.

"Safeguards Against the Theft and Diversion of Nuclear Materials." *Nuclear Safety* 15 (1974): 513.

Smalley, W. L. "The ERDA Radiological Assistance Program." *Nuclear Safety* 16 (1975): 345.

U.S., Atomic Energy Commission. *Environmental Survey of Transportation of Radioactive Materials to and from Nuclear Power Plants* (WASH-1238). Washington, D.C., 1972.

U.S., *Code of Federal Regulations*, Title 49. (Transportation). Washington, D.C. Radioactive Materials, Parts 173.369-173.399.

______. Title 10 (Energy). Washington, D.C.

Part 71. Packaging of Radioactive Material for Transport and Transportation of Radioactive Material Under Certain Conditions.

Part 73. Physical Protection of Plants and Materials.

U.S., Nuclear Regulatory Commission. *Environmental Survey of the Reprocessing and Waste Management Portions of the Fuel Cycle* (NUREG-0116), Sec. 4.9, Annex 4.10A. Washington, D.C., 1976.

______. *Materials and Plant Protection* (Regulatory Guides Div. 5). Washington, D.C.

______. Physical Protection of Irradiated Reactor Fuel - Interim Guidance (NUREG-0561). Washington, D.C., 1979.

Willrich, M., and Taylor, T. B. *Nuclear Theft: Risks and Safeguards.* Cambridge, Mass.: Ballinger Publishing Company, 1974.

12

Biological Effects of Condenser Cooling Systems

INTRODUCTION

General Considerations

An explanation was given in Chapter 1 of why in all steam-electric power plants, no matter whether they use fossil or nuclear fuel, waste heat must be removed by the cooling water in the turbine condenser. It was seen that the simplest way to dispose of this heat is to discharge the warmed water leaving the condenser in a once-through system. Where this is not feasible, for such reasons as the potential harmful effects on organisms living in the receiving water body, the limited supply of cooling water, or the inability to meet the water quality standards for the discharge, alternative (closed-cycle) systems with cooling towers or ponds are required (Chapter 13).

In a closed-cycle cooling system, the warmed condenser water is not discharged, as it is in a once-through system, but it is cooled and recirculated. Nevertheless, makeup water has to be supplied in a closed cycle to compensate for evaporative (and other) losses and for water discharged and replaced by fresh water to prevent accumulation of dissolved chemicals. Hence, regardless of the type of cooling system, a certain amount of water is taken in from a water body, passed through the condenser, and eventually discharged.

Various effects on aquatic organisms are associated with each of these aspects of the operation of a condenser-cooling system. Since the volumes of water involved are much greater for a once-through system than in a closed-cycle system, so also is the potential for biological damage. The basic principles, however, as discussed in this chapter, are the same in all cases. Biological effects that have received major attention are those associated with water temperature changes; therefore, they are treated here in some detail. It is shown how studies of these effects have led to the derivation of appropriate water quality criteria for the protection of aquatic organisms.

Sources of Biological Damage

An aquatic ecological system (or ecosystem) is highly complex and a disturbance usually leads to a number of interacting effects. It is not always possible, therefore, to associate a particular observed effect with a specific cause. Some understanding of the situation may be obtained, however, by considering the following individual factors that may cause damage to organisms living in a body of water that supplies cooling water to and receives the discharge from a power plant condenser:

1. changes in the water temperature which may have both direct and indirect effects (e.g., on metabolism, growth, disease, predation, etc.) on aquatic life forms.
2. mechanical, temperature, and other effects arising from impingement of larger organisms (e.g., fish) on water-intake screens and entrainment of smaller organisms that are not held back by the screens but pass through the condenser.
3. chemicals (e.g., chlorine) used as biocides to control slime formation on the condenser tubing and other chemicals from water treatment.
4. changes in the water quality (e.g., in the concentration of dissolved oxygen, chlorine, and salts in the water), arising directly or indirectly from the operation of the condenser-cooling system.
5. changes in the natural salt distribution and content (e.g., salinity) in the receiving water body.

These sources of potential damage and their effects are examined in later sections, but first it is necessary to explain the names used to describe and classify various aquatic organisms and their relationships to one another.

Classifications of Aquatic Organisms

A natural body of water is an ecosystem that is normally influenced by many factors, including climatic conditions (e.g., solar radiation intensity and water temperature) and the chemical composition of the water (e.g., salinity). The organisms present in an aquatic ecosystem may be classified in various ways, two of which are described here.

In the first place, the organisms may be classified according to their habitat (i.e., where they live) and mobility in the following broad categories: nekton, plankton, and benthos. The *nekton* are the free-swimming organisms of which fish (actually finfish) are the most important. The *plankton* consist of the animal and plant life forms that float on or near the surface or drift with the water current, whereas the *benthos* live on or in the bottom material.

The plant forms of the plankton, called *phytoplankton*, include various algae and bacteria; the animal forms, or *zooplankton*, include many minute organisms, such as protozoa (single cells), rotifers, copepods (small crustaceans), and the larval stages of finfish and shellfish (mollusks and crustaceans). The benthos is made up of plants, mollusks, larger crustaceans, worms, and insect larvae. Epibenthic organisms have some of the characteristics of both plankton and benthos; they spend the day on the bottom but may move up into the water layer at night.

Another classification of aquatic life forms is in terms of their functions in the *food cycle* or *food web*. The organisms are categorized as decomposers, primary producers, and consumers. The *decomposers* are mainly bacterial communities that decompose organic matter, such as dead organisms, and make the material available for use by the producers and consumers. The bacteria themselves also serve as food for much of the microscopic zooplankton. The decomposers are found in all parts of the aquatic ecosystem wherever dead organic material is available.

The main *primary producers* are the algae (phytoplankton), which live in the upper layers of the water that are penetrated by sunlight; they utilize solar energy to convert water, carbon dioxide, and minerals into complex organic compounds, such as sugars, proteins, etc. The *consumers* are the animals that must obtain their food from other living organisms; the most important consumers are zooplankton, benthic animals, and fish.

The smaller zooplankton consume phytoplankton, bacteria, and detritus; they convert them into food more readily utilized by the larger life forms, such as finfish and shellfish. Many fish also feed on the larger zooplankton, and larger fish often eat smaller fish. The detritus, consisting of organic matter (and associated bacteria) from inside and outside the water body, sinks to the bottom and provides food material for the benthos. Eventually all dead matter is decomposed by bacteria, thereby completing the food cycle (Fig. 12-1).

THERMAL EFFECTS

Introduction

Temperature is a particularly important factor governing the occurrence, metabolism, and behavior of aquatic organisms. It may not only affect the distribution of a single species, but may modify the species composition of an ecosystem. Aquatic organisms have upper and lower temperature tolerance limits, optimum growth temperatures, preferred temperatures (where temperature variations exist), and restricted temperature limits for migration, spawning, and development. Temperature also has indirect effects, such as those caused by changes in the dissolved oxygen concentration and in the toxicity of some chemicals.

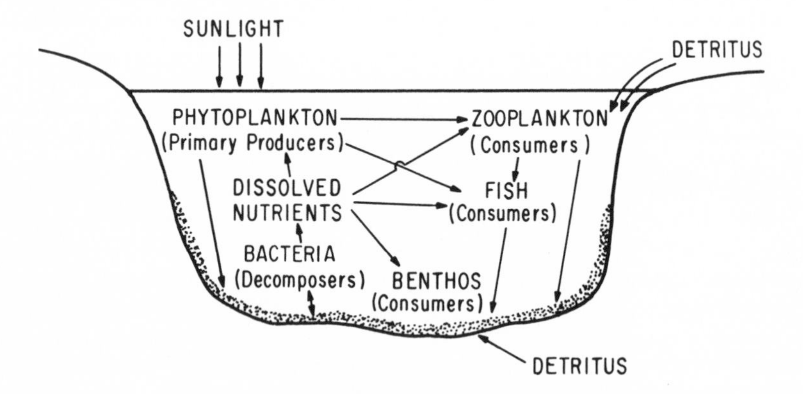

FIG. 12-1. Simplified outline of aquatic food web.

Numerous studies have been made of the effects of temperature on aquatic life, both in the laboratory and in the field. From the many laboratory observations, certain conclusions have been drawn concerning the expected effects of the discharge of heated water. These expectations have, however, not always been borne out in fact. The difficulty arises from trying to predict the effects in a complex natural environment from observations made under controlled laboratory conditions. Moreover, a distinction should be made between short-range (i.e., immediate) and long-range (i.e., later) effects. The short-range consequences, as judged from laboratory studies, may appear to be detrimental, but resulting ecological adjustments over the course of time may prove to be less harmful than expected. On the other hand, the short-range effects may appear to be insignificant, but over the years there may be serious changes.

Despite the many uncertainties, a number of general results have emerged from laboratory and field studies of the effects of temperature on aquatic organisms in various stages of their life cycles. Because of the complex interactions that occur in nature, the results described in the following sections are often modified. Nevertheless, they form the essential basis for the development of thermal criteria of water quality.

Thermal Effects on Decomposers

The temperature of most natural waters, even in summer, is below the optimum at which many bacteria multiply most rapidly. Hence, provided there is an adequate food supply, an increase in water temperature will usually increase the growth rate of decomposers. If the number of bacteria in the water body is less than the carrying capacity, as determined by the available food, the heated discharge from a power plant will tend to favor increased growth during most of the year.

For various reasons, such as depletion of oxygen, however, an increase in temperature may reduce the carrying capacity of the water, and this may result in a net decrease in the growth rate of decomposers. On the other hand, if the temperature increase results in the availability of more dead organic matter (e.g., fish or other organisms killed by power plant operation), the crop of bacteria could increase substantially. There may also be shifts in the species composition of the bacteria with changes in temperature, like those that normally occur with changing seasons. Limited observations indicate that, in the absence of chlorine, condenser cooling discharges have little overall effect on the crop of decomposers in the aquatic ecosystem.

Thermal Effects on Producers

Algae constitute the primary producers and under appropriate conditions each species has a particular temperature range of optimum growth, photosynthesis, and reproduction. This is shown in a generalized (not exact) manner for three freshwater algae groups in Fig. 12-2. The diatoms dominate at water temperatures up to about 27 °C (80 °F), whereas the green algae are most abundant between roughly 29 to 35 °C (85 to 95 °F). If other requirements are met, blue-green algae become the predominant type above 35 °C (95 °F). The latter, however, may be undesirable; some species are known to be poisonous to animals, but mass mortality of fish observed after decomposition of blue-green algae has probably been due to a reduction in dissolved oxygen.

The curves in Fig. 12-2 are based on laboratory observations of attached algae under controlled conditions. In nature, however, the environment is variable, and the temperature tolerance ranges may differ from those shown in the figure. For example, blue-green algae will dominate at lower temperatures in a lake rich in nutrients (eutrophication). Nevertheless, it is generally true that diatoms and green algae are the dominant producers at temperatures below about 29 °C (85 °F). It is only at higher water temperatures, which might be experienced in discharge canals or near the discharge point of the heated condenser water (and along shorelines) that the undesirable blue-green algae would be prevalent. Hence, as a rule, warm water discharges are likely to have only a localized effect on the algae population, provided the temperature of the receiving water body does not rise above 29 °C (85 °F).

Thermal Effects on Zooplankton and Benthos

Although animal plankton (i.e., zooplankton) include a wide variety of aquatic species, some general conclusions can be drawn concerning the effects of water temperature. There is an optimum temperature at which the rate of population growth is a maximum. Hence, an increase above a certain temperature will result in a depletion of the animal plankton; moderately high water temperatures, possibly above 30 °C (86 °F), are required for this to be significant. As a general rule, temperature has only a secondary effect on the zooplankton population. Animal plankton are often entrained with the condenser intake water and are subjected to both mechanical and thermal shock in passing through the condenser. The effect of entrainment is considered later.

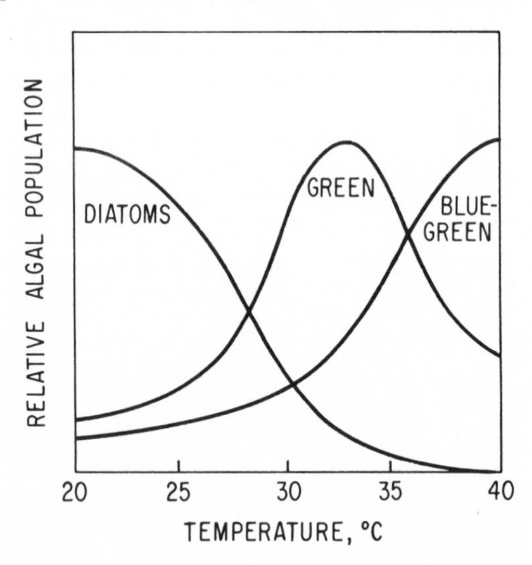

FIG. 12-2. Changes in algal populations with temperature. (Adapted from J. Cairns, *Industrial Wastes*, 1 [1956] 150.)

Some studies of thermal effects on the benthos have been made in connection with the condenser discharge from steam-electric power plants. During the summer and early autumn, when the normal water temperature is relatively high, the extra heat added near the discharge point may result in substantial reductions in the numbers and diversity of benthic organisms. At increasing distances downstream, as the temperature approaches normal, so also does the benthic population. During the cooler months, there is often a recovery in the areas affected by the condenser discharge in the summer.

Thermal Effects on Fish

Fish are among the important ultimate consumers and numerous studies have been made of the effects of water temperature on fish at various stages of their life cycle. Fish and other cold-blooded animals differ from warm-blooded animals in an important respect. The latter have various temperature-regulating mechanisms that permit their body temperature to remain essentially constant, regardless of the temperature of the surroundings. In nearly all cold-blooded animals, on the other hand, the body temperature does not remain steady but tends to approach that of the environment.

The metabolic processes involved in the building up and functioning of body cells are basically chemical reactions, commonly referred to as biochemical reactions. Like almost all chemical reactions, the rates at which the biochemical reactions occur are, within limits, related to the body temperature. In a warm-blooded animal, biochemical reactions take place at an essentially constant rate, since the body temperature remains almost constant. In fish, however, the rates of biochemical reactions increase with increasing temperature of the aquatic environment, provided this temperature is below the optimum for the species (see below).

It is reasonable to expect that there will be an upper limit to the rates of biochemical reactions as the temperature is raised. One reason is that enzymes, which play an important role in all life processes, become inactive above a certain temperature, and cell protoplasm tends to coagulate. In addition, with increasing temperature, hemoglobin in the blood becomes less effective in delivering oxygen to the tissues. There is consequently a limiting upper temperature at which fish can survive. There is also a lower temperature limit because biochemical reactions and blood flow may become so slow that life cannot be sustained.

In nature, fish can become acclimated to living at different temperatures within these limits; thus, the same fish species may be found at different seasons and in different climatic regions. For arctic and tropical (or semitropical) fish, which live in very cold and warm waters, respectively, the temperature range within which they can be acclimated is quite narrow. Many temperate freshwater and estuarine fish, however, are found to tolerate a wide range of temperature.

Observations in the laboratory have shown that acclimation of fish to living at a higher or lower temperature, within their tolerance limits, is possible when the temperature of the water is changed slowly. Rapid changes in temperature can, however, prove fatal.

Effect of Temperature on Fish Growth Rates

Within a certain temperature range, an increase in the water temperature generally results in an increase in the rate of growth of fish, provided sufficient food is available. At a certain temperature, which depends on the type of fish, the age, and several other variables, the growth rate attains a maximum. At increasingly higher temperatures, the fish grow more and more slowly as the temperature is raised. This growth pattern appears to be quite general for all species of fish that have been studied.

Figure 12-3 shows the effect of the water temperature on the fraction of the maximum size reached within a certain time for brook trout, northern pike, and largemouth bass. This fraction is a direct measure of the rate of growth at different temperatures. Both the growth rate and the maximum size are dependent on the fish species, but this does not affect the general conclusions. The optimum growth rates occur at temperatures of roughly 15 °C (59 °F), 21 °C (70 °F), and 27 °C (81 °F) for brook trout, northern pike, and largemouth bass, respectively. It is clear that brook trout grow better in cooler water, whereas largemouth bass increase in size more rapidly in warmer water.

It appears that, as a possible result of a substantial increase in the temperature of a stream, brook trout might be replaced by more thermally tolerant species, such as northern pike or varieties of bass, if the temperature were high enough. This assumes, of course, that stocks of these fish are available, that the conditions for reproduction and growth are satisfactory, and that the appropriate food is present. It has been reported that, with increasing temperature of the water, more desirable game (or sport) fish may be replaced by less desirable fish which tolerate higher temperatures. For example, carp and

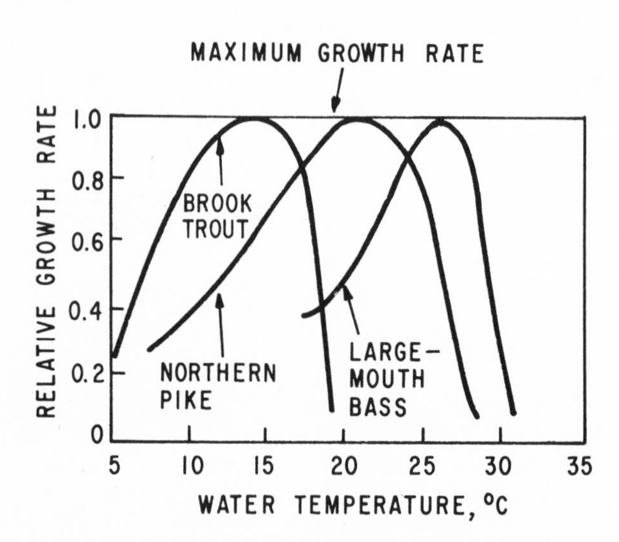

FIG. 12-3. Effect of water temperature on growth rate of brook trout, northern pike, and largemouth bass. (Adapted from D. I. Mount, *Environmental Effects of Producing Nuclear Power*, hearings before the Joint Committee on Atomic Energy, U.S. Congress, Washington, D.C. [1969].)

carpsucker are commonly found in warm ponds, lakes, and rivers. However, some sport fishermen may regard pike, carp, or bass, which thrive in warmer waters, as highly as others regard trout.

Although fish tend to grow faster within a certain temperature range, there are some limitations that must be borne in mind. In the first place, the observations have generally been made in the laboratory where ample quantities of food and oxygen are provided. In nature, this situation does not necessarily hold. Moreover, there are many variables in nature that do not arise under laboratory conditions.

Temperature Preference by Fish

The ability of fish to become acclimated to the temperature of the surroundings, within limits, is an example of protective reaction that is common among living organisms. Another instance of such behavior is the avoidance reaction or temperature selection. If fish have a choice of water at different temperatures, they will usually prefer a certain temperature and avoid others. (Lists of upper and lower avoidance and preferred temperatures for many fish species have been published.) There is thus an innate tendency for the fish to remain in water whose temperature is within their preferred range.

Many cases of migration based on temperature preference have been observed among fish. In estuarine and coastal waters, local and seasonal variation in temperature are quite large. Resident fish then migrate between colder and warmer regions according to circumstances. Similar migration occurs in lakes having layers of different temperature in summer (p. 347). Temperature preference (or avoidance) has been found to occur in nearly every fish species among the many—almost a hundred—that have been tested.

Fish are generally drawn to water of higher temperature during the cooler months, provided food and shelter are available and the current velocity is acceptable. An interesting example, among many, has been reported in connection with the discharge of heated water from the condensers of the coal-burning steam-electric power plant at New Johnsonville, Tennessee. The water is drawn from Kentucky Lake and is discharged into a harbor from which it eventually returns to the lake. During the winter the temperature of the water in the harbor was found to be about 5.5 °C (10 °F) above that of the lake water. Large numbers of shad were then attracted by the warmer water in the discharge harbor. After mid-March, when the water temperature approached 15.5 °C (60 °F), the shad became scarce but they were replaced by bluegill and channel catfish. In July, when the temperature in the harbor was above 27 °C (80 °F), the catfish returned to the lake, where the temperature was a few degrees lower.

Although fish normally exhibit a preference for a particular temperature range, to which they have become acclimated, they will voluntarily endure higher temperatures for a short period in special circumstances. In a natural lagoon in Pennsylvania, the temperatures in the summer range from 32 °C (89 °F) in the coolest parts to 41 °C (106 °F) in the warmest regions. Brown bullheads in the lagoon tended to stay in the zone at about 32 °C (89 °F). But when worms were thrown into the water where the temperature was 40 °C (104 °F), the fish would enter the warmer region to capture the worms, but would immediately return to the cooler water.

As a general rule, fish will not remain in a potentially harmful temperature environment if there is a chance to escape. Given the opportunity, most fish will successfully avoid excessive temperatures and will move into a region that is favorable to the particular species. They can do this by taking an appropriate escape route, which may be alongside or beneath a layer of heated water in a river, or by descending to a lower depth in a reservoir or lake. On the other hand, a sudden drop in the temperature of the artificially warmed water to that of the environment can have disastrous consequences even if the fish do escape (p. 324).

Thermal Effects on Reproduction

The preferred temperatures for spawning and egg development of many fish species are about 5.5 °C (10 °F) or more below those at which the fish grow and mature. This is illustrated by the fact that in the temperate zone fish commonly spawn in the autumn or in the late spring. However, this behavior is not universal; some fish species, such as the fathead minnow, which is an important source of food for larger fish, can apparently spawn throughout the summer. For fish that require lower temperatures for spawning, such temperatures must be available at the appropriate time of the year if the fish are to propagate.

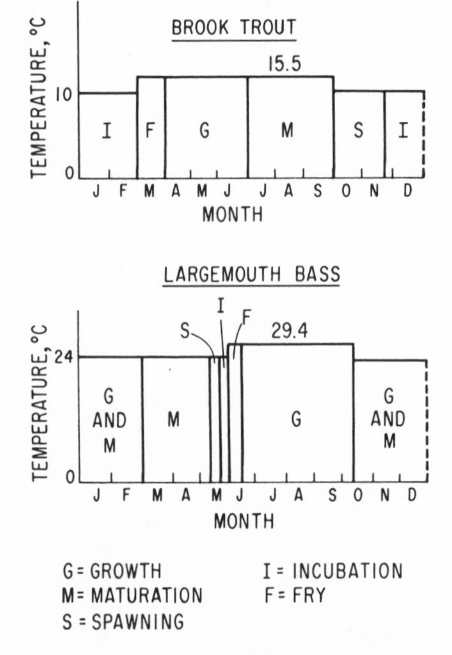

FIG. 12-4. Preferred temperatures for various life stages of brook trout and largemouth bass. (Adapted from D. I. Mount, *Environmental Effects of Producing Nuclear Power*, hearings before the Joint Committee on Atomic Energy, U.S. Congress, Washington, D.C. [1969].)

Brook trout spawn in the autumn, when water temperatures are generally falling. The metabolic processes in the eggs then occur fairly slowly and the incubation period is relatively long. The fry develop in March, and they grow during the spring and summer, when ample food becomes available. Yellow perch and largemouth bass exhibit different behavior; they spawn in the spring, and the incubation periods are quite short. Growth occurs in the summer and early autumn, when the natural food supply is most abundant.

The preferred temperatures for brook trout and largemouth bass during different stages of development are shown in Fig. 12-4. The actual water temperatures are generally lower in the winter and spring months and are sometimes higher in the summer. Brook trout require most of their food in late spring and early summer, while they are growing most rapidly. Largemouth bass (and yellow perch), however, normally grow fastest in the spring, summer, and early autumn. The different preferred temperatures and food requirements of different fish species account for the regions in which the fish are commonly found.

Within limits, fish can become acclimated to spawning at somewhat higher and lower temperatures than the preferred value. The range between the lower and upper limits, however, is small (p. 314), and the sequence of events involved in reproduction is considered to be the most thermally restrictive aspect of the fish life cycle. Outside the narrow limits, which are characteristic of each fish species, reproduction does not occur.

In the temperature range where spawning is possible, fish will often spawn prematurely in a warm condenser discharge. Furthermore, at the higher temperature, the eggs will incubate more rapidly than at lower temperatures. Consequently, the fry may appear at an earlier date when appropriate food sources have not yet developed. If this occurs, the ultimate effect of an increase in temperature would be the same as if the adult fish had been killed directly or had not spawned at all.

An indirect aspect of the effect of water temperature on spawning arises in connection with fish, such as salmon and shad, that ascend a river in early spring to spawn. If the migration is blocked by a zone of heated water, the fish would probably not move up to a preferred spawning area, thus reducing or eliminating a spawn.

Thermal Shock

Fish living in water at a certain temperature can become acclimated to higher or lower temperatures, within certain limits, provided the temperature changes occur slowly. For acclimation to occur, the heating rate should not exceed about 1.1 °C (2 °F) per day, and the cooling rate should be even less. The situations to be considered here are those in which the water temperature changes to a higher or lower temperature so rapidly that normal acclimation does not occur. The fish are then said to be subjected to *thermal shock*. In the absence of a refuge to which the fish can escape, their survival will depend to a great extent on the magnitude of the difference between the new temperature and the temperature to which the fish had been acclimated.

The general consequences of thermal shock are indicated diagrammatically in Fig. 12-5. It shows in a semiquantitative manner the response of fish to a fairly rapid transfer from a previous acclimation temperature to a new environmental (or exposure) temperature. The "fairly rapid" temperature change may occur over a period of up to a few hours without greatly affecting the conclusions.

The figure shows three main regions: a central *thermal tolerance zone* and upper and lower *thermal resistance zones.* (As used here, the term *resistance* is meant to imply the opposite of *tolerance.)* In the interior of the tolerance zone, the chances of fish surviving a temperature change are very good, but they decrease as the resistance zones are closely approached (and entered). The boundaries between the tolerance and resistance zones are called the upper and lower *incipient lethal temperatures.*

The incipient lethal temperature is defined as the environmental temperature at which 50 percent of a fish population will be killed within a short time if they are brought fairly rapidly to this temperature from a specified acclimation temperature. In the thermal tolerance zone, the chance of survival (or survival time) increases as the conditions move away from the incipient lethal temperature. Thus, within the tolerance zone, there is a smaller (i.e., more restrictive, rhombus-shaped region within which normal activity and growth will occur. An even smaller zone represents the conditions within which normal spawning is possible.

In the thermal resistance zone, the chance of survival (or survival time) decreases as the environmental temperature exceeds the incipient lethal value for the particular acclimation temperature. At a sufficiently high or low environmental temperature, death from thermal shock is virtually instantaneous.

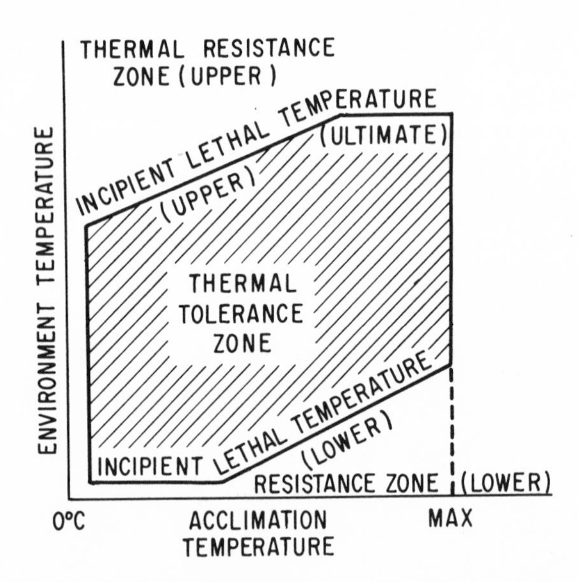

FIG. 12-5. Thermal tolerance and resistance zones for a typical (hypothetical) fish species. (Adapted from F. E. J. Fry *et al.*, *University of Toronto, Biological Series* 54 [1946].)

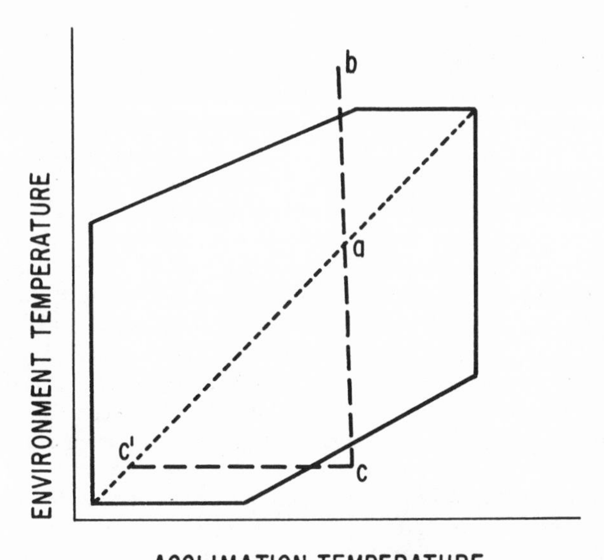

FIG. 12-6. Conditions for lethality or survival of fish as the result of a water temperature change.

Actual temperature values are not given in Fig. 12-5. One reason is that the temperatures at which various effects occur will depend on the fish species (and age), although the general consequences of a fairly rapid change in temperature appear to be similar for all species. The ultimate (or highest) upper incipient lethal temperature is always the same as the maximum acclimation temperature; observed values for common freshwater species range from 38 °C (100.4 °F) for channel catfish to 23.5 °C (74.3 °F) for brown trout. The ultimate (or lowest) lower incipient lethal temperature for most fish living in temperature latitudes is near the freezing point of the water (i.e., close to 0 °C [32 °F]). For more southerly latitudes, it may be a few degrees higher.

As a specific example of thermal shock, the results may be quoted of a laboratory study of brook trout. When the fish were acclimated to a temperature of 15 °C (59 °F), and the temperature of the environment was increased to 23 °C (73 °F) over a period of a few hours, 50 percent of the fish died within about a week. But if the environmental temperature was increased to 28 °C (82 °F) at about the same rate, 50 percent died within an hour. The ultimate upper incipient lethal temperature, beyond which these brook trout could not be acclimated, was found to be about 24 °C (75 °F). Almost instantaneous death occurred at a temperature of approximately 30 °C (86 °F).

Certain general conclusions can be drawn from Fig. 12-5, as shown in Fig. 12-6. At all points on the diagonal line the acclimation and environmental temperatures are equal; in other words, a water temperature on the diagonal means that the fish are acclimated to their environment. Consider a point, such as *a*, where this condition exists, and suppose there is a fairly rapid change in the environmental temperature that does not allow the fish to become acclimated.

The temperature change will then be represented by a vertical line starting from *a;* for example, *ab* represents a rapid increase and *ac* a rapid decrease in environmental temperature.

In the particular case under consideration, points *b* and *c* lie in the upper and lower thermal resistance zones, respectively; hence, more than 50 percent of the fish would be killed in a short time by *heat shock* or *cold shock*, respectively. If the new environmental temperature had been within the tolerance zone, there would have been few deaths, if any, as a result of thermal shock.

In the foregoing discussion it has been assumed that the environmental temperature changes rapidly. Suppose that, at the other extreme, the temperature changes so slowly that the fish become continuously acclimated to the changing environmental temperature. The conditions would then always be represented by a point on the diagonal line. Consider, for example, a very slow decrease in temperature from *a* to *c'*, where *c'* represents the same environmental temperature as *c*. All the fish would now be expected to survive the temperature change. An equivalent situation would arise from a slow increase of temperature.

The survival probability in a fish population subjected to a change in the environmental temperature thus depends not only on the magnitude of the change, but also, to a considerable extent, on the rate at which the change takes place. For a decrease in temperature from *a* to *c*, for example, the final state will lie somewhere along line *c'c*. The chances for survival are very good for a slow temperature decrease (final state close to *c'*), but may become smaller and smaller as the decrease is more and more rapid (final state approaches *c*).

Indirect Thermal Effects

Apart from the direct effects of temperature on aquatic life forms, there are some indirect effects that may be harmful. Fish often have a greater susceptibility to disease and to attack by parasites at higher temperatures, even when they have become acclimated to these temperatures. At the higher (and also lower) temperatures, fish are often less active in catching smaller fish on which they depend for food.

Adequate food and oxygen are, of course, essential to fish during their lifetime. If the water temperature is not too high, an increase in temperature can result in an increased metabolic rate and, hence, in an increased oxygen demand. In most circumstances, a moderate increase in the water temperature does not have any serious consequences in this respect. An exception arises, however, if the water is signifcantly polluted (e.g., by domestic sewage or other organic matter). The supplies of food and oxygen may then be limited, and an increase in temperature can cause harm to the fish population.

Temperature can also have a synergistic effect on the toxicity of chemicals, such as chlorine, heavy metals, and pesticides, present in water. These chemicals often become more toxic to aquatic organisms as the water temperature is increased.

Another indirect thermal effect on fish is *gas embolism*, commonly referred to as *gas-bubble disease*. The solubility of air in water decreases with increasing temperature (p. 320). But if water already saturated with air is heated rapidly, the excess dissolved air may not be released, and the water becomes supersaturated. When fish are exposed to this water, the air passes through the gills and is taken up by the blood. The excess dissolved air is then released as bubbles, mainly of nitrogen gas. The resulting embolisms (i.e., blockages of blood vessels by bubbles) cause the fish to die.

OTHER EFFECTS OF CONDENSER SYSTEMS

Effects of Impingement

Condenser cooling water intakes are provided with an arrangement of screens to prevent the entry of debris (which could cause damage to the pumps) and of fish, which likely would be killed. A typical system of the kind used in many nuclear power plants is depicted in Fig. 12-7. The outer screen, called the trash rack, is a stationary grating with bars about 8 to 10 centimeters (cm) (3 to 4 in.) apart. The debris collected on the trash racks is removed by a mechanical rake. The inner or traveling screen, consisting of a mesh with square openings, approximately 0.6 to 1 cm (1/4 to 3/8 in.) on each side, passes over upper and lower rollers. This screen is normally stationary, but it moves when a certain pressure differential builds up across it because of partial blockage. The screen is then backwashed with water so that the fish and debris collected on it are returned to the main water body.

The intake system just described is designed to prevent entrainment of all but the smallest fish, although in some instances larger fish have been killed by impingement on the screens. Once the fish are caught on the screens, they are often unable to escape and eventually succumb. The actual causes of death are not clear. In some cases, factors may exist that cause fish to be attracted to the intake or reduce their ability to avoid it. For example, in the winter, when the main water body is cold, fish may prefer the warmer water in the vicinity of the power plant. The pump suction then draws them to the screens.

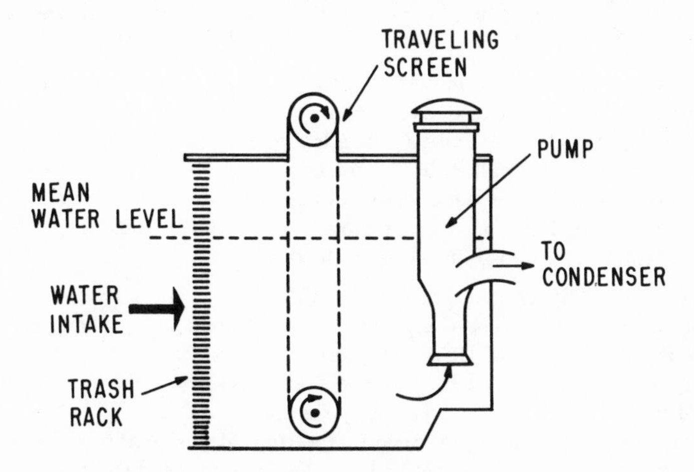

FIG. 12-7. Intake system for condenser cooling water.

Several methods have been proposed to minimize the problem, and the preferred solution seems to lie in the design and location of the water intake system. Reduction in the water intake velocity to less than about 15 cm (0.5 ft) per second, for example, permits the fish to escape before reaching the screens. Furthermore, the intake should be located out of the highly utilized shoreline zone. Various devices, including guidance mechanisms, such as louvers and low-pressure screen washers with return channels for live fish, have markedly reduced impingement damage.

Effects of Entrainment

Fish eggs, larvae, and small fish, as well as both plant and animal plankton, pass through the water intake screens, even if they have a small mesh, and hence they are drawn into the pumps and condenser. In once-through cooling systems, drawing millions of litres of cooling water per minute, very large numbers of organisms are entrained in this manner. These organisms may then be subjected to mechanical, thermal, and chemical effects. Mechanical damage can arise from abrasion, shear, impact, and the rapid change of pressure in passing through the pumps and condenser tubes. Thermal effects include (a) heat shock from the rapid increase in temperature within a few seconds when the organisms enter the condenser and (b) the sustained elevated temperature over a period of time between intake and final discharge of the water to the receiving body. Entrained organisms may also be affected by chemicals, especially chlorine, added to the condenser cooling water. The effects of entrainment on decomposers, producers, and consumers are considered in turn.

Decomposers. Bacteria (decomposers) are fairly tolerant to temperature changes and the increase in temperature in their passage through the condenser will have little effect on the population. Mechanical damage as a result of entrainment is also likely to be small. The presence of chlorine in the water during chlorination periods would be the only extensive source of bacterial mortality. However, bacteria reproduce so rapidly that any organisms lost are rapidly replaced.

Producers. The reported effects of entrainment on algae are not always in agreement, no doubt because several variables are involved. Experience has shown, however, that on the whole, the effect of entrainment on the local plant plankton population is small. It has been estimated that even if 20 percent of the population were killed, the net effect would be negligible. In the first place, the losses are confined to a small region; furthermore, the population of phytoplankton is restored by rapid reproduction.

Consumers. The animal plankton population is generally decreased by passage through the condenser, especially during the summer, when the intake water temperature is high. However, there is evidence that mechanical effects make a considerable contribution to the overall loss, possibly by causing the death of organisms already weakened by thermal shock.

The disturbance of the aquatic ecosystem from this loss of animal plankton is apparently not serious when the reproduction times for the species are short.

There are indications that if a substantial fraction of the animal plankton bypass the water intake, they may compensate for the loss by an increased reproduction rate in water that has been warmed by the condenser discharge. In addition, the organisms that have been killed (or weakened) may be readily consumed by fish, or they are decomposed by bacteria and thereby returned to the food cycle.

The most serious consequences of entrainment can occur when fish eggs, larvae, and juveniles and other plankton with long reproduction times are drawn into the condenser water intake. Even if they are not killed, young fish that survive thermal shock or physical damage may become prey to predators, or they may be unable to move out of water where the temperature is lethal. The remains of eggs, larvae, and fish that are killed return to the aquatic food cycle, but there may be a significant effect on the fish population because the losses cannot be regained. The overall consequences, however, cannot be stated in general terms because they are highly dependent on local conditions.

Effects of Chemicals

During the passage of the cooling water through the condenser, various small organisms tend to grow as slimy deposits on the interior walls of the tubes. These deposits can decrease the transfer of heat across the condenser walls. Consequently, a common practice is to chlorinate the water to kill the fouling organisms and prevent their accumulation. This is usually done by adding a "slug" of chlorine or sodium hypochlorite solution to the condenser intake water perhaps twice a week for a period of an hour or so. The residual chlorine in the condenser effluent must be kept at a very low concentration, generally from 0.1 to 0.5 milligram per litre (mg/l) of water at the discharge point.[a] This is then diluted further by mixing with the large volume of water in the receiving body. Lengthy exposure of aquatic organisms to chlorine at concentrations as low as 0.01 mg/l (or even less for especially sensitive species) can be toxic. For short exposure periods, however, substantially higher concentrations can be tolerated.

The regulations of the U.S. Environmental Protection Agency (EPA), as given in Title 40 of the *Code of Federal Regulations*, Part 423 (40CFR423), require that the maximum concentration of "free available" chlorine in the actual discharge from a steam-electric power plant should not at any time exceed 0.5 mg/l; the average over an extended period should not exceed 0.2 mg/l. (The concentration in the receiving water body should not be more than 0.002 mg/l for salmonids [e.g., salmon and trout] and 0.01 mg/l for other aquatic organisms.) Neither free available chlorine nor "total residual chlorine" may be discharged to a water body for more than 2 hours in any one day unless it can be established that the plant cannot operate under this condition.[b] If there is more

[a]Concentrations in milligrams per litre of water are essentially the same as parts per million by weight.

[b]The *free available* chlorine is the amount of chlorine present as molecular chlorine, hypochlorite, and hypochlorous acid. The *total residual chlorine* is the free available chlorine plus certain substances (e.g., chloramines) formed by the chemical action of chlorine or hypochlorite with impurities, chiefly nitrogen-containing compounds, present in the water.

than one plant at a particular location, discharge of water containing chlorine is permitted for only one plant at a time.

As implied by the EPA regulations, the chlorine hazard can be controlled by keeping the frequency of chlorination to the essential minimum, as required for efficient heat transfer across the condenser pipe walls. A number of alternative treatments, which are said to be less efficient than chlorine or hypochlorite, have been developed. One procedure is to flush the pipes periodically with hot water. Another approach is to make use of mechanical cleaning by brushing or by abrasion to remove the fouling organisms from the interior walls of the piping. For example, rubber balls coated with tungsten carbide are circulated through the cooling system. As a result of the low algal productivity, combined with the scouring action of sand drawn in with the condenser water, chlorination or other treatment has been found to be unnecessary at some locations on Lake Michigan.

Although sodium hypochlorite or chlorine is the only chemical added to the water entering the condenser, the effluent often serves to dilute other liquids discharged from a nuclear power plant. These may contain various chemicals used in the treatment of the water in the steam-generating system (e.g., chromate and phosphate) and for the regeneration of ion-exchange resins. Boric acid used for the control of pressurized water reactors is also discharged. In addition, the water may contain traces of nickel and copper compounds formed by corrosion of nickel-copper alloy condenser tubing. Of these chemicals, copper is the most significant hazard to aquatic organisms, especially in soft water (i.e., water low in dissolved calcium).

Dissolved Oxygen

Oxygen, dissolved in water, is essential to fish and other aquatic organisms; without a sufficient supply of oxygen, they cannot live. The saturation solubility of oxygen in water (i.e., the maximum amount of oxygen that a given volume of water can normally dissolve) increases with decreasing temperature. Thus, there is usually more oxygen dissolved in water in winter than in summer. As a general rule, the amount of dissolved oxygen in river water in the United States ranges from about 5 to 10 mg/l. These values may be compared with the saturation solubilities of oxygen gas in water in equilibrium with the atmosphere—namely, 6.6 mg/l at 38 °C (100 °F) and 10.4 mg/l at 15.5 °C (60 °F).

Some concern has been expressed that the heated water discharged from the condenser of a steam-electric power plant may contain insufficient dissolved oxygen to sustain fish life. Numerous measurements made at such plants have established, however, that there is usually little change in the amount of dissolved oxygen in the water that has passed through the condenser. In fact, in some instances an increase has been recorded, even when the temperature of the water has been raised by 5.6 °C (10 °F) or more. Such a result is not unreasonable if the intake water is not already saturated with oxygen. Since the rate—as distinct from the total amount—at which oxygen dissolves in water increases with rising temperature, additional oxygen can dissolve in the effluent water as it flows through exit pipes and discharge canals and over weirs. In general, the changes in the oxygen concentration in water arising from passage through a condenser are small compared to those that occur in most natural waters as a result of photosynthesis and respiration and of the oxidation of organic matter.

For most fish, an oxygen concentration below roughly 3 mg/l is lethal, although some species have survived at lower concentrations. The minimum oxygen concentrations specified in the water quality standards for most states are in the range of 4 to 6 mg/l. Except for waters in which the dissolved oxygen has been greatly reduced by sewage or by decomposing organic matter, the dissolved oxygen rarely falls to 3 mg/l. Furthermore, there is evidence that fish instinctively tend to avoid regions of low oxygen concentrations just as they do high temperatures.

In enclosed bodies of water, such as lakes and reservoirs, the oxygen content often decreases with increasing depth (e.g., in the late summer), so that the lower layers may contain very little oxygen (p. 348). Normally, fish tend to remain in the regions where the oxygen content is higher. However, the circumstances may be drastically changed if condenser water is taken in at a great depth, because of the lower temperature, and is discharged at the surface. The resulting decrease in the oxygen content of the upper layers of the lake (combined with the higher temperature) can have serious consequences for fish and other aquatic organisms.

Another harmful situation may arise in water heavily loaded with organic matter. The increased oxygen demand by decomposer organisms at the higher temperature of the discharge could result in a substantial decrease in the dissolved oxygen content of the water after it has left the power station.

Effects of Salinity Changes

Changes in salinity of the receiving water body, into which the condenser effluent is discharged, can occur in particular when the plant is located on an estuary or bay. Such changes could result if the condenser intake and discharge are in different water quality zones; these zones may occur in a horizontal direction along a river-estuary-ocean system or vertically in a salinity stratified estuary. The effects of salinity changes depend in a complex manner on the accompanying changes in temperature and dissolved oxygen content. In the following, the effects of salinity variations alone are considered.

A relatively few species migrate from the sea to fresh water (e.g., salmon and shad), from the sea to estuaries (e.g., herring), or from fresh water to the ocean (e.g., the eel) to spawn. Some fish migrate from the ocean to an estuary to feed. Otherwise, aquatic organisms that normally live in fresh water will not survive in seawater, and vice versa. Organisms living in estuarine or coastal waters become adapted to the ranges of salinity that they normally experience in their natural habitat. Furthermore, many species of estuarine organisms have complex life cycles, with different stages requiring different salinity characteristics.

Most fish (including shellfish) can tolerate salinities outside their normal range for a period of minutes or hours; however, more extended exposures may prove fatal. Species able to swim tend to avoid unfavorable salinity zones and seek regions of preferred salinity if available. Many shellfish, such as clams and oysters, are unable to move freely, but they can take protective action, temporarily at least, by closing their shells when exposed to abnormal salinities. Nevertheless, several instances have been reported of major losses of bivalve shellfish as a result of lowered salinity caused by fresh water from runoff of rain or melting snow.

Normal variations in salinity are important in the life histories of some species, especially in estuaries and bays; for example, the spawning of oysters may be triggered by such variations. Unusual variations in salinity can thus have adverse effects by causing oysters to spawn when other conditions are unfavorable. Larvae of fish and shellfish may survive only in marsh areas where the salinity is within a narrow range while the larvae are present. Furthermore, a small change in the water salinity may favor predators of shellfish and thus deplete the stock of these organisms.

FIELD OBSERVATIONS ON CONDENSER DISCHARGES

Introduction

The foregoing discussion has been concerned mainly with individual factors that have a bearing on the effects of condenser water discharges on aquatic life. As will be seen shortly, the results obtained have been used as the basis for formulating water quality temperature standards. In the field, however, the various factors can interact with one another in an unpredictable manner. It is therefore of interest to review some observed consequences of the discharge of warmed water in once-through condenser systems.

Although it is not the only criterion for assessing the effects of warmed condenser water discharge on the aquatic ecosystem, the conservation of a desirable fish species is of some practical interest. For several years, extensive studies have been made by government agencies and electric utilities of the effect on local fish populations of thermal discharges from steam-electric power installations, as well as from industrial plants. These studies were initiated and in progress long before nuclear power was of any significance. Some general observations are described first, and then more specific details are given of the effects of power plants on the fish populations of the Columbia, Tennessee Valley, and Connecticut Rivers, for which there is ample documentation.

A common experience has been the attraction of several fish species to the warmer water, in a discharge canal, river, lake, estuary, or reservoir, during the cooler months. It is not certain if it is merely the higher temperature of the water near the discharge that attracts the fish or if the greater availability of food is an important consideration. During the summer, the temperature avoidance reaction apparently becomes operative, and the fish, in most cases, leave the warmer water.

Essentially all bodies of water in which sport and commercial fish can live contain a certain proportion of less desirable "rough" fish. In some instances, an increase in the temperature of the water has been accompanied by an increase in the number of the less desirable fish, which compete with sport fish for food.

Although cases have been reported of a simultaneous increase in the latter species, there are others in which sport fish, such as trout, which prefer cooler water, have been almost eliminated by the warm condenser discharge.

The possibility has been raised that the presence of fish in warm water discharges does not necessarily mean that the conditions are either desirable or optimal. There is one recorded instance in which this is apparently the case. At the Connecticut Yankee Atomic Power Company's plant on the Connecticut River, the condenser water is discharged into a canal more than a mile long and then flows into the river. Large numbers of fish, especially brown bullhead and white catfish, have moved from the river into the canal. During the winter, these fish suffered a substantial loss in weight, significantly greater than that of catfish that had remained in the cooler river. There was apparently no shortage of food and of oxygen in the canal water that could account for the weight loss. The factors responsible for the weight loss may be the higher rate of metabolic processes in the warm water, an increased energy expenditure required to swim against the relatively rapid flow of water in the canal, and the stress of crowding.

Fish Kills

The destruction of a substantial number, from a few hundred to many thousands, of mature fish (as distinct from eggs, embryos, and juveniles) in a single event is commonly referred to as a *fish kill*. Some 3000 or more fish kills were reported in the United States during the 1960s, so that they are of relatively frequent occurrence. The great majority of these kills were due to causes completely unrelated to steam-electric power plants, conventional or nuclear. One of the worst recorded fish kills resulted from a thermal discharge by a manufacturing plant in Ohio in January 1967. Most lethal events arise from such causes as high summer water temperatures, depletion of dissolved oxygen, low river flow, parasitic infestation, or toxic industrial effluents.

About a dozen or so instances of thermal fish kills during the 10-year period could be attributed to electric power installations. However, there have probably been more kills resulting from impingement of fish on the screens in front of the condenser water intake than from the discharge of thermal effluents. Impingement has now been decreased, but not eliminated, by changes in the design of water intake systems.

A fish kill involving mostly larger fish occurred at the Indian Point Nuclear Generating Plant No. 1 on the Hudson River, near Peekskill, New York, in June 1965, as a result of impingement on the intake screens. This event coincided with the addition of chlorine (as sodium hypochlorite) to the water to inhibit the growth of organisms on the condenser tubes. When the chlorine inlet was removed from the river side to the pump side of the screens, the larger fish were no longer collected on the screens. Apparently the chlorine had previously either poisoned the larger fish or it had reduced their ability to avoid or escape from the water intake.

Other impingement fish kills at the Indian Point No. 1 plant, which is no longer in operation, were apparently an indirect effect of a temperature increase. In cold weather, fish were attracted to the warm discharge, which had been carried upstream, possibly by tidal action, toward the intake. The fish were then trapped on the intake screens.

In April 1973, several thousand adult menhaden, an Atlantic Ocean fish, were killed in the condenser discharge canal of the Pilgrim (nuclear power) Station near Plymouth, Massachusetts. The fish kill in this case was probably caused by gas-bubble disease (p. 316). Coastal waters are normally saturated with air, and rapid heating in the condenser resulted in supersaturation of the warmed discharge water. Fish in the canal thus experienced gas embolism and died.

The major emphasis on thermal effects of condenser discharges from power plants has been on the consequences of the higher than normal temperatures. It is possible, however, that potential fish kills from cold shock may be more significant than from heat shock. If the water temperature at a particular location is increased beyond the tolerance point, fish may be able to escape to a preferred environment where the temperature is lower. On the other hand, if the water temperature drops to that of the much colder surroundings, there is no higher temperature region available to the fish.

During the winter, fish are attracted by and become acclimated to the warmer water in the vicinity of the discharge. If the plant were to shut down, the temperature of the water would soon fall to the ambient value, and the fish may die of cold shock. Several cases of fish kills probably resulting from such circumstances have been reported for steam-electric power plants. The danger of cold shock can be minimized by appropriate design of the condenser discharge system so that fish are not attracted in winter to areas where the water temperature is much above ambient. This can be achieved, for example, with a submerged discharge in which there is rapid mixing with the surrounding water (p. 346). When two or more reactors are located at the same site, the risks of fish kills from cold shock are greatly reduced since there is only a small probability that all the reactors will be shut down at one time.

Observations on the Columbia River

Reactors for the production of plutonium, for use in nuclear weapons, were located near Hanford, Washington, on the Columbia River in order to utilize the river water for cooling purposes. The first reactor began operating late in 1944, and in 1964, as many as nine reactors were returning warmed water to the Columbia River over a distance of about 24 km (15 miles). Only one of the reactors, and the only one still in operation, produced electricity. The river water is used in the turbine condenser associated with this reactor. For the others, the water served as the reactor coolant in a once-through system, so that all the heat generated by fission was ultimately transferred to the Columbia River.

The water was discharged from large pipes directed upward from the bottom into the main flow of the river. Vertical mixing took place quite quickly, but the warmed water was transported many kilometers downstream before it had mixed throughout the width of the river. The natural temperature of the Columbia River ranges from about 2 °C (36 °F) to 20 °C (68 °F), and when most of the reactors were in operation, the summer temperatures were increased by about 2.5 °C (4.5 °F) over a distance of more than 80 km (50 miles). Much larger local increases in temperature were detected, but only in limited areas.

In view of the importance of salmon in the Columbia River to both commercial and sport fishing, special attention was paid to the effect of the reactor discharges on various stages in the life of these fish. Large numbers of Chinook salmon and other fish species swim up the river in spring, summer, and early autumn to spawn. By attaching sound-emitting tags to some of the salmon, their locations while migrating upstream could be determined. It was found that the speed of migration of the fish was not significantly affected by the warmer water, and there was no indication of any blockage. The salmon tended to travel near the shore on the opposite side of the river from that on which the reactors were located. But this preference may have resulted from factors not associated with reactor operation.

Normal temperature variations of the Columbia River water from year to year alter the spawning conditions. Hence, an increase in temperature of the river could have affected spawning adversely or beneficially, according to circumstances. However, less than 1 percent of the total run of Chinook salmon in the Columbia River spawned in places where the eggs could have been harmed by water discharged from the reactors. Consequently, the overall effect on salmon eggs in the river was probably very small. It is of interest that the number of "redds" (spawning nests) near the Hanford plants has increased substantially since 1960, but this is attributed to the influx of salmon that would have spawned in places now inundated by the erection of dams.

The return of young Chinook salmon from the Columbia River to the sea in late spring and summer was apparently not influenced by the warm water. Only in about 5 percent of the cross-sectional area of the river was the temperature high enough to injure the fish. Mathematical modeling and observations on caged fish showed that young salmon would rarely have remained long enough at an unfavorable temperature to be killed. Furthermore, the young Chinook salmon apparently swim close to the shore rather than in the main channel of the river, where the warm water is discharged.

The general conclusion drawn from a study of salmon in the Columbia River over a period of several years is that the biological effects of the warm water from the Hanford reactors have been inconsequential. The principal disruptive factor for Columbia River salmon has been hydroelectric power development. The population of these salmon is now sustained largely by an extensive hatchery system.

Observations on the Connecticut River

Six steam-electric power stations within a stretch of some 96 km (60 miles) have utilized water from the Connecticut River in once-through condenser cooling systems.[c] Of these plants, five burn conventional fuels, whereas the sixth, farthest downstream, is the Connecticut Yankee Power Company's nuclear in-

[c]The Vermont Yankee Nuclear Corporation's power plant, about 80 km (50 miles) farther upstream from the most northerly of these installations, commenced operation in 1972.

stallation at Haddam Neck. The effluent water from the condensers of the Haddam Neck station flows down a long canal before discharge to the river. The mouth of the canal is fan-shaped; this has the effect of decreasing the velocity of the heated discharge water, which tends to stay near the surface. The normal temperature of the Connecticut River near Haddam Neck ranges from 0 °C (32 °F) to 30 °C (86 °F), and the maximum increase in the surface temperature due to the power plant effluent is about 5.5 °C (10 °F), except near the mouth of the discharge canal. The river water below a depth of 3.7 m (12 ft) is heated by less than 1.1 °C (2 °F).

To investigate the possible effects of the thermal discharge from the Haddam Neck installation on aquatic life in the Connecticut River, a comprehensive study was initiated in January 1965, about 30 months before the nuclear reactor started operating at low power in 1967. It was continued for several years after full power, approaching 600 megawatts electric (MWe) was attained in January 1968. The American shad is the most valuable fish species in the river; consequently, considerable attention has been paid to it.

When the adult shad return from the sea to spawn in the late spring, they pass the nuclear power plant on their way to the main spawning grounds upstream. It has been observed, by means of tagged fish, that the shad follow the river channel which, in the Haddam Neck area, is closer to the bank opposite to the power plant. The shad pass alongside, under, or through the warmed water without apparent difficulty. The juvenile fish hatched upstream move down the river on their way to the sea from August to November. About 5 percent of the total were entrained by the cooling water, and most of these were killed by combined thermal and mechanical shocks. The adult shad population of the Connecticut River decreased after 1965, before the plant started operation, but it has since increased. This decrease and subsequent increase are attributed to natural causes.

As already seen, resident fish are attracted in winter by the warmer water near the mouth of the discharge canal. Some varieties of catfish, however, become emaciated as a result of wintering in the canal (p. 311). Other fish species are apparently not affected in this manner. The operation of the nuclear power plant at Haddam Neck does not appear to have had any noticeable effect on the catch of resident fish, such as catfish, perch, sunfish, and eels, in the Connecticut River in the vicinity of the plant. For obvious reasons, many fishermen prefer to fish in and near the mouth of the canal; in fact, the plant has stimulated winter fishing, which had not existed previously.

Operation of the Haddam Neck power station has caused some changes in the bottom-dwelling organisms (benthos) that are an important part of the aquatic food cycle. Around the plant intake, the water flow has washed away most of the silt and sand from the bottom of the river. The remaining gravel and stone are not suitable for the organisms that formerly lived on the bottom. Near the mouth of the discharge canal, in the warmer water, there was at first a marked increase both in the variety and abundance of small organisms. But the diversity and numbers have declined sharply on several occasions. Some three or four miles downstream, the benthos appears to be unaffected by the operation of the Haddam Neck plant.

Observations by the Tennessee Valley Authority

The Tennessee Valley Authority (TVA) operates several steam-electric power stations, most of which draw their cooling water either directly from the Tennessee River or from one of its tributaries. The warmed water from the condensers is then discharged into a reservoir (or lake) that is part of the main river flow. Although the TVA now has nuclear power plants in operation, the following experiences relating to the thermal effects of large installations using conventional fuels are of interest.

Except for the Paradise plant, to which reference is made below, the quantity of water used for condenser cooling is in each case not a large fraction of the total river flow. The effects on aquatic life are, therefore, expected to be minimal. In several instances, large numbers of fish have been attracted to the warmer waters in discharge areas, especially in winter and spring. Undesirable filamentous blue-green algae, which are predominant at higher water temperatures, have been observed, but the growth has not been of nuisance proportions.

The Paradise plant is located on the small Green River in Kentucky; it flows into the Ohio River just upstream from its junction with the Tennessee River. In this case, the total cooling water requirement represents a substantial part of the Green River flow. Soon after the first two units, with a total capacity of about 1400 MWe, began operating in 1963, a marked depletion was observed in the population of small zooplankton near the discharge area, but complete recovery occurred a few miles downstream. The total supply of food available for fish was considered to be substantially unchanged. The fish population was apparently decreased during the summer, but inventories near the warmer discharged water have been consistently larger than either upstream or downstream in the winter.

When a third (1150-MWe) unit of the Paradise plant was planned, the TVA recognized that the use of a once-through cooling system could have detrimental effects on the Green River. Consequently, a natural-draft cooling tower (Chapter 13) is used for the condensers of this unit. Cooling towers have also been built for the first two units of the plant, and they are put into operation when the mean river water temperature might otherwise be expected to exceed about 32 °C (90 °F).

Thermal Effects on Municipal Water Uses

Many water bodies that might be utilized by power plants for condenser cooling serve, after treatment, as sources of public water supplies, and they are also used for the disposal of organic wastes (e.g., in sewage). The possible effects of thermal discharges on water used by municipalities must therefore be examined.

An increase in the temperature, such as normally occurs in the summer, makes drinking water less palatable, at least to the American taste. This objection is partly overcome by the common practice of refrigerating or icing drinking water. But there is another factor that leads to a decrease in palatability. A higher temperature tends to favor the production of blue-green algae over other types of algae (see Fig. 12-2). Some of the blue-green algae cause the water to have an unpleasant taste and odor that are not easily removed. Thus, the discharge of warm condenser water could, in some circumstances, have a deleterious effect on municipal water supplies.

Before it is distributed for use by the public and by industry, water from a river or other source is treated for purification and disinfection and sometimes also to decrease the hardness. Some of the processes used in water treatment are accelerated at higher temperatures. Hence, the addition of warm condenser effluent to a water body might be of advantage in water treatment. More is said about this matter in connection with possible beneficial uses of warm water in the next chapter.

When organic wastes are discharged to a water body, bacteria utilize the dissolved oxygen to convert the obnoxious wastes into innocuous substances, such as carbon dioxide, nitrates, and water, by biochemical reactions. In this way the wastes are assimilated. The rates at which the processes take place increase with increasing temperature of the water, up to at least 32 °C (90 °F). As a result, the organic matter is consumed more rapidly, and so also is the dissolved oxygen.

Unless sufficient oxygen it taken up from the air, the dissolved oxygen content of the water will decrease. Ultimately, a point may be reached at which the aerobic bacteria (i.e., those that make use of dissolved oxygen) can no longer function. The decomposition of the organic matter is then taken over by anaerobic bacteria (which do not require dissolved oxygen). These bacteria generate products that give the water an unpleasant odor and color; furthermore, the rate of decomposition of waste matter is significantly decreased. The net effect of adding heat to a body of water could thus be to decrease its capacity for assimilating organic wastes. It is explained in Chapter 13, however, that if steps were taken to add oxygen to the water, the increase in temperature could be used to good purpose.

WATER QUALITY CRITERIA

The Federal Water Pollution Control Act

Although water quality standards had been adopted previously by some states and interstate bodies in the United States, a major impetus to the development and enforcement of such standards resulted from the passage by Congress of the Water Quality Act of 1965, which was an amendment to the Federal Water Pollution Control Act (FWPCA) of 1948. The Act encouraged the states to establish by June 30, 1967, water quality standards, subject to approval by the Department of the Interior, for interstate streams and coastal waters. In December 1970, authority for the approval of state water quality standards was transferred from the Department of the Interior to the EPA. If the standards proposed by a state were not found to be acceptable, the reviewing agency was authorized to set water quality standards for the particular waters involved.

In addition to confirming the foregoing requirements, the FWPCA amendments of 1972 (and 1977) authorized the EPA to establish "national standards of performance" for control of the discharge of pollutants (including heat) from "new sources." A new source is defined as any source, such as a steam-electric power plant, the construction of which is commenced after the publication of the proposed applicable national standard. States are authorized to develop and submit to the EPA for approval their own procedures for applying and enforcing the standards.

The amended FWPCA covers a wide variety of pollutants, but there is a special section (Sec. 316) dealing with thermal discharges. In effect, this section states that the national standards for thermal discharges may be relaxed in individual cases if it can be shown that less stringent limitations will be adequate "to assure the protection and propagation of a balanced, indigenous population of shellfish, fish, and wildlife in and on the body of water into which the (thermal) discharge is to be made." Relaxation of the standards in such cases is to be authorized by the EPA or, if appropriate, by the state in which the discharge occurs. Section 316 also stipulates that the "location, design, construction, and capacity of cooling water intake structures reflect the best technology available for minimizing adverse environmental impact."

When an activity requires a permit or license from a federal agency, the applicant must submit a certification from the state (or the EPA if the state lacks authority) that discharges (of all kinds) will comply with the appropriate provisions of the amended FWPCA of 1972 (and 1977). Thus, in accordance with Section 401 of the Act, an applicant for a permit to construct a nuclear power facility must provide such certification to the U.S. Nuclear Regulatory Commission (NRC). Upon receipt of the application, the NRC must inform the EPA of the application and certification. The administrator of the EPA is then required to take such action as is necessary to assure that a discharge originating in one state will not adversely affect the quality of the waters in any other state. Before the plant can be operated, the owner must obtain a discharge permit from the state, as required by Section 402 of the amended FWPCA. The FWPCA and its amendments are applicable to the discharge of a wide variety of pollutants into navigable waters. The text that follows, however, is concerned with thermal discharges.

The passage of the National Environmental Policy Act of 1969 required that a detailed environmental impact statement be prepared for each major federal action affecting the environment, including the licensing of nuclear power plants (Chapter 3). An important court decision in July 1971, in connection with the Calvert Cliffs Nuclear Power Plant, held that the U.S. Atomic Energy Commission, which was then responsible for the licensing of nuclear plants, must consider the thermal effects of condenser discharge, in addition to the effects of radioactive effluents, on the environment. Consequently, the water quality criteria developed for thermal discharges are now used by the NRC in assessing the environmental impact of a proposed nuclear power plant before issuing a construction permit or an operating license.

Development of Water Temperature Standards

Fish and other aquatic life occurring naturally in each body of water are species or varieties that are competing with each other with varying degrees of success and have, over the years, reached an ecological equilibrium (or balance). The number and variety of aquatic life forms in the state of equilibrium depend on the temperature of the water as well as on other conditions in the environment. If an appreciable quantity of heat is added to a body of water, the equilibrium is disturbed. As a result, there may be significant changes in the aquatic life. A new ecological equilibrium will eventually be attained, but it may possibly be less desirable to human interests than the original equilibrium.

There are many difficulties in the way of setting temperature standards that will minimize the harm to an aquatic ecosystem. In the first place, the temperatures of surface waters in the United States range from about 0 °C (32 °F) to 40 °C (104 °F), depending on the latitude, altitude, season, time of day, water flow, depth, and other factors. Because of the many variables, it seems unlikely that two bodies of water, even at the same latitude, will have exactly the same thermal characteristics. Moreover, the habitats of a given aquatic species often have different thermal characteristics.

It has been seen that, within limits, fish can be acclimated to a changed temperature environment, although the growth, spawning, and survival rates may be affected, either beneficially or adversely. The temperature range over which such acclimation is possible varies, however, with the nature of the fish. Furthermore, consideration must be given to the normal seasonal and daily variations in the temperature of the water body. At some times, the normal (or natural) temperature of the water will be nearer to the upper acclimation temperature than at other times.

In view of the many variables involved, especially the different temperature tolerances of different fish species and natural variations of water temperature, it is clearly impossible to establish a single temperature standard for the whole country or even for one state. The requirements must be closely related to each water body, to its characteristic important aquatic life forms, and to the season of the year. A possible approach to the establishment of standards is to propose a permissible increase in temperature above the normal value for the specific water body, with the proviso that certain temperatures not be exceeded.

To provide a basis for the review of water quality standards proposed by the states, in accordance with the Water Quality Act of 1965, the Secretary of the Interior appointed a National Technical Advisory Committee (NTAC) on Water Quality Criteria. (Water quality criteria are not standards, but are used to develop acceptable local standards.) One part of the NTAC report, issued in 1968, dealt with the temperature requirements of bodies of water that might be used for cooling systems by electric utilities and other industries. In 1971, at the request of the EPA, the National Academy of Sciences and the National Academy of Engineering (NAS-NAE) undertook a revision of the NTAC criteria. This comprehensive revision was published under the title *Water Quality Criteria 1972.*

The FWPCA amendments of 1972 require the EPA to develop and publish from time to time criteria for water quality reflecting the latest applicable scientific knowledge. The first such publication, entitled *Criteria for Water Quality* was issued in 1976. The criteria for thermal effects are essentially identical to the 1972 NAS-NAE recommendations.

Federal approval (by the Department of the Interior, and subsequently by the EPA) of most existing water temperature (and other) standards proposed by the states was based on the guidelines of the NTAC report of 1968. Consequently, the following discussion deals first with the suggestions contained in this report.

NTAC Guides for Temperature Standards

As a general guide, the NTAC recommended that, during any month of the year, heat should not be added to a stream in excess of the amount that would raise the temperature of the water by 2.8 °C (5 °F), at the expected minimum daily flow for that month.[d] In lakes and reservoirs, the temperatures of the upper layers should not be raised by more than 1.6 °C (3 °F) in those areas where important aquatic organisms might be affected adversely. For estuarine waters, the temperature increase should not exceed 2.2 °C (4 °F) in the autumn, winter, and spring, or 0.83 °C (1.5 °F) in the summer north of Long Island Sound on the East Coast and north of California on the West Coast. The permissible temperature increases for reservoirs, lakes, and estuarine waters should be based on the monthly average of maximum daily temperatures before the addition of heat. By using a base temperature corresponding to the monthly average minimum flow of a river or the monthly average maximum daily temperature in other cases, an attempt is made to allow for natural temperature variations.

Because of the large number of salmon and trout waters that have been made marginal or unproductive, the NTAC recommended that inland trout streams, headwaters of salmon streams, trout and salmon lakes, and reservoirs containing such fish not be warmed at all. Moreover, no heated effluents should be discharged in the vicinity of spawning areas.

To provide guidelines for the establishment of water quality temperature standards, the NTAC report included Table 12-I, which gives the provisional maximum temperatures that were thought to be compatible with the well-being of various species of fish at different stages of their life cycle. Note that the suggested maximum for temperate-water fish, such as catfish, gar, shad, etc., is 34 °C (93 °F), whereas for salmon and trout, which are cold-water fish, the maximum is 20 °C (68 °F). In harmony with what has been said earlier in this chapter, the proposed maximum temperatures for spawning and egg development are lower than those required for subsequent growth.

States' Water Quality Standards

The standards proposed by individual states were not expected to be identical with the recommendations of the NTAC in order to obtain federal approval.[e] All that was required was that the proposed criteria be compatible with the committee's suggestions. As a general rule, maximum temperature increases

[d]In this and the following sections, the distinction between temperature change and actual temperature must be borne in mind, although they are both expressed in the same units, °C or °F.

[e]For a summary of the states' temperature (and some other) standards, see *Nuclear Safety* 13 (1972): 268.

TABLE 12-I
PROVISIONAL MAXIMUM ACCEPTABLE TEMPERATURES

	°C	°F
Growth of catfish, gar, white, yellow, and spotted bass, buffalo, carpsucker, and threadfin and gizzard shad	33.9	93
Growth of largemouth bass, drum, bluegill, and crappie	32.2	90
Growth of pike, perch, walleye, smallmouth bass, and sauger	28.9	84
Spawning and egg development of catfish, buffalo, and threadfin and gizzard shad	26.7	80
Spawning and egg development of largemouth, white, yellow, and spotted bass	23.9	75
Growth and migration of salmon and trout; egg development of perch and smallmouth bass	20.0	68
Spawning and egg development of salmon and trout (other than lake trout)	12.8	55
Spawning and egg development of lake trout, walleye, northern pike, sauger, and Atlantic salmon	8.9	48

and maximum permissible temperature are specified. Some states have also specified maximum rates of change of temperature. The purpose of this restriction is to minimize the possibility of thermal (hot or cold) shock and to permit acclimation of the fish to the changed water temperature. It is recalled from Chapter 3 that the Technical Specifications, which must be approved by the NRC and which become part of the operating license for a nuclear power plant, include limits on the rate of temperature change (increase or decrease) of the discharged condenser water. These limits apply to normal operation and are compatible with the standards of the state in which the plant is located.[f]

Most states have established 20 to 21 °C (68 to 70 °F) as the maximum temperature and from 0 to 2.8 °C (0 to 5 °F) as the permissible increase above normal for streams with cold-water fish. For warm-water fish, the maximum temperature is usually from 28 to 34 °C (82 to 93 °F) with a maximum permissible increase or decrease of 2.2 to 2.8 °C (4 or 5 °F). The temperature maxima frequently vary with the season (or month) of the year. Where the criteria include maximum rates of temperature change, they are usually 0.56 to 1.1 °C (1 to 2 °F) per hour. For coastal waters, the standards are generally a maximum increase of 0.83 °C (1.5 °F) in the summer and 2.2 °C (4 °F) in other seasons. In some cases, the requirements of certain fish have made it necessary to allow no change in temperature from natural conditions; obviously, such waters cannot be used for once-through condenser-cooling systems. Furthermore, unusual situations may arise in individual states that require special consideration. A difficulty in establishing meaningful temperature standards is that in some water bodies the natural temperature in summer may exceed the stipulated maximum.

In addition to the temperature standards for the preservation of aquatic life forms, many states have imposed limitations on the temperature of water used

[f]In the event of an emergency shutdown of the reactor (see Chapter 4), the water temperature may decrease more rapidly than permitted by the Technical Specifications.

for recreational purposes (e.g., swimming), for drinking water, for irrigation of agricultural land, and for industrial processes. However, these standards are not necessarily different from those applicable primarily to fish. Where they are different, standards for the protection of aquatic organisms are usually the more stringent.

NAS-NAE and EPA Criteria: Temperature Maxima for Prolonged Exposures

The thermal water quality criteria recommended in the 1972 NAS-NAE report, and which were incorporated in the EPA criteria of 1976, fall into two main categories: (a) average maximum acceptable temperatures for prolonged exposures (e.g., a week or more), and (b) maximum temperatures for short-term exposures that depend on the exposure time. The average maximum temperatures for long exposures of freshwater fish, which are considered in this section, must be defined for warm and cold seasons for aqueous species that are important in the water body; additional, overriding criteria are applicable during the reproduction and development periods for these species.

Criteria for Warmer Months. These criteria are for mid-April through mid-October in the Northern states and for March through November in the South. Their objective is to maintain the growth of important aquatic organisms necessary to sustain actively growing and reproducing populations. The criterion for a particular location would be determined by the most sensitive life stage of an important species likely to be present in that location at that time.

For each fish species (and stage of development), there is, on the one hand, an optimum temperature and, on the other hand, an ultimate upper incipient lethal temperature (p. 314 *et seq.*). Values of these temperatures for a number of common freshwater fish species are given in Table 12-II; the optimum temperatures quoted are those for rate of growth (see Fig. 12-3). The NAS-NAE recommendation is that, in the warmer months, the maximum average weekly

TABLE 12-II
CHARACTERISTIC WATER TEMPERATURES FOR SOME COMMON FRESHWATER FISH SPECIES

Species	Optimum		Ultimate Upper Temperature		Criterion for Maximum	
	°C	°F	°C	°F	°C	°F
Channel catfish	30	86	38	100.4	32.7	90.8
Largemouth bass (fry)	27.5	81.5	36.4	97.5	30.5	86.7
Smallmouth bass (average)	27.3	81.1	35	95	29.9	85.8
White sucker	27	80.6	29.3	84.7	27.8	82
Bluegill	22	71.6	33.8	92.8	25.9	78.6
Winter flounder	18	64.4	29.1	84.4	21.8	71.2
Cisco	16	60.8	25.7	78.3	19.2	66.6
Sockeye salmon	15	59	25	77	18.3	64.9
Brook trout (average)	14.5	58.1	25.5	77.9	18.2	64.8
Brown trout (average)	12.5	54.5	23.5	74.3	16.2	61.2

temperature of the water for long exposure periods should not exceed the optimum temperature plus one-third of the difference between the ultimate upper incipient lethal temperature and the optimum temperature (in °C). The maxima derived in this matter (and the corresponding values in °F) are given in the last columns of Table 12-II.

The temperature criterion for the warmer months may thus be written as follows:

Maximum weekly average temperature shall not exceed
Optimum temperature + 1/3 (Ultimate upper incipient lethal temperature
− Optimum temperature)

for the aquatic species under consideration. For example, for channel catfish, the optimum growth temperature is 30 °C, whereas the ultimate upper incipient lethal temperature is 38 °C. Hence, where channel catfish are an important species, the maximum average weekly temperature of the water during the summer months should not exceed 30 + 1/3 (38 − 30) = 32.7 °C (90.8 °F). This is slightly less than the provisional NTAC acceptable maximum temperature in Table 12-I.

Criteria for Colder Months. Criteria for the colder months apply from mid-October through mid-April in the North and from December through February in the South. (In some regions, these criteria may also be applicable in the summer.) Their purpose is to protect against fish kills due to cold shock. Although the criteria are especially applicable during the colder months, they may be important in the warmer months for lakes, when rapid changes in the water layers, which are described in Chapter 13, may cause a sharp decrease in the surface temperature.

The NAS-NAE and EPA recommendation for the maximum average weeklytemperature for an important species during the colder months first is stated below and then explained in subsequent paragraphs. Let T_n be the normal temperature of the bulk of the water body in the cold season. The acclimation temperature, T_a, corresponding to a value of T_n as the lower incipient lethal temperature, is then obtained from the lower part of Fig. 12-5 for the fish species of interest. In other words, if the fish had been living in water at temperature T_a, the corresponding lower incipient lethal temperature would be equal to T_n. The recommendation is that the temperature in the condenser discharge water in the colder months shall not exceed T_a − 2 °C. The 2 °C (3.6 °F) constitutes a factor of safety that provides assurance that very few fish will be killed.

To understand what this means, consider, for example, a case in which the normal winter temperature, T_n, of the water body is 5 °C (41 °F), and bluegill is the important aquatic species. For bluegill, the acclimation temperature, T_a, corresponding to a lower incipient lethal temperature of 5 °C (41 °F) is found to be 20 °C (68 °F), as represented qualitatively in Fig. 12-8. If the safety factor of 2 °C (3.6 °F) is subtracted, the result is 18 °C (64.4 °F). The NAS-NAE and EPA criteria are therefore that the average weekly water discharge temperature should not exceed 18 °C (64.4 °F) in the colder months if the bluegill are to survive.

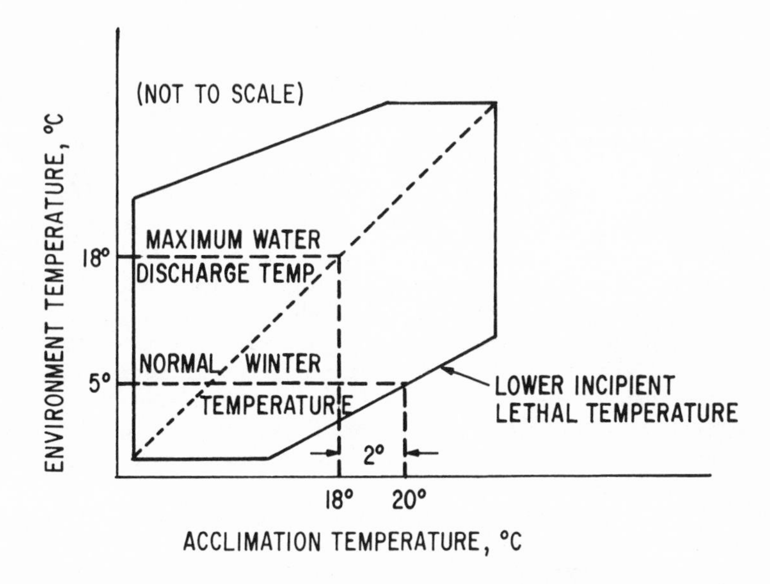

FIG. 12-8. Estimation of maximum permissible average weekly water discharge temperature in winter (not to scale).

Suppose the power plant were shut down fairly rapidly after operating for some time during the colder months. The temperature at (or near) the point of the condenser discharge would drop from 18 °C (64.4 °F) or less to the normal water temperature of 5 °C (41 °F). The latter temperature is the lower incipient lethal temperature for bluegill that have been acclimated at 20 °C (68 °F). Since the fish have been living in (and have been acclimated to) the discharge at a temperature of 18 °C (64.4 °F) or less, the incipient lethal temperature will be a few degrees below 5 °C (41 °F). Hence, the bluegill should survive if the water temperature falls to 5 °C (41 °F).

Criteria for Reproduction and Development Periods. Although some temperature changes can be tolerated during the reproduction and development stages of freshwater fish, these stages are much more sensitive to temperature than are adult fish (p. 313). The information on this subject, however, is limited in extent, and the NAS-NAE report recommends that high priority be assigned to summarizing existing data and to obtaining what is lacking.

Since it is not possible to propose quantitative criteria, it has been recommended that, during reproduction seasons (generally April to June and September to October in the North, and March to May and October to November in the South), the water temperatures should meet the specific site requirements for successful migration, spawning, egg incubation, fry rearing, and other reproductive functions of important species. These temperature requirements should supersede all others during the times of the year when they apply at a particular location.

Other Criteria. The NAS-NAE report called attention to the possibility that significant changes in temperature or in thermal patterns over a period of time could cause some change in the composition of aquatic communities (i.e., the species present and the number of individuals in each species), as noted on page 333. A significant change in the community structure may be detrimental, even though species of direct importance to man are not eliminated. The water temperature requirements at each plant site should therefore be such as to preserve the normal diversity of the aquatic species.

Another aspect of the addition of heat to a water body is that it may occasionally result in growths of nuisance organisms (e.g., blue-green algae), provided that other environmental conditions essential to such growths, such as nutrients, exist. For example, dense algal growths, which periodically break loose and release undesirable organic matter to the receiving water, have been observed in the condenser discharge channels of steam-electric power plants. Nuisance algal growths also sometimes occur in stratified lakes as a result of temperature changes induced by altered circulation patterns. There is not sufficient evidence to indicate the temperature increase that will necessarily result in an increase in nuisance organisms; hence, careful evaluation of the local conditions is required in each case.

Short-Term Exposure Criteria. Aquatic organisms entrained with the condenser intake water and carried through the condenser are exposed to an elevated temperature for some time, especially if the discharge occurs into a long canal. Even if not entrained, organisms may enter the heated discharge before it mixes with the receiving water body. To protect aquatic life, it is required to know the length of time each important organism can survive at temperatures that exceed the upper incipient lethal temperature appropriate to the existing acclimation temperature. The NAS-NAE report describes a method for estimating this time for many different fish species.[g] For example, it is found that juvenile largemouth bass could survive a rise in temperature from 21.1 °C (70 °F) to 32.2 °C (90 °F) for about 7 minutes. As is to be expected, the survival times decrease as the temperature rise increases, and vice versa.

Where once-through cooling is used and largemough bass are an important species, it is evident that the condenser water should not be discharged into a long canal where entrained juveniles would be exposed to the elevated temperature for hours. Most of the juvenile bass would be expected to survive passage through the condenser, but they would be killed by the long exposure in the discharge canal. In these circumstances, the condenser effluent should be discharged in such a way that it enters and mixes with the receiving water body soon after leaving the condenser. However, the fish might not survive passage through the mixing zone described in the next section. The foregoing considerations refer only to thermal effects and do not include damage that might arise from mechanical factors.

Mixing Zones

When the condenser water is discharged to a receiving water body, the discharge spreads out to form a *plume* in which the temperature decreases

[g]An extensive tabulation of the data required for making the calculations for different species is given in Appendix II-C of the NAS-NAE report *Water Quality Criteria 1972*.

steadily from that at the discharge point to the normal ambient water temperature. The shape, volume, and temperature characteristics of the plume depend on the conditions of the discharge and the receiving body, as described in the next chapter.

An important part of the plume is the *mixing zone.* This expression is applied to all pollutants including heat; the discussion here, however, is restricted to the *thermal mixing zone* (Fig. 12-9). It is then defined as a region in which a discharge with a higher temperature than that of the receiving water is in transit. In this region the water characteristics necessary for the protection of aquatic life are based on the short-term criteria described above. At the boundary of the mixing zone, the temperatures are such as to satisfy the long-term requirements for the protection of aquatic life. In other words, the mixing zone may be thought of as that part of the plume in which the temperatures exceed the maximum permissible for a prolonged exposure (i.e., a week or more).

The NAS-NAE Committee on Water Quality Criteria made a number of recommendations concerning mixing zones, some of which are applicable to thermal discharges. Nonmobile benthic and sessile (i.e., attached) organisms in mixing zones may experience long or intermittent exposures to temperatures exceeding the values recommended for prolonged exposures; as a result, the populations of the organisms will suffer damage. Such damage may be minimized by suitable location and design of the discharge system (see Chapter 13) so as to maintain a sufficiently low temperature at the bottom of the receiving water body.

Organisms entrained in the condenser water will inevitably be discharged into the mixing zone. In addition, weak swimmers and drifting organisms may

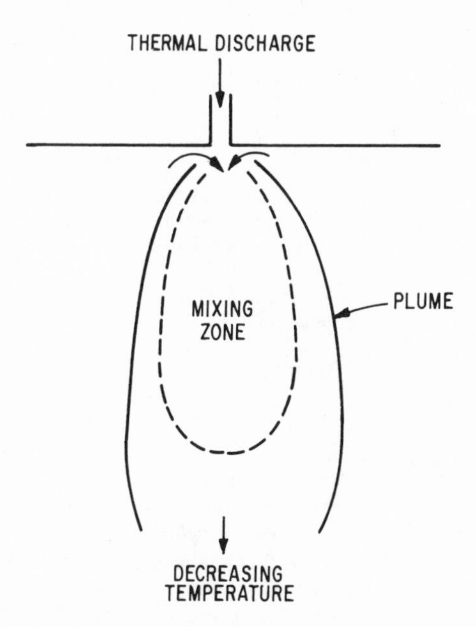

FIG. 12-9. Simplified (idealized) representation of thermal plume and mixing zone on the water surface (for more detail, see Chapter 13).

be drawn into the mixing zone by the discharged water flow. The exposure time in the mixing zone is, however, relatively short because of the motion of the water. The dimensions and temperature distribution in the mixing zone should be such as to satisfy the criteria for short-term exposure, with due allowance for the times spent in the different temperature regions. Strong swimmers can avoid or move out of a mixing zone, but migrating fish may swim through such a zone. Water quality temperature standards for the protection of drifting and entrained organisms should also protect migrating fish.

The dimensions of the mixing zone relative to that of the receiving water body should be such that the natural upstream and downstream migration patterns of various fish species are not blocked. A passage zone, alongside or below the mixing zone, should be available to provide favorable conditions for migration. A continuous stretch of water bordered by the same bank of the receiving body should be available to permit passage of a portion, at least, of swimming and drifting species. A shoreline region is also required to provide spawning, nursery, and feeding areas. As a guideline, the NAS-NAE report recommended that no more than two-thirds of the width of a receiving water body be devoted to mixing zones, leaving at least one-third as a passage zone.

It is evident that there can be no simple criteria that would be applicable to mixing zones in all cases. Each situation must be evaluated separately in light of the existing local conditions. The shape and dimensions of the mixing zone will thus vary with the location, width, character, and flow rate of the receiving water body, and the nature of the aquatic organisms living therein or migrating through it. A few states have defined acceptable mixing zones, but in most instances the mixing zone characteristics for proposed power plants are determined on a case-by-case basis.

Bibliography—Chapter 12

Clark, J. "Thermal Pollution and Aquatic Life." *Scientific American* 220, no. 3 (1969): 18.

Coutant, C. C. "Biological Aspects of Thermal Pollution: Part I — Entrainment and Discharge Canal Effects"; and "Part II — Scientific Basis for Water Temperature Standards at Power Plants." *CRC Critical Reviews of Environmental Control*, 1 (1970): 341; and 3 (1972): 1.

______. "Cold Shock to Aquatic Organisms: Guidance for Power-Plant Siting, Design, and Operation." *Nuclear Safety* 18 (1977): 329.

______. "Compilation of Temperature Preference Data." *Journal of the Fisheries Research Board of Canada* 34 (1977): 739.

______. "Effects on Organisms of Entrainment in Cooling Water." *Nuclear Safety* 12 (1971): 600.

______. "Temperature Selection by Fish—A Factor in Power-Plant Impact Assessments." *Proceedings of the Symposium on the Environmental Effects of Cooling Systems at Nuclear Power Plants*. Vienna: International Atomic Energy Agency, 1975.

______. "Thermal Effects." *Journal of Water Pollution Control Federation* (1976): 1486.

Davis, J. P. "The New Federal Water Pollution Control Act and Its Impact on Nuclear Power Plants." *Nuclear Safety* 15 (1974): 262; *Nuclear Safety* 16 (1975): 421.

Eisenbud, M., and Gleason, G., eds. *Electric Power and Thermal Discharges*. New York: Gordon and Breach, Science Publishers, Inc., 1969.

Goodyear, C. P.; Coutant, C. C.; and Trabalka, J. R. *Sources of Potential Biological Damage from Once-Through Cooling Systems of Nuclear Power Plants* (ORNL-TM-4180). Oak Ridge, Tenn.: Oak Ridge National Laboratory, 1974.

International Atomic Energy Agency. *Proceedings of the Symposium on the Environmental Effects of Cooling Systems at Nuclear Power Plants.* Vienna: International Atomic Energy Agency, 1975.

Merriman, D. "The Calefaction of a River." *Scientific American* 224, no. 5 (1970): 42.

______. "Does Industrial Calefaction Jeopardize the Ecosystem of a Long Tidal River." *Proceedings of the Symposium on the Environmental Aspects of Nuclear Power Stations*, 50. Vienna: International Atomic Energy Agency, 1971.

Mount, D. I. "Thermal Effects Statement." *Hearings before the U.S. Congress, Joing Committee on Atomic Energy, on the Environmental Effects of Producing Electric Power*, Part I, p. 356. Washington, D.C., 1969.

Nakatani, R. E.; Miller, D.; and Tokar, J. V. "Thermal Effects and Nuclear Power Stations in the U.S.A." *Proceedings of the Symposium on the Environmental Aspects of Nuclear Power Stations*, p. 561. Vienna: International Atomic Energy Agency, 1971.

Parker, F. L., and Krenkel, P. A., eds. *Biological Aspects of Thermal Pollution.* Nashville, Tenn.: Vanderbilt University Press, 1969.

______. *Thermal Pollution: Status of the Art* (Report No. 3). Nashville, Tenn.: Vanderbilt University, Department of Environmental and Water Quality Engineering, 1969.

U.S., Atomic Energy Commission. *Proceedings of the Symposium on Thermal Ecology* (CONF-730505). Washington, D.C., 1974.

______. *Thermal Effects and U.S. Nuclear Power Stations* (WASH-1169). Washington, D.C., 1971.

______. *Toxicity of Power Plant Chemicals on Aquatic Life* (WASH-1249). Washington, D.C., 1973.

U.S., *Code of Federal Regulations*, Title 40 (Protection of the Environment).
Chap. 1. Environmental Protection Agency.
Part 122. Thermal Discharges (Subchap. N. Effluent Guidelines and Standards);
Part 423. Steam Electric Power Generating Point-Source Category.

U.S., Congress, House of Representatives, Committee on Public Works. *Laws of the United States Relating to Water Pollution Control and Environmental Quality.* Washington, D.C., 1973.

U.S. Department of the Interior. *Water Quality Criteria.* Report of the National Technical Advisory Committee, Federal Water Pollution Control Administration. Washington, D.C., 1968.

U.S., Energy Research and Development Administration. *Proceedings of the Symposium on Thermal Ecology*—II (CONF-750425). Washington, D.C., 1976.

U.S., Environmental Protection Agency. National Academy of Sciences—National Academy of Engineering Report. *Water Quality Criteria 1972.* Washington, D.C., 1973.

______. *Quality Criteria for Water.* Washington, D.C., 1976.

13

The Disposal of Waste Heat

INTRODUCTION

Cooling Water Requirements

The rate at which waste heat must be removed in the turbine condenser of a steam-electric power plant depends on the operating power of the plant and the efficiency with which the heat is converted into electrical energy. The efficiency is often expressed by a quantity called the *heat rate*; it is the amount of heat input, in British thermal units (Btu), used to generate 1 kilowatt-hour (kWh) of electricity in the given plant.[a] The thermal energy equivalent of 1 kWh is 3413 Btu. Hence, the thermal efficiency of the power plant is equal to 3413 Btu divided by the heat rate in Btu; if the result is multiplied by 100, the efficiency is obtained in percent, as it is usually stated (see Chapter 1).

For nuclear power plants using light-water reactors (LWRs), such as are common in the United States (see Chapters 2 and 4), the heat rate is about 10 400 Btu/kWh. The efficiency for the conversion of heat into electricity is thus about $(3413/10\,400) \times 100 = 33$ percent. In a nuclear plant, approximately 5 percent of the heat supplied is dissipated in the plant, leaving 62 percent to be removed in the condenser water (p. 3). Hence, for every kilowatt-hour of electrical energy generated, $0.62 \times 10\,400 = 6500$ (approximately) Btu must be removed. This result may be put in another form: for each kilowatt of electric power that is generated, heat must be removed from the condenser at a rate of 6500 Btu per hour. Consequently, for an LWR plant with a generating capacity of 1000 megawatts electric (MWe) (i.e., 1 million kilowatts), the rate of heat removal would have to be 6500 million (i.e., 6.5×10^9) Btu per hour.

An indication of the quantity of condenser cooling water required to remove heat at this rate can be obtained in the following manner. The rate (in Btu/h) at which water removes heat from a condenser is equal to the flow rate of the water (in lb/h) multiplied by the increase in temperature (in °F) of the water in passing through the condenser; that is to say,

$$\underset{(\text{Btu/h})}{\text{Rate of heat removal}} = \underset{(\text{lb/h})}{\text{Rate of water flow}} \times \underset{(°\text{F})}{\text{Temperature increase.}}$$

[a]The Btu is defined as the amount of heat required to increase the temperature of 1 lb of water by 1 °F.

For nuclear power plants in the United States, the temperature increase of the condenser water in most cases is in the range of 6.3 to 19 °C (10 to 30 °F), with an annual average of 10 °C (18 °F), when the plant is operating at full (or rated) capacity (see Appendix, Table A-I). The average rate of water flow through the condensers in a 1000-MWe plant, assuming a temperature increase of 18 °F can then be obtained by writing

$$6.5 \times 10^9 \text{ Btu/h} = \underset{(\text{lb/h})}{\text{Rate of water flow}} \times 18\ ^\circ\text{F},$$

so that

$$\text{Rate of water flow} = 360 \text{ million lb/h}.$$

Water flow rates are commonly expressed in cubic feet per second (ft^3/s) or gallons per minute (gal/min). At normal temperature, the weight of a cubic foot of water is about 62.3 lb and that of a gallon is 8.33 lb. It follows, therefore, that when operating at its rated capacity, the plant would require, on the average, water flow rates of roughly 1600 ft^3/s or 720 000 gal/min to remove the heat from the condenser. These numbers apply to the hypothetical average case in which the water temperature increase is 10 °C (18 °F). If the temperature rise is more than 18 °F, the quantity of water required would be decreased proportionately, and vice versa.[b]

The condenser water flow rate is determined in metric units by

$$\underset{(\text{W})}{\text{Rate of heat removal}} = \underset{(\text{kg/s})}{\text{Rate of water flow}}$$

$$\times \underset{(^\circ\text{C})}{\text{Temperature increase}} \times 4.186 \times 10^3.$$

The 1000-MWe plant has a generating efficiency of 33 percent, whereas 62 percent of the heat supplied is removed by the condenser water. Hence, the rate of heat removal is $1000 \times (62/33) \approx 1900$ MW (or 1900×10^6 W). For an assumed water temperature increase of 10 °C,

$$1900 \times 10^6 \text{ W} = \underset{(\text{kg/s})}{\text{Rate of water flow}} \times 10\ ^\circ\text{C} \times 4.186 \times 10^3,$$

so that

$$\text{Rate of water flow} = 45 \times 10^3 \text{ kg/s}.$$

The volume of 10^3 kg of water is 1 m^3; hence, the required water flow rate is 45 m^3/s, which is equivalent to 2.7 million litres/minute (l/min).[c]

[b]For a modern fossil fuel plant, the water flow rate for the same temperature rise would be roughly two-thirds of that for an LWR nuclear plant of the same electrical output (Chapter 1).

[c]One cubic meter is equal to 1000 litres.

Open, Closed, and Variable Cycles

Clearly, large volumes of cooling water are needed for the condensers of a nuclear power plant. In a once-through cooling system, the water must be drawn continuously at the required high rate from (and discharged to) an adjacent water body. At the same time, the temperature conditions in the receiving water must comply with the existing standards, as described in Chapter 12. Hence, large power plants with once-through cooling are possible only on major rivers, large lakes, or the ocean, where the water supply is adequate.

In many locations, however, once-through cooling is not possible. For example, where the water required would represent a large proportion of the available water supply, especially during periods of low flow (e.g., summer or drought conditions). In other cases, the plant design does not permit operation of a once-through system while meeting the local water quality (temperature) standards. Another situation in which once-through cooling is not permissible, even if the temperature standards can be met, is when the damage by entrainment of fish eggs, larvae, and juveniles would cause a significant decrease in the population of an important aquatic species. In such cases, an alternative method of treating the condenser effluent, for example, by the use of cooling towers, ponds, etc., is mandatory.

The extremes of condenser heat discharge are represented, on the one hand, by once-through (or open-cycle) cooling and, on the other hand, by a closed-cycle system, in which the warmed water is cooled in a tower or pond and recirculated through the condenser. Between these extremes are variable-cycle systems. For example, where the water supply is adequate to meet the temperature standards with once-through cooling except at certain times, a "helper" tower or pond can serve to cool all or part of the water at these times prior to discharge. Another possibility is a combination of open- and closed-cycle systems in variable proportions, as may be dictated by the available flow of fresh cooling water and the temperature conditions.

A listing of the cooling systems for nuclear power plants operating, under construction, or on order in the United States is given in the Appendix. Since these lists were compiled, a number of plants on order have been cancelled.

Environmental Protection Agency Regulations

In the Federal Water Pollution Control Act (FWPCA) amendments of 1972 (discussed in Chapter 12), Section 301 states that, not later than July 1, 1983, all sources of pollution "shall require application of the best available technology economically achievable. . . which will result in progress toward the national goal of eliminating the discharge of pollutants." To meet this objective, the U.S. Environmental Protection Agency (EPA) has proposed in the *Code of Federal Regulations*, Title 40 Part 423 (40CFR423) regulations for effluents of all kinds from steam-electric power plants. As far as condenser water is concerned, the EPA has concluded that the "best available technology economically achievable" would be the use of cooling towers or other water recirculation system.

By limiting thermal discharges to the relatively small quantities of the blowdown (p. 351) from cooling towers or ponds, the EPA regulations preclude the use of once-through cooling for power plants of more than 500-MWe capacity placed in service after January 1, 1970. However, as seen on page 329 of this book, Section 316 of the amended FWPCA provides for exemptions (or exceptions) from the regulations in special cases. For example, when it can be shown that finfish, shellfish, and wildlife would not be adversely affected by a once-through system. Exceptions might also be granted where sufficient land is not available for cooling towers or where unavoidable salt drift or water vapor plumes from these towers would be a serious problem. Nevertheless, it is expected that, in the future, most nuclear power plants will have to employ some form of closed-cycle cooling for the condensers.

Transfer of Heat to the Atmosphere

Regardless of the type of cycle employed, the waste heat removed by the condenser cooling water is eventually dissipated to the atmosphere. The transfer of heat from the warmed water to the air takes place by evaporation, radiation, and convection. Of these three processes, evaporation of the water is the most important. When a liquid is converted into a vapor it absorbs a certain quantity of heat, called the *latent heat* of vaporization. When this heat is taken up from the liquid itself, the temperature must fall. Thus, heat can be transferred from the water to the atmosphere by vaporization.

In heat transfer by radiation, heat energy passes from one body (or material) to another without the necessity for direct contact between them. For convective heat transfer, the materials must be in contact; heat transfer by convection occurs only when one material is a fluid, such as air, and it moves relative to the other, which is the warmed water in the situation being considered. No matter how the heat is transferred from the condenser water to the atmosphere, the greater the temperature difference between the water and the air, the greater is the rate of heat loss from the water.

WARM WATER DISCHARGE SYSTEMS

Methods of Discharge

In a once-through cooling system, all the warm water leaving the condenser is discharged to the receiving water body. On the other hand, in a closed-cycle system, the water is recirculated; nevertheless, substantial volumes of water, called the blowdown (p. 351), are discharged from time to time (or continuously) to prevent excessive accumulation of dissolved salts in the cooling water. The method of discharging the water is important, but it is especially so for a once-through system because of the much larger volume of water discharged than in a closed cycle. In any case, the particular discharge technique used in a given situation is dictated by the need to minimize the biological impact on the aquatic environment by meeting the applicable water quality standards.

Many different types of discharge are possible, but two extreme situations may be considered: namely, (a) complete stratification and (b) complete mixing. In complete stratification, the condenser water is discharged slowly at the surface so that it forms a relatively shallow layer on the receiving water body.

There is essentially no mixing, and heat dissipation to the atmosphere occurs at a maximum rate for the given temperature conditions. At the other extreme of complete mixing, the condenser water is discharged rapidly through nozzles or ports located at a depth, often at or near the bottom, in a flowing water body. Nearly all the heat is thus stored in the receiving water, and heat dissipation from the surface to the atmosphere is relatively slow over a very large area.

Practical discharge systems generally lie between complete stratification from a slow surface discharge to complete mixing from a rapid submerged discharge. The maximum temperature increase above ambient prescribed in thermal regulations would generally make complete stratification unacceptable. However, if the discharge occurs at a moderate depth below the surface or if the rate of discharge is increased (or both), there will be some mixing with the ambient water, and the temperature increase at the surface would be lowered. In submerged discharge, the mixing rate can be varied by the nature of the ports or diffusers through which the condenser water enters the receiving body. Most new nuclear power plants have multiport subsurface diffusers.

In the design analysis of a proposed discharge system, the thermal plume (p. 336) is often considered as being made up of two zones: the near- and far-field regions. In the near-field region, the temperature distribution is determined by the mixing of the discharged water with the ambient water due to the momentum of the discharge and to buoyancy effects. In the far-field region, however, the normal conditions in the receiving water body are dominant; these include natural turbulence, temperature stratification, and heat dissipation from the water surface. From the engineering standpoint, the near-field region is important because the temperature distribution can be controlled by the design of the discharge structure and the rate of discharge of the water.

The near-field surface temperature distributions for a relatively slow surface discharge and a fairly rapid submerged discharge have been determined by computer modeling. The results for a particular case are depicted in Fig. 13-1 for discharge from a condenser in which the temperature of the water has been increased by 10 °C (18 °F); the temperatures indicated are the values at the surface in excess of the intake water temperature. The situations shown in the figures represent the simple (ideal) case in which there is no movement of the water as a result of natural flow or wind. In actual practice, however, the plume would be distorted by water currents, wind, bottom topography, and other variables (see Fig. 13-2).

Because of the rapid mixing in a submerged discharge, the mixing zone area (p. 337) is smaller than for a surface (or near-surface) discharge. The effects on aquatic life forms, especially organisms entrained in the condenser water are thus different. Another difference in biological impact arises from the fact that in a surface discharge, the plume surface often hugs the nearby shore and organisms in the littoral (near-shore) zone could be harmed. On the other hand, a deep discharge could adversely affect bottom (benthic) organisms, including fish eggs, and larvae.

In the following sections, typical surface and submerged discharges are discussed in more detail. The cases considered approach the extremes mentioned earlier, and various modifications are possible, as already noted.

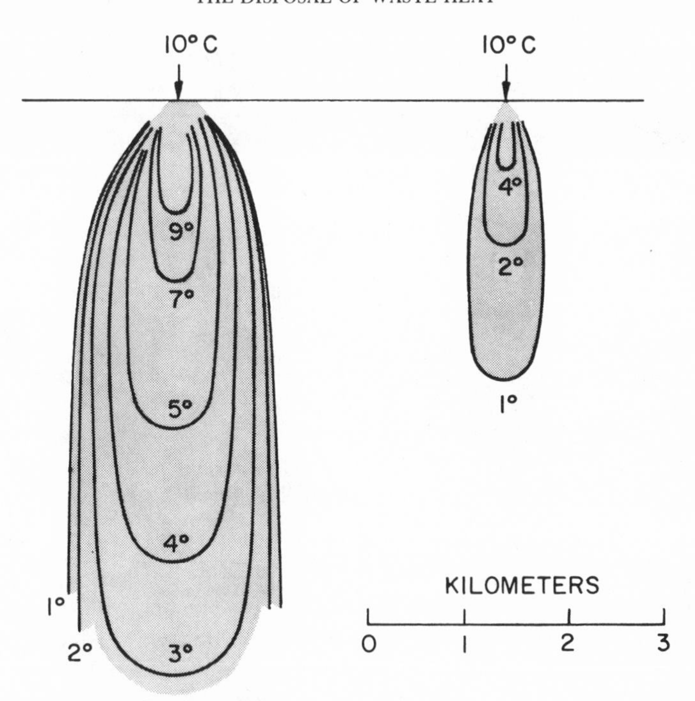

FIG. 13-1. Ideal representations of temperature increases and dimensions in a surface discharge with gradual mixing (left) and rapid mixing (right).

Surface Discharge

For surface (or near-surface) discharge, the condenser effluent usually enters the receiving water body by way of a shallow canal that is generally widened at the exit end. The discharged water flows on and just below the surface of the receiving water. Because of its higher temperature, the discharge has a lower density than the main body of water, and it therefore tends to remain near the surface, extending in the direction of stream flow and spreading sideways; in this way, a warm plume of large area is formed. There is some downward mixing and entrainment of cooler water as a result of turbulent motion and shear, but this does not usually extend to a depth of more than 3 to 4.6 m (10 to 15 ft).

A general representation of the temperatures in excess of the intake water temperature in the plume from a surface discharge in a flowing river is shown in Fig. 13-2 in the form of top and side views. In this case, the plume may extend for several kilometres and be carried to the near shore by the water currents (top view). The maximum depth of the plume is about 5 m (16 ft) below the bottom of the discharge canal (side view).

As the plume extends downstream, the additional heat present near the surface of the water is dissipated to the atmosphere by vaporization, convection, and radiation. The rate of dissipation increases with an increase in the temperature difference between the water and the air. Hence, other conditions being equal, a larger quantity of heat will be dissipated within a given distance

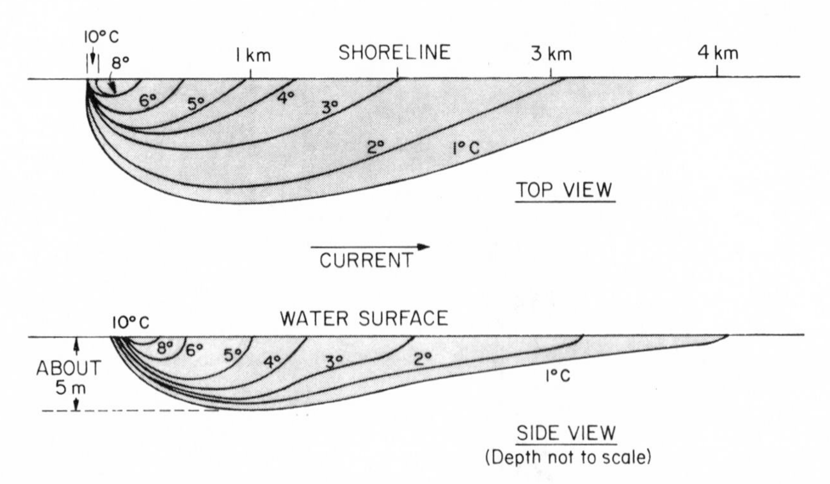

FIG. 13-2. Temperature increases in near-field plume of slow surface discharge in flowing water (hypothetical). (Adapted from *Engineering for Resolution of the Energy-Environment Dilemma*, National Academy of Engineering [1972].)

if the surface is warmer than if it is cooler. A low flow rate of the discharge would be advantageous in this respect. However, to comply with local temperature standards, it may be necessary to increase the flow rate with some loss in the rate of heat dissipation. Surface cooling and mixing eventually cause the plume to lose its identity; the whole receiving body (e.g., a river) will thus be warmed up slightly for a downstream distance of many kilometers.

A difficulty may arise with surface discharge as a result of the warmer water backing up from the discharge point to the intake upstream. The entry of warmer water into the condenser would decrease the turbine efficiency. Some upstream spread of the plume is a common occurrence, especially if the wind is blowing in that direction. When the water is discharged into a river estuary, the upstream spread of the plume is greatly accentuated during flood tide.

One solution to the recirculation problem is to have a sufficient distance between intake and discharge points. This procedure has the added benefit of avoiding the attraction of fish to the water intake by the presence of warmer water. Another measure is to erect a vertical 'skimmer" wall near the intake, so that water is drawn in from a short distance below the surface. Even if there is no backup of warmed effluent, the deeper water is often cooler than that at the surface.

Submerged Discharge

In a submerged discharge, the condenser effluent is carried by pipes to a depth in the receiving water body and is expelled through jets or multiport diffusers, either parallel or perpendicular to the direction of flow. The discharged water can mix rapidly with the bulk of the water in the receiving body and moves on downstream. In this way it is possible to limit the extent of the mixing zone, although the entire plume of water peaked a small amount above ambient may extend over a large area. A schematic representation of the excess temperature distribution in the near-field part of the plume is given in a side view in Fig. 13-3.

The heat present in the condenser effluent is now distributed through a large volume of water, and the temperature is rapidly decreased; this reduces the rate of heat loss to the atmosphere. Hence, although the temperature falls off rapidly with distance near the point of a submerged discharge, the volume of water at a temperature slightly above normal is greater than it would be for an equivalent surface discharge. In situations where there is a stream flow that is large compared to the discharge flow, so that the increase in temperature throughout the volume of the receiving body is small, the thermal effects in the bulk of the water will be minimal. The possible damage caused by heat and scouring on the bottom organisms should, however, not be overlooked.

When the Indian Point No. 1 nuclear power plant (p. 323) first started operation, the condenser water was discharged to a canal, from which the effluent spread over the surface of the Hudson River. At certain times of the year, however, it was found impossible to meet the New York State 1969 standard of a maximum surface temperature of 32 °C (90 °F). Consequently, a submerged discharge system was installed in 1970 which is different from that described above. It consists of a number of rectangular ports, 1.2 m (4 ft) high, 4.6 m (15 ft)wide, and 0.9 m (3 ft) apart. The centers are 3.7 m (12 ft) below the U.S. Coast and Geodetic Survey Sea Level Datum. The effluent is emitted in a horizontal direction at a velocity of 3 m (10 ft) per second. The heated water tends to rise and reaches the surface about 38 m (125 ft) from the discharge. Mixing and diffusion with the bulk of the river water serve to maintain temperatures that meet the New York State standards.

Discharge into Lakes and Reservoirs

Discharge of heated water to a lake or a large reservoir may involve some special thermal problems. Many deep lakes become stratified in temperature (and density) during the summer into three more or less distinct layers. In the top layer *(epilimnion)*, which is warmed by the sun and mixed by the wind, the

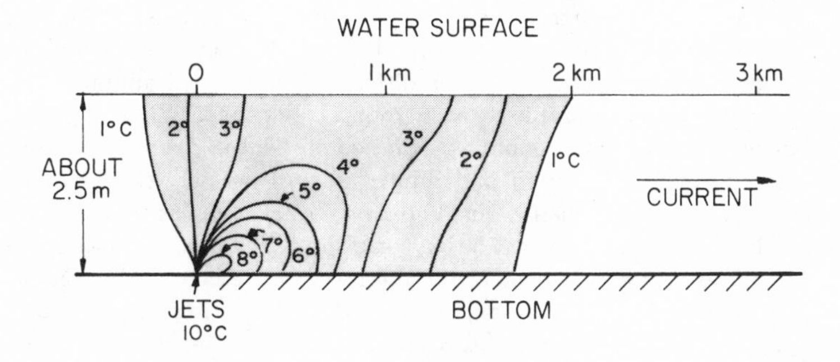

FIG. 13-3. Temperature increases in near-field plume of rapid submerged discharge in flowing water (hypothetical). (Adapted from *Engineering for Resolution of the Energy-Environment Dilemma*, National Academy of Engineering [1972].)

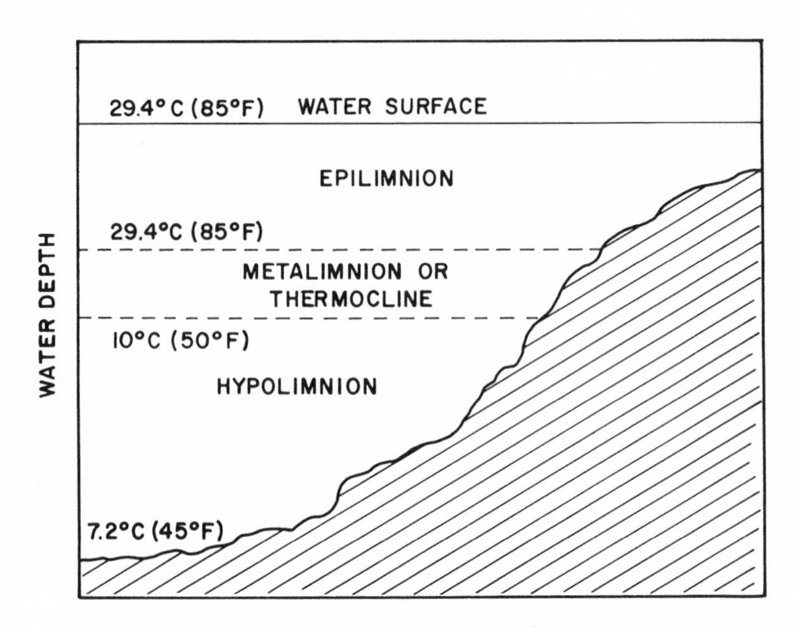

FIG. 13-4. Typical stratification of a lake (or reservoir) in the summer (temperatures are approximate and vary with location).

temperature is essentially the same throughout and approaches that of the atmosphere. Below this is an intermediate layer *(thermocline,* or *metalimnion),* in which the temperature drops sharply, by perhaps as much as 17 to 19 °C (30 to 35 °F), with increasing depth. At the lowest level in the lake there is the third layer *(hypolimnion),* in which only a small further drop in temperature occurs. The general situation in summer is depicted in Fig. 13-4. It is seen that the temperature in the bottom layer is substantially less than at the top of the lake.[d]

During the late autumn and in the winter in temperate latitudes, a so-called "turnover" generally occurs. The surface water cools and tends to sink, thereby causing the lower layers to move upward; eventually, there is complete mixing. In the winter there are no separate layers, and the temperature of a large lake or reservoir is nearly uniform throughout its depth. The temperature at the bottom of a deep lake in winter is not very much different from that in summer. This makes the bottom water desirable for year-round condenser cooling.

The bottom layer of water contains a limited amount of oxygen, taken up from the air during the autumn and winter mixing period. After summer stratification has occurred, the intermediate layer blocks the entry of any further oxygen into the bottom layer. The oxygen is gradually used up both by the aquatic organisms living in the cooler bottom layer of the lake in the summer and also by the decay of dead plankton and other organic matter descending from the top layer. In most small lakes there is a substantial decrease in the quantity of oxygen present in the bottom layer, and the variety and numbers of aquatic species living there is diminished.

[d]For simplicity, the technical names of the three layers are not used subsequently; the layers are described as top, intermediate, and bottom, respectively.

As a general rule, when a steam-electric power plant is located on a lake, cold water is withdrawn at a substantial depth and is discharged at a somewhat higher temperature near the surface. In view of the natural temperature difference between the bottom and top layers in the summer, the temperature of the effluent will be closer to the normal surface temperature of the lake. Of course, this would not be the case in winter when stratification is lost as a result of turnover.

Some possible drawbacks to deep intake and surface discharge are the following. Transfer of water from the bottom results in a decrease in the amount of oxygen available for aquatic life near the surface. Furthermore, the addition of heat to the top layer of water could retard the turnover that normally occurs when the weather becomes colder. Such a delay might possibly affect the ecological balance of the lake.

The National Technical Advisory Committee on Water Quality Criteria (Chapter 12) recommended that warm effluent not be discharged to the bottom layer of a lake unless a special study shows that this may be desirable in a particular local situation. Normally, the cool bottom water provides the required environment for cold-water fish in the summer. An increase in temperature produced by a thermal discharge could thus have serious consequences.

Discharge into Estuaries and the Ocean

Because of the large volumes of water in estuaries and in the ocean, these waters would appear, at first sight, to be promising for use in once-through cooling systems. There are, however, a number of problems that must be considered. In the first place, many aquatic organisms are sensitive to the salinity of the water (p. 321). It is of prime importance, therefore, that the water quality, especially the salinity, at the condenser intake should not differ greatly from that in the receiving water body.

Estuarine waters are frequently the breeding areas for several varieties of ocean fish. The population of these fish could be severely affected by damage caused by entrainment of eggs, larvae, and juveniles in the condenser water. The effects of entrainment can be so severe as to preclude the use of estuarine waters for once-through cooling in some locations.

In many coastal areas there exists a delicate ecological balance, which could be upset by an increase in temperature (and salinity) of the water. Consequently, the water quality criteria of states with coastal waters generally permit a temperature increase not to exceed 0.83 °C (1.5 °F) in the summer months. However, it is pointed out later in this chapter that, under controlled conditions, the warm condenser effluent may possibly serve to increase the harvest of shellfish in coastal waters.

When the ocean is used as a source of once-through cooling, care must be taken to ensure effective mixing of the discharge with the ocean water. Submerged discharge is commonly employed, but surface discharge is possible where mixing is favored by wave action. The outlet should not, however, be located near an estuary where tidal currents might carry the warm water inland toward spawning or migration areas.

The use of cooling water from estuaries or from the ocean requires that the condenser tubing, as well as the intake and discharge pipes, be made of a corrosion-resistant material. Since shellfish and other marine organisms are adversely affected by copper, the use of alloys containing copper may not be acceptable. It would then be necessary to employ tubing of more expensive stainless steel or other copper-free alloy.

CLOSED-CYCLE SYSTEMS

Introduction

For one or another of the reasons discussed earlier, many steam-electric plants, both nuclear and fossil-fuel, are using or will use closed-cycle (or recirculation) cooling systems. Various types of cooling ponds and towers are employed to dissipate the waste heat from power stations. Closed-cycle cooling systems add to the cost of the electricity generated, but the increase is an acceptable price for minimizing damage to existing aquatic ecological systems.

Although in once-through cooling the waste heat is discharged to a water body, the heat is eventually dissipated to the atmosphere over a wide area. With cooling ponds and towers, on the other hand, the heat from the condensers is discharged directly to the atmosphere in the general vicinity of the pond or tower. As a result, there are a number of atmospheric environmental effects that must be considered in locating the power plants.

In comparing closed-cycle systems with one another and with once-through cooling, two aspects of water use are of interest. One is the rate at which water is continuously withdrawn from an adjacent water body, and the other is the *consumptive use*—that is, the net rate of loss of water from the body. In closed-cycle cooling, the withdrawal rate, which is required to make up for evaporative and other losses, is not more than 2 or 3 percent of the rate in once-through systems. An important consequence is a great decrease in the damage to aquatic organisms from impingement on screens and entrainment because of the low rate of water withdrawal in closed-cycle cooling (Chapter 12).

In once-through (or open-cycle) systems, essentially the only loss of water is from evaporation. In closed-cycle systems of the wet type, however, there are other losses, as described shortly. Hence, the consumptive use of water is roughly 50 to 100 percent larger than in once-through cooling. In dry cooling systems (p. 361), there is essentially no consumptive use of water.

Cooling Ponds and Canals

In a cooling pond[e] the heated condenser water is commonly discharged at one end, usually at the surface, and the cool water is withdrawn at the other end of the flow path. Cooling ponds are often constructed with internal dikes (or guides) to direct the water flow and extend the path length. Several days are usually required for the water to circulate from one end to the other.

The heated water loses heat to the atmosphere mainly by vaporization and also by radiation and convection. The rate of heat loss (for a specified water sur-

[e]The word *pond* in this context implies a natural or man-made enclosed body of water, which may be a lake or a reservoir.

face) depends on the temperatures of the water and the ambient air, the wind speed, and the relative humidity.[f] For given atmospheric conditions, the heat loss increases in proportion to the area of the water surface. An area of about 400 to 1200 hectares (1000 to 3000 acres), depending on climatic conditions, is required for a nuclear power plant with an electrical capacity of 1000 MW. In a region of high relative humidity, where the air normally approaches saturation with water vapor, the water evaporates (and cools) fairly slowly. A larger pond area would then be required than in a region of low humidity. Wind facilitates cooling, partly by blowing fresh air across the water surface and partly by causing waves that increase the area of water exposed to the air.

As a rule, circulation of water in a cooling pond does not extend below a depth of some 3 to 4.6 m (10 to 15 ft). Hence, the maximum depth of water, at the condenser intake, does not need to exceed this amount; the average depth of the pond can then be less. A rough rule-of-thumb is that the minimum capacity of a cooling pond, assuming optimum water use, is equal to the volume of water passing through the condensers in a 24-hour period. The bottom of the pond or the ground below should be fairly impervious to water, although some seepage (and absorption) is usually tolerable.

An arrangement of canals can serve the same purpose as a cooling pond, as is the case in the closed recirculation system for the condenser water at the Turkey Point power plants in Florida. There are 15 canals, side by side, each 60 m (200 ft) wide and 1.5 m (5 ft) deep. The total length is about 270 km (168 miles), and a complete circulation of the water requires about two days. This system of canals is expected to serve two nuclear and two fossil-fuel plants.

As a result of continued evaporation of the pond (or canal) water, chemicals used for water treatment and minerals normally present in the water tend to become concentrated. Unless some corrective action is taken, the concentration might reach a point at which scale would deposit on the condenser tubes and thus decrease the cooling effectiveness of the water. To prevent this deposition, part of the water is discharged, either continuously or periodically. The discharged water is called *blowdown*.[g] The rate of blowdown depends on the amount of dissolved chemicals and minerals in the water. An average figure for a 1000-MWe nuclear plant would be roughly 11 000 to 15 000 litres (3000 to 4000 gal) per minute.

As a general rule, the blowdown is discharged to a flowing water body. The discharge must comply with both state and EPA water quality standards, especially those concerned with chlorine content and temperature (Chapter 12). Furthermore, the EPA regulations (40CFR423) require that the "temperature at which the blowdown is discharged . . . [should] . . . not exceed . . . the lowest temperature of the recirculated cooling water prior to the addition of make-up water." At the Pebble Springs Nuclear Plant in Oregon, it is planned to use part of the blowdown for irrigating agricultural land.

[f]The relative humidity, as commonly given in the weather reports, is the ratio, expressed in percent, of the water vapor actually present in the air to the amount the air would hold if it were saturated at the existing temperature.

[g]The same term is used in a somewhat related sense in connection with the water in a steam generator (p. 244).

Spray Ponds and Canals

The cooling efficiency of a cooling pond or canal system can be enhanced, and hence its area greatly decreased, by utilizing spray cooling. The water is broken up into small droplets by pumping it through nozzles to form sprays about 6 m (20 ft) high. Various types of spray nozzles have been designed; examples are fixed, tethered floating, and rotating forms. About 200 modules, each with a motor-driven pump supplying four nozzles, may be used for a 1000-MWe plant. The increase in the water surface produced by the sprays results in a more rapid loss of heat by evaporation and by convective heat transfer across the air-water surface. Because of their relatively small land requirements, there is an increasing interest in the use of spray canals and spray ponds for the dissipation of waste heat from nuclear power plants.

Makeup Water and Consumptive Use

Cooling ponds and canals of all types require a certain amount of makeup water to allow for evaporative (and other) losses and for blowdown. For a cooling pond, the makeup water may be provided by a stream flowing through the pond; otherwise, water may be pumped from a nearby river or other water body. In some locations, a cooling pond must be large enough to permit it to function effectively without the addition of makeup water at times of low flow of an adjacent river.

The makeup water requirements for cooling ponds and canals with sprays are much the same. They are somewhat greater for ponds and canals without sprays because the larger area results in increased losses from evaporation (due to solar heating) and seepage. Typically, makeup requirements for cooling ponds and canals with or without sprays is between 57 000 to 76 000 litres (15 000 and 20 000 gal) per minute for a 1000-MWe nuclear power plant.

Environmental Aspects of Ponds and Canals

An environmental problem arising from the use of cooling ponds and canals, with or without sprays, is that mist or fog may form over and near the surface in cold weather. The situation is especially acute for spray ponds and canals because of the downwind movement of the spray. Fogging and icing of nearby roads, power lines, and trees can occur for a distance of a few hundred metres. Such adverse effects must be taken into consideration in locating spray ponds and canals. On the other hand, large cooling ponds (or lakes) without sprays can have beneficial use to the public for recreational purposes. For example, cooling lakes have been developed as exceptional sport fisheries for such fish as largemouth bass, which readily tolerate the warmer water.

Types of Cooling Towers

There are three general types of cooling towers, commonly referred to as *wet*, *dry*, and *wet/dry*. In the wet towers, cooling is caused mainly by evaporation (and to a lesser extent by direct heat transfer) from drops or films of water to a flow of air. The basic principle is similar to that of a spray system in a cooling pond or canal; the main difference lies in the manner in which the water and air are brought into contact. Since some 60 to 90 percent, depending on the season of the year, of the cooling is due to evaporation of the water, wet towers are frequently called *evaporative cooling* towers.

In dry towers, by contrast, the water to be cooled and the air are not in direct contact and there is no evaporation. The water (or in some cases steam) passes through a series of metal pipes, and heat is transferred, mainly by convection, to air flowing over the pipes. Wet/dry towers are a combination of wet and dry systems; they usually consist of two sections, the lower section acting as a wet tower and the upper as a dry tower.

In all types of towers, the heat is removed by a flow of air. The flow may be caused by mechanical means (e.g., by a fan) or by making use of natural draft. In particular, there are two common forms of wet towers, *mechanical-draft towers* and *natural-draft towers*, which differ in the manner in which the air flow (or draft) is produced.

Wet Towers: General Principles

Regardless of the type of draft, there are certain general characteristics applicable to all wet towers. The water in these towers flows downward over a packing, or *fill*, consisting usually of long, narrow slats set either horizontally or vertically, according to the particular tower design. The slats are made of wood, plastic, or asbestos-cement. Warm water from the condenser discharge is pumped to the top of the tower and runs down over the fill. When the slats are horizontal, the water splashes down from one layer to the next and so breaks up into many small droplets. Towers in which the fill is set vertically are said to use *film packing*, because the water flows downward in the form of thin films.

In either droplet or film form, the water offers a large exposed surface from which evaporation and direct heat transfer can occur. The resulting rapid removal of heat by the surrounding air flow in the tower causes the temperature of the water to decrease. The cooled water collects in a basin at the base of the tower from which it is returned to the condenser at a temperature that may be up to 17 °C (30 °F) lower than when the water was discharged.

The lowest water temperature that is theoretically attainable by evaporative cooling is the *wet-bulb* temperature of the ambient atmosphere. The wet-bulb temperature is the lowest temperature indicated by a thermometer with its bulb covered by a moist wick (or other fabric) when exposed to the ambient air. The wet-bulb temperature depends on the actual (or *dry-bulb*) temperature of the air and on its relative humidity; the lower the temperature and the humidity, the lower the wet-bulb temperature. In practice, the temperature of the water leaving a wet cooling tower is higher than the theoretical minimum (i.e., the existing wet-bulb temperature). The difference between these two temperatures is called the *approach* to the wet-bulb temperature. Under good conditions (e.g., in summer), the approach might be from 4 to 8 °C (7 to 15 °F), but it is generally more, especially in winter.

The temperatures given in Table 13-I are values estimated for the Cherokee Nuclear Station, South Carolina, which will use mechanical-draft wet towers for condenser water cooling.

TABLE 13-I
VALUES ESTIMATED FOR THE CHEROKEE NUCLEAR STATION, SOUTH CAROLINA

	Winter		Summer	
	°C	°F	°C	°F
Dry-bulb temperature	8.9	48	33.3	92
Wet-bulb temperature	4.4	40	24.2	76
Approach	16.4	29.5	6.3	11.3
Water exit temperature	20.8	69.5	30.7	87.3
Temperature decrease	13.3	24	13.3	24

Drift has been a problem with wet towers in the past, but the situation has now been greatly improved. In addition to the danger of icing of nearby roads, trees, and utility lines in winter, the drift is troublesome with cooling towers because the water contains chemicals used to prevent biological fouling, corrosion, and structural deterioration, in addition to the salts normally present. These chemicals could cause harm to the plant life when the drift settles on the ground.

In order to reduce the drift, wet towers have *drift eliminators;* they force the air to change its flow direction sharply as it leaves the fill. The effect is to cause the entrained water droplets to separate from the air and remain within the tower. Drift eliminators also serve to equalize the flow of air through the tower. As a result of recent improvements in the design of drift eliminators, drift losses have been decreased to 0.005 percent of the total water flow rate in mechanical-draft wet towers and to 0.0025 percent in natural-draft towers.

Water losses from wet cooling towers arise mainly from evaporation and blowdown; the latter is required for the same reason as in a cooling pond or canal—namely, to prevent the concentration of dissolved salts from becoming high enough to cause scale to deposit on the condenser tubes. The drift loss for wet towers of current design is small in comparison with the evaporative loss.

The rate of water use in a cooling tower depends on local atmospheric and other conditions; hence, only very rough average values can be given here. Fairly typical loss rates for a 1000-MWe nuclear plant would be as follows:

Evaporation and drift: 45 000 to 57 000 litres/min
(12 000 to 15 000 gal/min)

Blowdown: 11 000 to 15 000 litres/min)[h]
(3000 to 4000 gal/min).

These amounts are roughly the same for both mechanical-draft and natural-draft towers.

Makeup water for wet cooling towers is thus required at the rate of approximately 57 000 to 72 000 litres (15 000 to 19 000 gal) per minute for an elec-

[h]For salt-water cooling towers, the blowdown would be much larger, perhaps 76 000 litres (20 000 gal)/per minute.

trical capacity of 1000 MW. This is roughly equal to the water requirements of cooling ponds and canals. Makeup water is usually supplied by pumping from an adjacent water body. In some cases a reservoir is needed to provide makeup water at times of low flow of a river from which the water is normally drawn. Since the blowdown is usually discharged to the water body, the consumptive use of water is represented by the evaporative and drift losses only.

In modern cooling towers, drift losses are relatively small, as may be seen from the following considerations. When cooling towers are used, the temperature increase of the water in its passage through the condensers is commonly about 14 °C (25 °F) or more; this is somewhat larger than is usually the case for once-through cooling. If the temperature increase is taken to be 14 °C, the flow rate of the condenser cooling water for a 1000-MWe installation is calculated to be about 2.3 million litres (600 000 gal) per minute. On the basis of a drift loss of 0.01 percent for a mechanical-draft tower and 0.0025 percent for a natural-draft tower, the drift losses would be 230 litres (60 gal) and 58 litres (15 gal) per minute, respectively. For wet towers of older design, the drift rates are larger.

Except for the blowdown, all the water lost from an evaporative cooling tower is transferred to the atmosphere within a limited area. Under certain temperature and other conditions, this could result in the formation of fog, leading to reduced visibility. In cold weather, ice could deposit on roads, trees, and power lines in the vicinity even when the drift is small. The dangers of fogging and icing are sometimes cited as drawbacks to the use of wet towers for cooling condenser water. The available evidence indicates that such dangers are not general and may be limited to a few days of high relative humidity in winter. As described shortly, natural-draft towers are much taller than mechanical-draft towers. Because they discharge the moist air at high elevations, they are less likely to produce fog and ice near the ground.

Mechanical-Draft Wet Towers

The principles described above apply to evaporative cooling towers of all types. Differences arise, however, in the method employed to move the air into which the water vaporizes. In mechanical-draft towers, the air flow is produced by means of a fan (or fans), whereas in natural-draft towers no mechanical aid is used to cause the air flow. Each type has its advantages and disadvantages.

In mechanical-draft towers, fans may either push (*forced draft*) or pull (*induced draft*) the air over and around the wetted fill. For large power plants, when considerable volumes of water have to be cooled, the induced-draft type is preferred. The subsequent discussion therefore refers only to wet towers that utilize induced draft to produce the required flow of air.

Contact between the water and the air is achieved either by *counterflow* or *crossflow* techniques. In counterflow towers, the air is drawn in through louvers located near the base. The air thus flows from the bottom upward, whereas the water to be cooled, which is sprayed in at the top, flows downward (i.e., counter to the air flow). In the crossflow towers, the air is sucked in at the sides. The flow is thus mainly at right angles to (i.e., across) the direction of the water flow from top to bottom. In the United States, crossflow has generally been

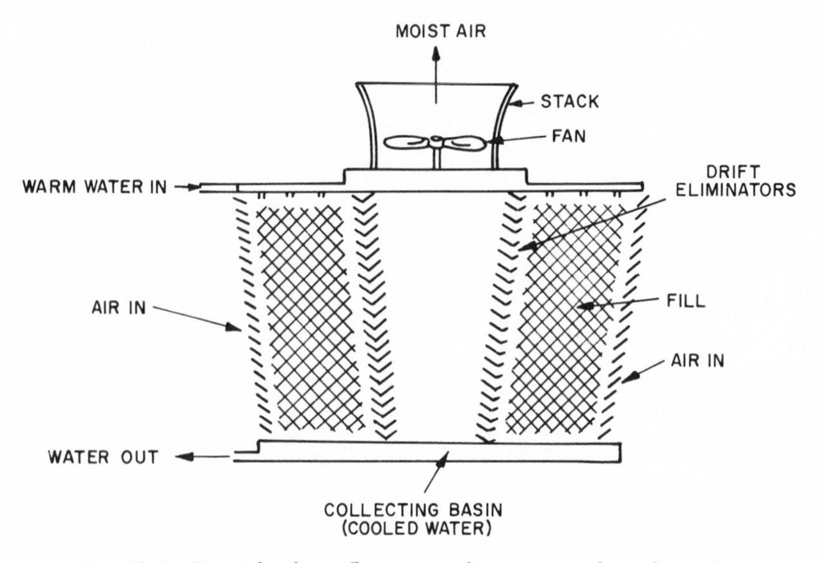

FIG. 13-5. Principle of crossflow wet cooling tower with mechanical draft.

preferred over counterflow for mechanical-draft wet towers for use in connection with large power plants.

Mechanical-draft wet cooling towers consist of a number of individual units. A cross-sectional diagram indicating the main features of such a unit of the crossflow type is shown in Fig. 13-5. Typically, each unit is a rectangular framed structure made of wood, galvanized steel, or concrete; it covers a ground area of about 230 m² (2500 ft²) and is some 12 to 18 m (40 to 60 ft) high. At the top of each unit is a fan roughly 8.5 m (28 ft) in diameter operated by a 150 kW (200-hp) electric motor. The fan is closely surrounded by a flared stack; this particular shape facilitates the movement away from the tower of the exhaust air, which is usually saturated with water vapor. The fresh air is drawn in through louvers on two opposite sides of the tower. The spacing, width, and slope of the louvers are designed to produce a uniform air flow with a minimum resistance and loss of water due to splashing.

From 8 to 20 units of the rectangular type just described are generally combined into one cooling tower structure, with a single collecting basin. A nuclear power plant operating at a full electrical capacity of 1000 MW would probably require about 30 to 40 individual units combined into three of four structures. The operation of this number of large fans is accompanied by considerable noise.

A new development in mechanical-draft wet towers is a circular type made of concrete. In one design, the circular towers have a diameter of 82 m (270 ft) and are 23 m (74 ft) in overall height; each tower has 13 fans, 8.5 m (28 ft) in diameter, arranged within an interior circle of 52 m (170 ft) diameter (Fig. 13-6). The outer 15 m (50 ft) of the radius of the structure contains the fill. Three such towers will be needed for a nuclear power plant with a capacity of 1280 MWe. Advantages of circular towers over rectangular mechanical-draft towers of the same cooling capacity are the smaller land area required, the lower cost, and the same plume buoyancy as an equivalent natural-draft tower (see below).

Natural-Draft Wet Towers

In natural-draft cooling towers, the flow of air is produced in the same manner as in a chimney. The difference in density between the less dense, warmer moist air inside the tower and the cooler air outside causes an updraft of the air in the interior. To produce sufficient draft, natural-draft cooling towers have to be very high, usually from 90 to 165 m (300 to 540 ft). This height must be borne in mind when comparing the accompanying illustration (Fig. 13-7) of a natural-draft tower with a mechanical-draft crossflow tower given earlier. At the bottom of the natural-draft tower is the region where vaporization and cooling of the water occurs. Both crossflow and counterflow designs have been used in natural-draft towers.

Natural-draft towers have a profile that is approximately hyperbolic in shape, as seen in the figure. Consequently, they are sometimes referred to as hyperbolic cooling towers. This particular shape provides additional strength to the reinforced-concrete structure. It is also more economical in the use of material than an equivalent straight-sided cylindrical tower. Hyperbolic towers are capable of withstanding winds of hurricane speed (i.e., well over 160 km [100 miles] per hour).

Natural-draft cooling towers are more expensive to build than those of the mechanical-draft type, but since they have no fans, the operating and maintenance costs are much less. They are also much quieter in operation. The exhaust air containing large amounts of water vapor leaves at the top of the tall tower and is carried even higher by the updraft and buoyancy of the moist air. A

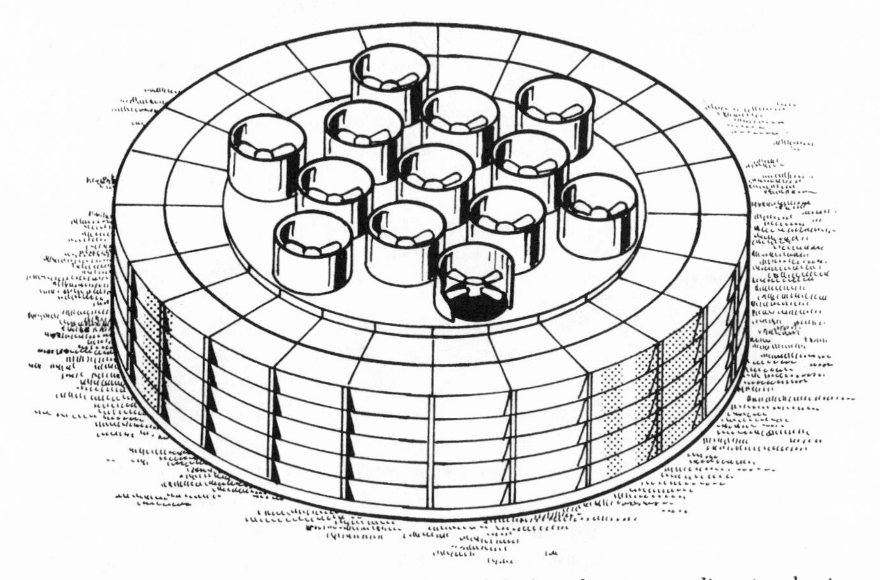

FIG. 13-6. Circular arrangement of mechanical-draft cooling towers; diameter about 82 m (270 ft) and 23 m (74 ft) high.

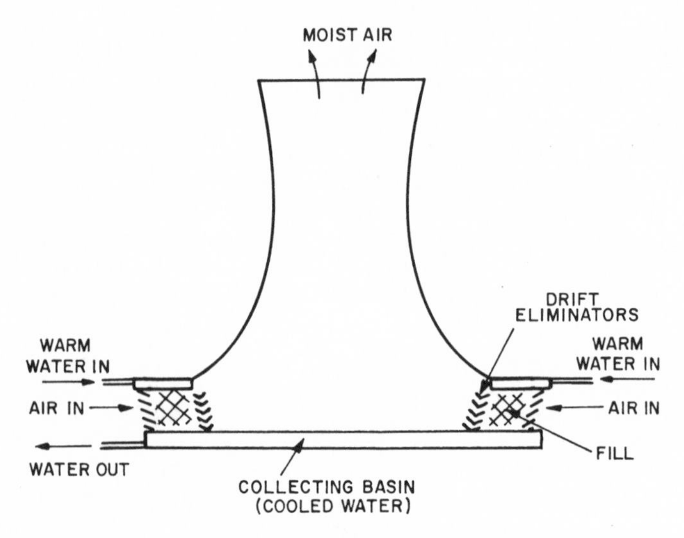

FIG. 13-7. Principle of hyperbolic natural-draft wet cooling tower.

rising visible plume (or cloud) of condensed water droplets is a common feature of natural-draft wet towers. The possibility of the formation of fog or ice on the ground is, however, less than that for a mechanical-draft tower.

Natural-draft cooling towers have been used extensively in Europe for some time, but in the United States the first one was constructed as recently as 1962. Since then, they have come into increasing use for cooling condenser water for electric power plants. In the past, towers have commonly been 75 to 90 m (250 to 300 ft) in diameter at the base and 90 to 105 m (300 to 350 ft) high. Larger natural-draft towers have been constructed, and it is estimated that a height of 165 m (540 ft) with a base diameter of perhaps 185 m (600 ft) is about as large as is practical at present. A single tower some 120 m (400 ft) in diameter and 150 m (500 ft) high is considered to be adequate to satisfy the cooling requirements of a 1000-MWe nuclear power plant. Many people find the great size of these towers esthetically displeasing, especially in areas of natural beauty or in the vicinity of historical sites or landmarks.

Hybrid Towers

Hybrid (or fan-assisted, natural-draft) towers constitute a new concept designed to combine the favorable features of both mechanical- and natural-draft systems. These towers have a reinforced-concrete structure, similar in shape to a natural-draft tower, but somewhat smaller—namely, about two-thirds the diameter and half the height. In one type, motor-driven fans, oriented vertically around the circumference at the tower base, force air over the water in a counterflow arrangement. Hybrid towers have also been designed with the fan at the top, as in mechanical-draft towers. By controlling the fans, the relative contributions of mechanical and natural draft can be adjusted to suit the ambient atmospheric conditions.

Some advantages claimed for hybrid towers are smaller size and less visual impact than natural-draft towers. Moreover, fogging, icing, and noise would be less than for mechanical-draft towers. Construction costs would be greater than for the latter, but operating and maintenance cost would be lower.

Chemicals in Wet Cooling Towers

Chemicals are added to the water in wet cooling towers for various purposes. Nearly all of these chemicals are eventually discharged to an adjacent water body in the blowdown. The remainder, about 0.0025 to 0.01 percent of the total in towers of current design, escapes with the small droplets of water in the drift.

Mechanical-draft towers are often constructed of wood, and many wet towers use wooden slats as the fill. Bacteria can then cause deterioration of the wood. In addition, microbial and fungal growths accumulate as slimy masses that can foul or even plug the condenser piping. To reduce these effects, a biocide is added to the water. Frequently, sodium hypochlorite (or chlorine) serves this purpose, as in many once-through cooling systems, but the discharges must meet EPA and state standards. More expensive biocides, such as ozone or certain organic compounds of sulfur, seem to be possible alternatives to hypochlorite or chlorine; they are much less toxic to aquatic life, both plant and animal.

To prevent corrosion of structural metal components of the tower, corrosion inhibitors are often added to the circulating water. Other additives may be included to reduce the formation of scale deposits in various parts of the cooling system.[i] A useful inhibitor is sodium chromate plus a zinc salt, sometimes with an antiscaling additive. The zinc compound is itself a weak inhibitor, but it exerts a synergistic effect with the chromate. Unfortunately, chromium is very toxic to living organisms, and when chromate is used as an inhibitor, it may be necessary to reduce the chromium concentration before the blowdown is discharged. An alternative corrosion and scaling inhibitor is an organic or inorganic phosphate plus a zinc salt. But phosphate may be undesirable especially in a lake because of its contribution to algal growth. Natural-draft towers are constructed of concrete, and it is usually not necessary to add a corrosion inhibitor to the water.

The chemicals present in the tower blowdown are, of course, also present in the drift. The escaping water droplets, carried by the wind, eventually either fall to the ground or on vegetation, or they may evaporate while still in the air and descend as solid particles. In any event, the dissolved solids, called the *fallout* from cooling tower drift, will accumulate on surfaces in the vicinity of the power plant. As a result of the great improvement in drift elimination, the fallout is now only a minor problem, extending a hundred metres or so downwind from a tower. In fact, it is claimed that salt water can be used in wet cooling towers without causing significant environmental damage from fallout.

[i]Formation of scale, consisting mainly of insoluble or slightly soluble compounds of calcium and magnesium, is not a problem in once-through systems because the salts do not become concentrated as they do as a result of evaporation in a wet cooling tower.

Comparison of Wet Closed-Cycle Cooling Systems

The choice of a wet, closed-cycle cooling system—i.e., a cooling pond or canal, with or without sprays, or a mechanical-draft or natural-draft wet tower—depends on the available land, the location of the plant, the climatic conditions, and economic considerations. A cooling pond is often the least expensive alternative, but it requires a large amount of land of suitable topography to permit formation of a pond or small lake. Where water must be stored for use in periods of low water flow, a cooling pond may be the most desirable means of heat dissipation, provided land is available. A cooling canal requires about the same land area as a pond for equal electrical capacity, but it has the advantage of providing a long flow path for the water. In a lake, the flow may be short circuited unless internal dikes are constructed to prevent this.

Spray ponds and canals are advantageous in that they require only about 5 percent (or less) of the area of a pond or canal without sprays. On the other hand, the possibility of icing caused by drift from the sprays must not be overlooked in locating a spray pond or canal. Moreover, the cost of installing, operating, and maintaining the motors and pumps to produce the sprays is an important consideration. Additional experience is required to determine if spray ponds and canals are effective for large power plants.

Mechanical-draft wet towers are less expensive to construct than natural-draft towers with the same cooling capacity; the former are, however, more costly to operate and are noisier. Mechanical-draft towers are preferred where hot, dry weather may be experienced because the natural draft is then relatively weak. On the other hand, natural-draft towers would be advantageous where fogging and icing in cold weather or salt drift may cause problems. The use of wet towers (and other recirculating water cooling methods) increases the generation cost per unit (e.g., kilowatt-hour) of electrical energy by roughly 5 percent over that for a once-through cooling system.

Despite the large size of a natural-draft tower, it requires a substantially smaller total ground area than equivalent conventional, rectangular mechancial-draft towers. One reason is that the mechanical-draft tower modules must be spaced to prevent moist air exhausted from one module being drawn (i.e., recirculated) into an adjacent one. A 1000-MWe nuclear power plant, for example, might need about 2 hectares (5 acres) for a natural-draft tower and its equipment or 14 hectares (35 acres) for mechanical-draft towers; however, circular towers of the latter type, as described earlier, can be accommodated within a smaller area because they are less subject than separate modules to air recirculation.

It must be remembered that even in so-called closed-cycle wet systems makeup water is required to compensate for losses. Although this is usually not more than about 3 percent of the volume of the cooling water passing through the condenser, the total annual makeup requirement is substantial. For a 1000-MWe power plant, the amount is roughly 30 billion litres or 30 million m^3 (8 billion gal) per year; this is equivalent to 180 000 acre-feet of water, which must be available from a nearby water body.

Dry Cooling Towers

The wet cooling methods described above greatly reduce the impact of the waste heat from power plant condensers on aquatic ecological systems; nevertheless, they require not an inconsiderable volume of makeup water. Moreover, they present problems associated with possible fogging and icing, and the blowdown discharged must comply with federal and state regulations. These drawbacks do not arise with dry cooling towers. They are called dry because there is no actual contact between the air and the water to be cooled. There is, then, no evaporation of the water, no blowdown, and no drift, and makeup water is needed mainly to replenish losses by leakage. On the other hand, the absence of vaporization means that the theoretical minimum temperature to which the water can be cooled is not the wet-bulb temperature, but the actual atmospheric (or dry-bulb) temperature, which is generally higher (p. 353). This has some adverse consequences, as noted later.

The water from a condenser discharge could be pumped through a closed circuit of metal pipes and air drawn past the pipes, thereby transferring heat from the water to the air. This is the same technique as is used to cool the water in an automobile radiator. The pipes carrying the water are finned in order to increase the heat removal area. The air flow could be achieved by mechanical or natural draft, just as in wet towers. Heat transfer would occur by convection to the moving air. This procedure has not been used in the past to any extent for cooling condenser water from power plants, but it may find application in the future as heat transfer techniques are improved.

The commonly used dry cooling towers function in a different manner. They operate by condensing the exhaust steam directly or indirectly, using the ambient air as coolant; hence, they are sometimes called *air-cooled condensers.* In wet cooling systems, the waste heat in the exhaust steam is first transferred to the condenser water and then dissipated to the atmosphere. But when dry cooling is used, the waste heat is generally transferred to the ambient air directly without the use of an intermediate cooling water system. At the same time the steam is condensed.

In the *direct cycle,* the exhaust steam simply goes to finned condenser pipes over which air is drawn by mechanical or natural draft. The air removes heat from the steam so that it is condensed, and the water is returned as feedwater to the steam generator (Fig. 13-8). A problem with the direct air-cooled system is that very large volumes of steam, at low pressure, must pass through the condenser. This can be done by dividing the steam flow among a large number of parallel channels.

In the *indirect cycle* (or Heller) system, the steam is condensed by bringing it into contact with jets (or sprays) of cool water from previously condensed steam. The resulting warmed water is then cooled by pumping it through finned pipes over which air is drawn. Part of the cooled water is then recirculated to the condenser sprays, and the remainder serves as feedwater for the steam generator. A simplified schematic representation of an indirect, dry cooling cycle is shown in Fig. 13-9. The volume of water circulated through the cooling tower in the indirect system is much less than that of the steam in the direct cycle.

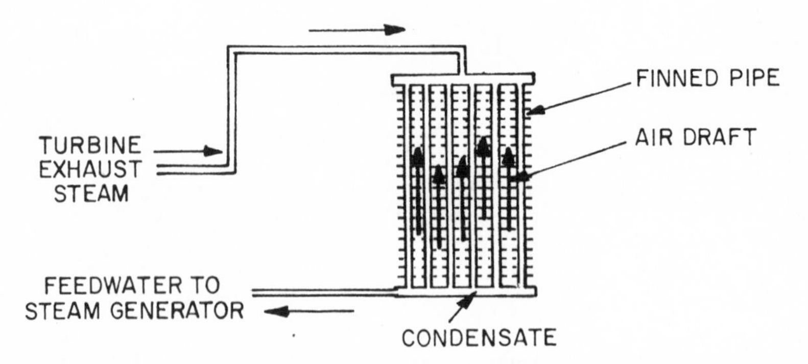

FIG. 13-8. Schematic outline of direct-cycle, air-cooled condenser.

Dry cooling systems, especially those of the Heller type, have been in use in Europe and elsewhere for several years for small power plants. The largest such plants are reported to have electrical capacities of 200 to 250 MW. Until the 1960s, there was little interest in dry cooling in the United States. A few coal-burning installations of low capacity in Wyoming, where water is not readily available, have used indirect-cycle air-cooled condensers. Their successful operation has led to the construction of a power plant in Wyodak, Wyoming, with an electrical capacity of 330 MW, which employs this system for disposing of waste heat.

The major advantage of dry cooling systems, and the one that may eventually override the drawbacks, is that they require little makeup water. Consequently, decisions concerning a power plant site can be based on the best land use and location of the load center, regardless of water supplies and the need for disposing of blowdown. A secondary benefit is the absence of drift, fogging, and icing.

In addition to the high cost of construction, an important disadvantage of dry cooling is the decrease in efficiency of the conversion of heat into electricity in the turbine. This is the case because the condensate (from the exhaust steam) may be 11 to 17 °C (20 to 30 °F) higher in temperature than for a wet-type cooling system. For the same reason, the pressure of the exhaust steam is also higher, and turbine design will have to be modified to operate effectively at a

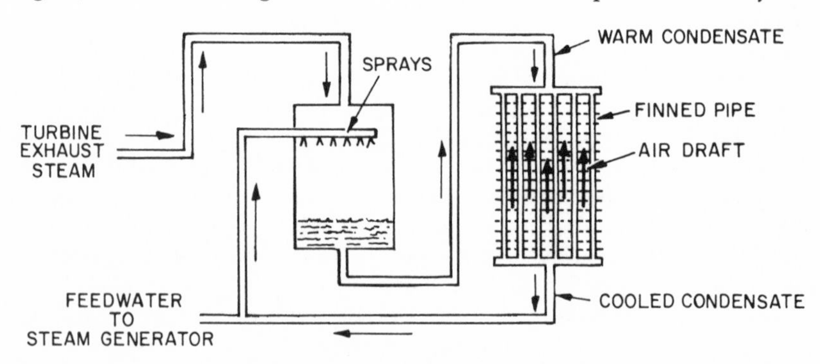

FIG. 13-9. Schematic outline of indirect-cycle, air-cooled condenser.

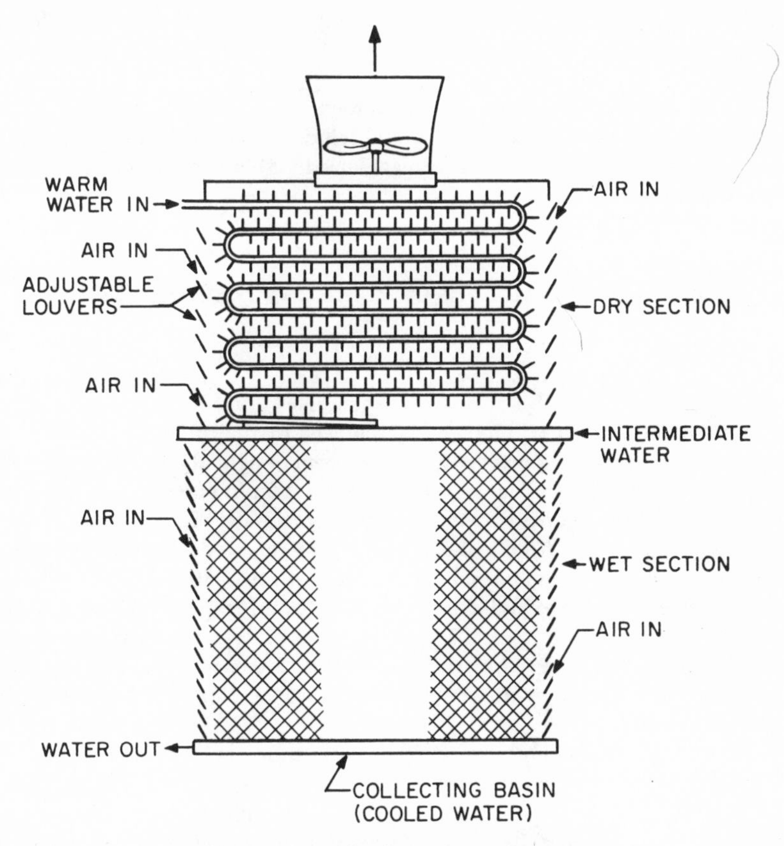

FIG. 13-10. Principle of wet/dry cooling tower
(air flowing through the dry section is adjusted by louvers).

higher back pressure. The use of dry cooling towers is expected to increase the cost of generating electricity by about 10 percent (on the average) above the cost with wet towers or roughly 15 percent above the cost with once-through condenser cooling.

Wet/Dry Towers

Wet/dry towers, capable of using either wet or dry cooling or both in different proportions, are a relatively recent development, which is attracting interest. Various combinations of wet and dry systems have been proposed. The one described here is designed to minimize the consumption of water and reduce the visible plume, with its associated fogging and icing in cold weather. This type of wet/dry tower consists of two regions: a lower wet tower and an upper dry tower (Fig. 13-10). The exhaust steam is condensed by cooling water in the same way as for any wet-type system. The warmed condenser discharge is first air-cooled by passage through finned pipes in the upper (dry) section; it then flows down over the fill in the lower (wet) section for further cooling. Air is drawn by a fan in parallel paths through both dry and wet parts of the tower, and the streams are mixed before discharge to the atmosphere. The air flow between the dry and wet sections can be adjusted to give preference to one or the other.

In winter, when the air is cold, advantage can be taken of dry cooling, with a marked decrease in local fogging and icing. At the same time, there is less vaporization of the water and a decreased need for blowdown, thereby reducing the makeup water requirement.

BENEFICIAL USES OF THERMAL DISCHARGES

Introduction

As long as steam turbines are used to generate electricity from heat energy, it is inevitable that heat will have to be removed by the condenser water. In the more efficient fossil-fuel power stations of recent design, about 1.32 kWh of heat are removed by the condenser water for every 1 kWh of electricity generated. For nuclear plants presently in operation, approximately 2 kWh of heat must be dissipated per kilowatt-hour of electricity. Reactors under development, using gas, liquid metal, or molten salt as core coolant, will discharge only about as much waste heat to the condensers as modern fossil-fuel plants. Thus, regardless of the type of installation, fossil-fuel or nuclear, substantially more of the energy contained in the fuel must be dissipated as heat in the condenser coolant than is converted into electricity. It would be desirable, therefore, to make some use of this heat energy.

The potential usefulness of the heat energy in the condenser effluent increases with the temperature. Since the temperature is not very high, the heat content of the water has limited use. Furthermore, in the winter, when the heat would have most value, the condenser water temperature is low. For example, with once-through cooling, the water discharge temperature might be as low as 13 °C (55 °F). With cooling towers, the winter temperature would be higher, but still of limited usefulness. Because of the large quantities of condenser discharge from power plants, its low temperature, and cost of delivery to the point of use, it is unlikely that any significant fraction of the waste heat will be utilized. However, whatever amount of such heat that is put to beneficial use will save an equivalent amount of fuel.

Water at a higher temperature (or low-pressure steam) would be useful for space heating and cooling and for desalination of salt or brackish water. A power station could be operated to provide condenser water at a higher temperature, but its electrical generating efficiency would be decreased. A dual-purpose plant could be designed specifically to provide a desirable balance of heated water (or steam) and electricity. The decrease in efficiency (and increased cost) of generating electricity might then be compensated for by the value of the heat in the condenser water (or steam). Such a dual-purpose plant would be attractive from the standpoint of fuel conservation and reduced thermal pollution. To minimize heat losses and pumping costs, there would have to be prospective customers for the hot water or steam in the vicinity of the power plant. This requirement presents a problem where nuclear plants are concerned.

The discussion of the possible beneficial uses of thermal discharges from turbine condensers are presented below under four headings: namely, space heating and cooling, agriculture, aquaculture, and miscellaneous. Investigations, sponsored by U.S. government agencies and by electric utilities, are in progress in all of these areas. But even if the results show promise, it must be realized that many years must elapse before large-scale applications become feasible.

Space Heating and Cooling

Warm water from a central source has been used for several years for space heating in Europe. In Iceland, homes and offices in the capital city of Reykjavik (population about 80 000) are heated by hot water (temperature around 85 °C [180 °F]) of natural origin some 16 km (10 miles) away. Artificially produced hot water is used for a similar purpose in several towns in the Soviet Union, West Germany, Sweden, Denmark, Finland, and other countries.

To be useful for space heating and cooling, the temperature of the water should be at least 93 °C (200 °F) and preferably higher; for example, one proposal envisions a water temperature of 149 °C (300 °F). But if the condenser water is discharged at 93 °C (200 °F), instead of 30 °C (85 °F), the efficiency of the turbine is diminished by about 8 percent (i.e., from about 32 to 24 percent) for water-cooled reactors. Whether or not the value of the waste heat used would compensate for the increase in power generating costs would depend on the design of the power plant and local conditions.

Although the use of hot water for space heating is obvious, it is less apparent that water at temperatures above 93 °C (200 °F) can be employed to produce cooling. A commercial system that will do this is presently available. The hot water serves to evaporate and concentrate a solution of lithium bromide in water. The more concentrated solution then takes up vapor from a pool of water, and the evaporation causes the latter to be cooled in the usual manner. The cooled water is circulated through radiators, just as hot water is in a heating system. The operating principle, known as *absorption refrigeration*, is used in gas refrigerators with ammonia and water as the component materials.

Large apartment houses, office buildings, and factories might be equipped with individual heating-cooling units using hot water piped from a power plant. An alternative, and perhaps more economical, possibility, which would be required for single-family residences in any event, would be to establish a number of substations throughout a city. The water, hot or cold according to the local temperature, would be distributed to the various users from the substations. The degree of heating or cooling desired by any individual could then be regulated by opening or closing valves.

During the summer, the heat in the water would be discharged to the atmosphere by way of the refrigeration equipment at the substations, but in the winter the heat would be released where it was used. The general result would be a wider distribution of the waste heat from the power station than is possible at present, with condenser water discharged into a body of water or cooled at or near the plant.

If the hot water effluent from a steam-electric installation were to be utilized in the foregoing manner, a serious problem would arise from the great variability of the demand. The requirements for heating or cooling change markedly as the seasons change; they even vary during the course of a day and from one day to the next. Clearly, what is needed is an alternative use for the hot water (or steam) that could be turned on or off at short notice.

It is unlikely that the heated water from an electric power plant will provide heating and cooling for an existing city, because of the large installation costs. But in building a new city or a large housing development, the pipes required for the heating-cooling water could be laid at the same time as those for ordinary water. An essential condition would be that a power station of appropriate electrical capacity could be located within a distance of 16 to 19 km (10 or 12 miles).

Warm Water in Agriculture

There are several ways in which warm water might be used to increase food supplies. Water at a temperature of 21 to 43 °C (70 to 110 °F) could be employed both to heat greenhouses in the winter and to keep them cool, if necessary, during the summer. A simple film-type evaporative cooler has been developed in which fans blow air past fiber pads kept saturated with the warmed water. Such cooling is most effective when the humidity of the air is low.[j]

It has been claimed that, with an adequate supply of moisture, fertilizer, and carbon dioxide, two or more crops of tomatoes, and other fresh vegetables, could be grown annually in some parts of the United States. The used, cooler water could be returned to the condensers, but this action would be dependent on several factors, such as the volume of water, distance to the power plant, and type of cooling system.

There is evidence that young chickens (broilers) gain more weight from a given quantity of feed or require less feed for the same weight gain if kept within a certain temperature range. One reason for this difference is that at the higher temperature less of the feed is required to maintain the chicken's body temperature. If the temperature of the environment is too high, however, the rate of growth of the chickens declines. There also appear to be optimum temperature ranges for the production of eggs by laying chickens and for the growth of hogs. A controlled environment attained by means of the thermal discharge from a power plant could thus be used advantageously in raising chickens and hogs and in producing eggs.

A five-year project on 68 hectares (170 acres) of land in Oregon has been sponsored by the Eugene (Oregon) Water and Electric Board in conjunction with the EPA to study some possible agricultural uses of warm water. The water is not from a power station but from a paper plant, with a discharge temperature of 32 to 54 °C (90 to 130 °F); this is somewhat higher than usual for the condenser effluent from a steam-electric installation. The warm water has been used for protecting fruit trees from frost by means of overhead sprays and for irrigation purposes. By spraying peach orchards, production of full

[j]Evaporative coolers operating on the same principle have long been used in the southwestern United States, where the relative humidity is very low, to provide cool air for homes.

crops was possible at times when unprotected orchards suffered serious losses from frost. Furthermore, when warm water was used for irrigation and soil warming, the growth rates of asparagus, lettuce, and cucumber were increased.

Following the encouraging results of a small-scale test in growing tomatoes and cucumbers at Muscle Shoals, Alabama, the Tennessee Valley Authority (TVA) has constructed several large greenhouses on land near the Browns Ferry, Alabama, nuclear power facility. The warm water discharged from the plant is used to heat greenhouses in which various flowers and vegetables are being grown. Among other uses of warm condenser water under consideration by the TVA are a controlled environment for growing livestock and warming the soil in open fields.

Another way in which heated water might be utilized in agriculture is for desalting (i.e., purifying) salt or brackish water by distillation. The warm condenser discharge would be used to heat the saline water under reduced pressure, thereby producing vapor (steam) which is condensed to yield purified water. The higher the temperature of the warm water, the more efficient the distillation process and the lower the cost of the equipment. The temperature of the condenser discharge would have to be at least about 82 °C (180 °F) to provide a balance between the opposite effects of temperature on the efficiencies of generating electricity, on the one hand, and of desalting water, on the other hand. Several large-scale processes have been developed for the purification of saline water by distillation at low pressures.

Warm Water in Aquaculture

The warm water discharge from power plants is being used to some extent in fish farming. The culture of channel catfish for commercial purposes is conducted at many localities in the southern United States. These fish show optimum growth at a temperature of about 32 °C (90 °F); they are then able to gain 0.45 kg (1 lb) in weight for the consumption of 0.57 kg (1.25 lb) of a high-protein (soybean, alfalfa, and fish meal) feed. Under natural conditions, catfish grow mainly during warm weather. But it appears that year-round growth can be realized if the temperature is maintained within the optimum range for growth and if ample food and oxygen are available. Since lower temperatures are required for spawning and egg development (Chapter 12), fingerlings are raised in a hatchery and transferred to the warmed water. In this way, the annual yield of fish per hectare of a water body can be greatly increased.

The TVA has been highly successful in the intensive culture of catfish using condenser discharge water from the Gallatin (fossil-fuel) power plant in Tennessee. In the United Kingdom, plaice (European flounder) and sole have been grown in the warm effluent from the Hunterston nuclear power station in Scotland. More than five years of experience have demonstrated the feasibility of the project.

Several edible marine shellfish, such as shrimp, oyster, and lobster, have also been found to exhibit maximum growth when the water is maintained at a suitable temperature. In Japan, harvests of 2240 kg of shrimp per hectare (2000 lb/acre) have been obtained from seawater warmed by the discharge from a power plant, and several investigations of shellfish culture are proceeding in the

United States. For example, an experiment on shrimp farming is being conducted in connection with the Turkey Point power station in Florida, and waste heat from the Long Island Light Company's power plant at Northport, New York, is being used for commercial oyster culture. Furthermore, the Maine Power and Light Company is sponsoring an attempt to increase the growth rate of lobsters by means of warmed water. In England, prawns are being cultivated at the Hinkley Point nuclear power station.

A different form of aquaculture using warm water from a power plant is the growth of green algae; when dried, the product can be used as feed for chickens, animals, and fish. Annual yields as high as 33 600 kg of algae, before drying, have been realized per hectare (30 000 lb/acre) in water at a controlled temperature of 29 to 32 °C (85 to 90 °F), with nutrients derived from animal wastes or sewage effluent. Part of the TVA's field experiment in connection with the condenser discharge from the Browns Ferry nuclear power plant, mentioned in the preceding section, will be concerned with the cultivation of algae.

There are a number of problems in connection with the use of discharges, especially in the culture of shellfish and other fish. One is the possible effect of various chemicals, such as chlorine and heavy metals that may be present in the condenser cooling water. Either the amounts added will have to be limited, or the water would have to be diverted temporarily after chemicals are added. There is virtually no chance that any significant amount of radioactivity will appear in the condenser discharge. Because of the low pressure on the turbine side, a leak in the condenser piping would cause water to be drawn inward rather than possible radioactive gases or steam to escape.

Another problem would be associated with shutdown of a nuclear power plant. The warm condenser water would then cease to be available, and the effect on aquatic life, which had become adapted to a higher temperature environment, could be disasterous. Consequently, some provision would have to be made for maintaining the temperature when the reactor is shut down either deliberately or in an emergency. A less obvious problem arises from the possible organic pollution from the fish culture system. Because of the presence of fish metabolic wastes and unused food, it may be necessary to treat the effluent water before it can be discharged without violating EPA regulations.

Miscellaneous Uses of Warm Water

The addition of heat to a stream generally reduces its ability to assimilate the organic wastes in sewage because of the depletion in the amount of dissolved oxygen. However, if the warmed water were to be deliberately oxygenated, the rate of consumption of organic matter would be increased. The treatment capacity of a sewage plant might thus be significantly enhanced. Incidentally, in some installations sewage is already being oxygenated to improve the operation at normal temperatures. More needs to be known about the effects of temperature on sewage operations before it can be determined whether thermal discharges could be utilized or not.

In the treatment of municipal water supplies, iron or aluminum salts (or both) are added to cause small suspended particles to form flocculent masses, or *floc*. The floc is then removed by sedimentation and filtration. It has been found that an increase in the temperature of the water results in more effective floc formation and an improvement in filtration efficiency. Estimates indicate that from 30 to 50 cents could be saved in the cost of chemicals per million gallons of water treated for a 5.6 °C (10 °F) increase in the temperature. Possibly use could be made of condenser effluent in connection with water treatment, but a heat exchanger would be required to transfer heat to the water, thus adding to the overall costs.

The possibility of using condenser effluent for defogging and deicing airport runways has been considered. The defogging would be achieved by evaporating the droplets of moisture in the fog by means of air warmed by the discharged water. For deicing, it would probably be necessary to circulate the water through buried pipes. It is doubtful, however, if the large capital costs of the necessary installations would be justified in view of the few times when they might be required. Furthermore, occasional use of the condenser discharge does not really solve the waste-heat disposal problem.

Bibliography—Chapter 13

Eisenbud, M., and Gleason, G., eds. *Electric Power and Thermal Discharges.* New York: Gordon and Breach, Science Publishers, Inc., 1969.

Elliott, T. C. "Cooling Towers." *Power* 3 (Special Report) (1973).

Furlong, D. "The Cooling Tower Business Today." *Environmental Sciences and Engineering Notes* 8 (1974): 712.

Karheck, J.; Powell, J.; and Beardsworth, E. "Prospects for District Heating in the United States." *Science* 195 (1977): 948.

Kolflat, T. D. "Cooling Tower Practices." *Power Engineering* 78, no. 1 (1974): 32.

Lee, S. S., and Sengupta, S., eds. *Proceedings of the Conference on Waste Heat Management and Utilization.* Miami, Fla.: University of Miami, 1977.

Mathur, S. P., and Steward, R., eds. *Proceedings of the Conference on Beneficial Uses of Thermal Discharges.* New York: New York State Department of Environmental Conservation, 1971.

Miliaras, E. S. *Power Plants with Air-Cooled Condensing Systems.* Cambridge, Mass.: Massachusetts Institute of Technology Press, 1974.

National Academy of Engineering. "Engineering Alternatives for Heat Dissipation Schemes." *Engineering for Resolution of the Energy-Environment Dilemma,* p. 81, 1972.

Parker, F. L., and Krenkel, P. A., eds. *Engineering Aspects of Thermal Pollution.* Nashville, Tenn.: Vanderbilt University Press, 1969.

Reynolds, J. Z. "Power Plant Cooling Systems: Policy Alternatives." *Science* 207 (1980): 367.

Rimberg, D. *Utilization of Waste Heat from Power Plants.* Park Ridge, N.J.: Noyes Data Corp., 1974.

U.S., *Code of Federal Regulations,* Title 40 (Protection of the Environment). Chap. 1. Environmental Protection Agency.
Part 122. Thermal Discharges (Subchap. N. Effluent Guidelines and Standards);
Part 423. Steam Electric Power Generating Point-Source Category.

U.S., Energy Research and Development Administration. *Proceedings of the Symposium on Cooling Tower Environment—1974* (CONF-740302). Washington, D.C., 1976.

U.S., Environmental Protection Agency. *Analysis of Engineering Alternatives for Environmental Protection from Thermal Discharges* (EPA-42-73-161). Washington, D.C., 1973.

U.S., Federal Power Commission Staff Study. *Problems in Disposal of Waste Heat from Steam-Electric Plants.* Washington, D.C., 1969.

Woodson, W. D. "Cooling Towers." *Scientific American* 224, no. 5 (1971): 70.

APPENDIX TO CHAPTER 13*

TABLE A-I

TEMPERATURE INCREASE OF CONDENSER WATER, °C (°F)

PLANT	ONCE-THROUGH	TOWERS		COOLING POND OR SPRAY CANAL
		Mechanical	Natural	
	High Values			
Nine Mile Point	18 (32)			
Fitzpatrick	18 (32)			
Seabrook	24 (44)			
New England Power	20 (37)			
Tyrone		21 (38)		
Yellow Creek		19 (34)		
Watts Bar			21 (38)	
Bellefonte			20 (36)	
Hartsville			20 (36)	
Skagit			22 (39)	
Trojan			25 (45)	
Wolf Creek				17 (30)
	Low Values			
Indian Point	7.0 (12.6)			
Oyster Creek	7.8 (14)			
Salem	7.6 (13.6)			
Surry	7.8 (14)			
Calvert Cliffs	5.6 (10)			
Bailly	7.8 (14)			
Vermont Yankee		11 (20)		
Hatch		11 (20)		
Farley		11 (20)		
Three Mile Island			11 (20)	
Fermi			10 (18)	
Zimmer			11 (20)	
Turkey Point				8.9 (16)
Robinson				10 (18)
Allens Creek				11 (19.5)
South Texas				10.5 (19)

*Adapted from F. A. Heddleson, *Summary Data for Commercial Nuclear Power Plants in the United States*, ORNL-NSIC-141, Nuclear Safety Information Center Report, Oak Ridge National Laboratory (April 1978).

TABLE A-II
ONCE-THROUGH SYSTEMS

Rivers

Dresden 1	Kankakee River
Yankee Rowe	Deerfield River
Connecticut Yankee	Connecticut River
Quad Cities 1	Mississippi River
Fort Calhoun 1 and 2	Missouri River
Cooper	Missouri River
Waterford	Mississippi River
LaCrosse	Mississippi River

Great Lakes

Big Rock Point	Lake Michigan
Nine Mile Point	Lake Ontario
Ginna	Lake Ontario
Point Beach	Lake Michigan
Zion	Lake Michigan
Kewaunee	Lake Michigan
Cook	Lake Michigan
Fitzpatrick	Lake Ontario
Bailly	Lake Michigan
Sterling	Lake Ontario

Oceans, Bays, etc.

Humbolt Bay	Pacific Ocean
San Onofre	Pacific Ocean
Millstone[a]	Long Island Sound
Diablo Canyon	Pacific Ocean
Crystal River	Gulf of Mexico
Calvert Cliffs	Chesapeake Bay
Shoreham	Long Island Sound
St. Lucie	Atlantic Ocean
Seabrook	Atlantic Ocean
North Coast	Atlantic Ocean
Mendocino	Pacific Ocean
Pilgrim	Cape Cod Bay
Atlantic	Atlantic Ocean
Jamesport	Long Island Sound
New England Power	Atlantic Ocean

TABLE A-II — CONTINUED
ONCE-THROUGH SYSTEMS

Estuaries

Plant	Water Body
Indian Point	Hudson River
Oyster Creek	Barnegat Bay
Salem	Delaware River
Surry	James River
Maine Yankee	Back River
Brunswick	Cape Fear River
Hope Creek	Delaware River

Lakes and Reservoirs

Plant	Water Body
Peach Bottom 1	Conowingo Pond, Susquehanna River
Robinson[a]	Lake Robinson
Oconee	Lake Keowee
Arkansas	Dardanelle Reservoir, Arkansas River
Bell	Lake Cayuga
North Anna	North Anna Reservoir, North Anna River
McGuire	Lake Norman, Catawba River
Summer[a]	Lake Monticello
Harris	Special Reservoir
Commanche Peak	Squaw Creek Reservoir
Clinton	Salt Creek Reservoir

[a]Also see Table A-VII.

TABLE A-III

SPRAY CANALS AND COOLING PONDS

Plant	*Makeup Water*
Dresden 2 and 3	Kankakee River
Turkey Point	Biscayne Bay
Quad Cities	Mississippi River
Midland	Tittabawassee River
LaSalle	Illinois River
Surry 3 and 4	James River
Greenwood	Lake Huron
Braidwood	Kankakee River
Allens Creek	Brazos River
Wolf Creek	Wolf Creek
South Texas	Colorado River
Pebble Springs	Columbia River

TABLE A-IV

MECHANICAL-DRAFT COOLING TOWERS

Plant	*Makeup Water*
Palisades	Lake Michigan
Ft. St. Vrain	South Platte River
Hatch	Altamaha River
Arnold	Cedar Rapids River
Farley	Woodruff Reservoir, Chattahoochie River
WPPSS 2	Columbia River
Catawba	Lake Wylie
River Bend	Mississippi River
WPPSS 1 and 4	Columbia River
Tyrone	Chippewa River
Perkins	Yadkin River
Cherokee	Broad River
Blue Hills	Toledo Bend Reservoir, Sabine River
Palo Verde[a]	City of Phoenix sewage system
Clinch River	Clinch River
Marble Hill	Ohio River
Black Fox	Verdigris River
Yellow Creek	Pickwick Reservoir, Tennessee River
Sundesert	Palo Verde Drain, Colorado River

[a]Also see Table A-VII.

TABLE A-V

NATURAL-DRAFT COOLING TOWERS

Plant	*Makeup Water*
Three Mile Island	Susquehanna River
Rancho Seco	Folsom Canal
Fermi (with pond)[a]	Lake Erie
Trojan	Columbia River
Davis-Besse	Lake Erie
Limerick	Schuykill River
Zimmer	Ohio River
Forked River	Barnegat Bay
Arkansas 2	Dardanelle Reservoir, Arkansas River
Susquehanna	Susquehanna River
Watts Bar	Chickamauga Reservoir, Tennessee River
Beaver Valley 2	Ohio River
Grand Gulf	Mississippi River
Vogtle	Savannah River
Bellefonte	Gunterville Reservoir, Tennessee River
Perry	Lake Erie
Douglas Point	Potomac River
Summit[a]	Chesapeake and Delaware Canal
Byron	Rock River
Fulton	Conowingo Pond, Susquehanna River
Quanicassee	Weadock Canal
Callaway	Missouri River
Montague	Connecticut River
Davis-Besse	Lake Erie
Koshkonong (Haven)	Lake Michigan
WPPSS 3 and 5	Ranney Wells, Chehalis River
Hartsville	Old Hickory Reservoir, Cumberland River
Skagit	Ranney Wells, Skagit River
Barton	Coosa River
Greene County	Hudson River
Phipps Bend	Holston River
Erie	Lake Erie

[a]Also see Table A-VII.

TABLE A-VI

VARIABLE-CYCLE OR HELPER-CYCLE SYSTEMS

Mechanical-Draft Towers

Plant	*Makeup Water*
Browns Ferry	Wheeler Lake, Tennessee River
Monticello	Mississippi River
Vermont Yankee	Connecticut River
Peach Bottom 2 and 3	Conowingo Pond, Susquehanna River
Prairie Island	Mississippi River

Natural-Draft Towers

Plant	*Makeup Water*
Sequoyah	Chickamauga Reservoir, Tennessee River
Beaver Valley	Ohio River

TABLE A-VII

SPECIAL SITUATIONS

Plant	*Circulating Water System*
Millstone	Once-through system discharging through a quarry.
Robinson	2250-acre reservoir used as a cooling lake.
Fermi	Natural-draft towers discharging to a 50-acre cooling pond in a closed cycle.
Summer	The 7000-acre Lake Monticello is used for once-through cooling. Lake Monticello is a pumped storage lake off Parr Reservoir, with 480-MWe generating capacity.
Palo Verde	Sewage water from Phoenix, Arizona, provides makeup water for mechanical-draft towers.
Summit	Fan-assisted natural-draft towers, high-salinity makeup water.

14

Nuclear Power Plant Site Assessment

INTRODUCTION

Basic Requirements

The selection by an electric utility of a site for a proposed power plant, regardless of whether it is to use nuclear or fossil fuel, is determined in the first place by the following considerations:

1. availability of sufficient land area
2. compatibility of the proposed plant with the existing uses of adjacent land
3. availability of surface and ground water for condenser cooling and other purposes
4. proximity to a load center (or centers), where there is a substantial demand for electric power
5. accessibility to an electrical transmission system
6. accessibility to railroad, highway, or water transportation.

If the plant is to utilize nuclear energy, a number of other factors must be considered in selecting a suitable site. This chapter is mainly concerned with these factors.

As explained in Chapter 3, the U.S. Nuclear Regulatory Commission (NRC) is responsible for licensing and regulating nuclear facilities from the standpoint of public health and safety. Furthermore, the NRC is required to prepare a detailed environmental statement of its licensing and regulating actions that may affect the quality of the human environment. In fulfullment of these responsibilities, the NRC assesses the suitability of a proposed site for a nuclear power station by evaluating the public health and safety and environmental aspects of the planned site-plant combination.

The major criteria on which site assessment is based are included in Title 10 of the *Code of Federal Regulations*, Part 51 (10CFR51), "Licensing and Regulatory Policy for Environmental Protection," and Part 100 (10CFR100), "Reactor Site Criteria." In addition, some of the items in 10CFR50, Appendix A, "General Design Criteria for Nuclear Power Plants," have a bearing on site-related safety considerations (see Chapter 3).

The NRC Regulatory Guides describe acceptable, but not necessarily exclusive, methods for implementing specific parts of the regulations. Of special interest for site selection are Regulatory Guides 4.7, "General Site Suitability for Nuclear Power Stations," and 4.2 "Preparation of Environmental Reports for Nuclear Power Plants."

Preliminary Site Selection

The Environmental Report submitted, in accordance with 10CFR51, by an applicant for a permit to construct a nuclear facility includes an analysis of alternative plant sites. In this analysis, the applicant is required to present an initial survey of site availability, using a screening process to eliminate sites whose less desirable characteristics are readily recognizable. The purpose of the screening process is to select a reasonable number of realistic siting options. This preliminary evaluation can generally be based on existing information or on information obtained from reconnaissance by specialists familiar with the site area. State, regional, or local power plant siting laws, if there are any, must be taken into account in assessing potential sites.

As a result of the preliminary analysis, the number of suitable areas will have been reduced so that the investigation of a realistic set of alternative site and plant combinations is possible. The plant alternatives selected are based on the energy source options, such as fossil fuels, hydroelectric, etc., as well as nuclear. The criteria to be used in selecting among the potential site-plant possibilities are the health and safety of the public, the potential environmental effects, and the economic and social aspects of constructing and operating the plant. Analysis in depth will be necessary at this stage, since the relative merits (or drawbacks) of certain combinations of sites and plants may be less obvious than for the sites alone. A final selection is made on the basis of a cost-benefit analysis of the more promising site-plant combinations. The preferred combination is the one that provides the best balance between, on the one hand, economic and environmental costs, and, on the other hand, benefits.

Assessment of Site Suitability

When the procedures described above have resulted in the selection of a preferred site, a detailed analysis of numerous factors must be made to allow the NRC to determine the acceptability of the site for a nuclear power plant. Safety considerations are of primary significance in evaluating the suitability of a particular site. These matters are treated mainly in the Safety Analysis Report submitted with the application for a permit to construct a nuclear plant on the site. The safety issues include geologic (and seismic), hydrologic, and atmospheric characteristics (or meteorology) of the site; population distribution and densities in the vicinity as they relate to protection of the public from potential radiation hazards of normal operation and of postulated accidents; and possible effects of accidents associated with nearby activities.

The environmental impact of the proposed plant-site combination is discussed in the Environmental Report, mentioned above. It is inevitable that construction and operation of a nuclear plant will have an environmental impact, as would be the case for any major industrial facility. There must be assurance, however, that the unavoidable adverse effects on the environment would be balanced by the benefits derived from the proposed power plant. Furthermore, the adverse effects should be less than those of available alternative plant-site combinations.

The potential benefits of a power plant are the availability of electric power for industry, commerce, and the home; an increase in the local tax base and employment; and, possibly, new recreational facilities. These must be balanced against such environmental factors as impacts on biota and ecological systems, water supply and quality, land use and esthetics, and the costs of social and community services.

In the remainder of this chapter, the criteria that may be used in assessing the suitability of a proposed site for a nuclear power plant are described. For convenience, the topics are treated in two parts: (a) safety-related considerations and (b) environmental (including socioeconomic) considerations. Some of the topics (e.g., water supply) have both safety and environmental aspects and are therefore discussed in both parts.

SAFETY-RELATED CRITERIA IN SITE ASSESSMENT

Geology and Seismology

In assessing the suitability of a site from the geologic and seismic aspects, the factors considered are surface faulting, ground motion, and foundation conditions (e.g., liquefaction, subsidence, and landslide potential). The nature of the investigations required to provide information in these respects are provided in Appendix A, "Seismic and Geologic Siting Criteria for Nuclear Power Plants," to 10CFR100.

Sites that include "capable faults"[a] are not suited to nuclear power plants, since it is not practicable to design such plants with assurance that their safety-related features will remain intact if the fault should become active. Furthermore, sites within 8 kilometers (km) (5 miles) of a capable fault greater than 305 meters (m) (1000 ft) in length are presently regarded as being unacceptable for nuclear plant sites. In any case, extensive geologic and seismic field studies and analyses are required for a proposed site. These studies are less extensive for sites for which adequate geologic data exist to determine fault "capability." Conservative design is required in safety-related features of the plant when geologic and seismic information is not conclusive (NRC Regulatory Guide 1.29, "Seismic Design Classification").

The seismic history (and potential) of a site is used to specify the design loads of the plant components; hence, it has an important bearing on the overall costs. The loads are determined from the Operating Basis Earthquake (OBE) and the

[a]A fault is a displacement of rock layers, often apparent as a linear disturbance at the surface. In simple terms, a capable fault is one that is capable of becoming active (e.g., by slippage). According to 10CFR100, Appendix A, a capable fault is a fault that has one or more of the following characteristics: (a) movement at or near the ground surface at least once within the past 35 000 years or movement of a recurring nature within the past 500 000 years; (b) instrumental records of macroseismicity (i.e., appreciable ground motion) of sufficient precision to demonstrate a relationship with the fault; and (c) a structural relationship to another fault that is capable according to characteristics (a) and (b) such that movement on one could be reasonably expected to be accompanied by movement on the other.

Safe Shutdown Earthquake (SSE), also referred to as the Design Basis Earthquake, which are based on regional and local geology and seismology and on specific characteristics of the subsurface material. The OBE is the earthquake that could reasonably affect the site during the normal operating life of the plant, whereas the SSE is the maximum potential earthquake that might occur at the site. The plant must be designed so that it will continue to operate safely when subjected to an OBE and can be shut down safely in the event of an SSE.

Investigations are required to determine the static and dynamic engineering properties of the ground material underlying the site. Suitable foundation conditions for a nuclear plant are found in sites with unfractured bedrock. In regions where there are few or no sites of this type, it is advisable to select a site in an area known to have a low liquefaction potential and an absence of cavities resulting from mining or extensive withdrawal of petroleum or groundwater.

Hydrology

As a safety feature, a highly dependable source (or sources) of water should be available to provide for emergency and shutdown cooling and to supply service water for fire protection and for emergency heat exchangers. These sources should not be susceptible to damage by earthquakes and should not be subject to depletion.

Sites located in river valleys, on flood plains, or along coastlines may be subject to flooding as a result of high tides or weather-related conditions. In addition, there may be floods induced by seismic activity, such as tsunamis (i.e., so-called "tidal waves") at shoreline sites, or seiches at lakeside sites. An earthquake may also cause flooding as a result of dam failure, river blockage, or diversion of a waterway.

In many cases, the damage that might be caused by flooding can be controlled by engineering design or by protection of safety-related structures, systems, and components. Such protection may, however, add substantially to the cost of the plant at the proposed site. Acceptable methods for determining the design-basis (p. 39) flood at a particular location are described in NRC Regulatory Guide 1.59, "Design Basis Floods for Nuclear Power Plants."

Another aspect of hydrology that is related to the safety of the general public can arise from the presence of aquifers beneath the plant site. If such aquifers are (or may be) used by large populations for domestic, industrial, or irrigation water supplies, they could provide potential pathways for radioactive material to man in the event of an accident. It is necessary, therefore, to make a study of the regional groundwater systems; the results are included in the applicant's Safety Analysis Reports.

Meteorology

The atmospheric conditions at a site are an important consideration in evaluating the dispersion of airborne radioactivity from routine operations and from postulated accidents and the dissipation of heat from condenser cooling water. In addition to complying with the NRC regulations for radiation levels, as given in 10CFR20 and 10CFR50, Appendix I (see Chapters 5 and 6), the nuclear plant must meet the requirements of the Clean Air Act (and its amendments). These latter requirements are not likely to present any problems unless (a) a proposed site is in an area where the existing air quality is near (or already exceeds) the Clean Air Act limits or (b) there is a potential for interaction of a cooling tower plume with a plume containing noxious or toxic substances from a nearby facility.

Extensive meteorological data are required to assess the potential for dispersion of radioactive material from normal operation and from design-basis accidents. If the meteorology of the proposed site is unfavorable in this respect, the exclusion area (p. 34), may have to be unusually large. In some circumstances, compliance with 10CFR100, which sets limits on the radiation levels following a hypothetical major accident of very low probability, may necessitate the inclusion of adequate compensating engineered safety features in the plant design (Chapter 4).

The meteorological conditions at a proposed site may differ from those representative of the area. For example, topographical features, such as hills, mountain ranges, canyons, deep valleys, and lake or ocean shorelines, can affect local meteorology. The dispersion characteristics of the atmosphere may then not be the same as those in the general region of the plant site. If they are less favorable, a larger exclusion area, lower effluent limits, or more stringent plant design would be required.

Population Distribution Constraints

The site must meet the requirements of providing an exclusion area, a low-population zone, and a population-center distance that will satisfy the criteria of 10CFR100. These requirements have already been described in Chapter 3. It may be noted here, however, that the dimensions of the exclusion area and the low-population zone are determined by the limits on the radiation doses at the boundaries that might result from a hypothetical severe accident. These dimensions are, therefore, related to the meteorological and topographical conditions at the site and to the characteristics of the engineered safety features.

From an analysis of approved sites, it appears that a minimum exclusion distance of 640 m (0.4 mile), even with unfavorable atmospheric dispersion characteristics, usually provides assurance that the engineered safety features can be designed to comply with 10CFR100. Assuming the reactor is located at the center of the exclusion area, the latter would occupy at least 154 hectares (350 acres). If the minimum exclusion distance is less than 640 m it may be necessary to place special conditions on the plant design before the site can be considered as acceptable. Furthermore, a distance of 4.8 km (3 miles) from the reactor to the outer boundary of the low-population zone is usually adequate.

As seen in Chapter 3, present regulations require the population-center distance to be at least one and one-third times the distance from the reactor to the outer boundary of the low-population zone. However, the actual population-center distance may have to be much greater if the proposed site is near a large city. Nuclear power plants, preferably, should be located in areas of low population density. If the population density at a proposed site is not acceptably low, the applicant for site approval will be required to consider alternative sites with lower population densities.

Experience, especially the accident at Three Mile Island (p. 36), has indicated the need for more precise (and more stringent) criteria relating to population distribution around a nuclear plant than those outlined above. It is expected, therefore, that the NRC will modify 10CFR100, possibly in accordance with the recommendations of the Siting Policy Force (p. 36). In assessing the suitability of a site for a nuclear power plant, emphasis will probably be placed on the actual population distribution, rather than on engineered safety features and calculated radiation doses.

Potentially Hazardous Activities

Potential accidents associated with existing or projected nearby industrial, military, and transportation activities could affect the safety of the nuclear power plant. A site should not be selected if, in the event of such an accident, the plant could not be shut down safely. For example, an explosion or fire at a nearby chemical plant, oil refinery, or petroleum product storage facility may produce missiles, shock waves, flammable vapor clouds, toxic chemicals, or incendiary fragments. These may affect the plant itself or plant operations in a way that jeopardizes safety.

Nearby military installations, such as munitions storage areas and ordnance test ranges, may threaten plant safety. The acceptability of a site will then depend on establishing that the nuclear plant can be designed so that its safety will not be affected by an accident at the military installation. An otherwise unacceptable site may become acceptable, however, if there is an agreement under which the military authorities make changes that will reduce the likelihood and severity of potential accidents.

An accident during the transportation of hazardous materials, by air, waterway, railroad, highway, or pipeline, may generate shock waves, missiles, or toxic or corrosive gases that can affect the safe operation of a nearby nuclear plant. The consequences of the potential accident will depend on the distance of the transportation link from the site and on the nature and maximum quantity of the hazardous material per shipment. Unless a firm and enforceable agreement can be reached to limit the transport of hazardous materials or unless the transportation link can be relocated, the proposed site would not be acceptable. Airports pose a special hazard to nuclear power plants; possible threats arise from the aircraft itself as a missile and from secondary effects of a crash, such as fire. On the other hand, a tall natural-draft cooling tower and its plume of water droplets could be a danger to aircraft.

If a preliminary examination indicates that potentially hazardous activities within 8 km (5 miles) of a proposed site could affect the safety of the nuclear plant, a detailed evaluation of the potential hazard is required. For sites in the general vicinity of an airport, a specific analysis must be made of such factors as frequency and type of aircraft movement, flight patterns, and local meteorology and topography. The objective of the analysis is to determine the probability of an accident that could result in the significant release of radioactive material from the plant. The analysis should demonstrate that the probability of any potential accident affecting the plant in such a way as to cause the release of radioactivity in excess of the 10CFR100 guidelines is less than about 10^{-7} per year (i.e., once in more than 10 million years).

ENVIRONMENTAL CRITERIA IN SITE ASSESSMENT

Preservation of Important Habitats

Important habitats are those that are essential to maintaining the reproductive capacity and vitality of populations of "important species" including the harvestable crop of economically valuable species.[b] Such habitats include breeding areas (e.g., nesting and spawning sites), nursery, feeding, resting, and wintering areas, or other areas of seasonally high concentrations of individuals of important species. Some species utilize a given breeding and nursery area every year, and if the characteristics of that area are changed, breeding success may be substantially affected.

The construction and operation of a nuclear power station, including new transmission lines, roads, etc., can result in the destruction or alteration of habitats of important species. If endangered or threatened species are present at a proposed site, potential adverse effects should be estimated. The impact of these effects should then be evaluated relative to the local population and the estimated population in the entire range of the species. The reproductive capacity of populations of important species (including the harvestable crop where appropriate) must be maintained, unless justification for proposed or probable changes can be provided. In general, a detailed justification is required when the destruction or significant alteration of important habitats is likely to exceed a few percent.

Some sites may contain, be adjacent to, or have an impact on ecological systems or habitats that are unique, limited in extent, or necessary to the productivity of populations of important species. Examples of such special areas are wetlands and estuaries. These sites cannot be evaluated as to suitability for a nuclear power plant until sufficient data are available to permit a reliable assessment of the impact of the plant on the area. Design characteristics of the facility that would satisfactorily mitigate the potential ecological impacts must be defined.

Movements of Aquatic Organisms

Seasonal or daily migrations in water bodies are essential to maintaining the reproductive capacity of several important fish species. Disruption of migratory patterns can occur in various ways; for nuclear power plants such disruption can arise from partial or complete blockage of a waterway by the discharge of heated water or chemicals or the construction or placement of structures. A

[b]Important species, in this respect, are defined as animals or plants that (a) are commercially or recreationally valuable, (b) are endangered or threatened (as defined in the Endangered Species Act of 1973), (c) have, as a species or specific population, important or unique esthetic or scientific value, or (d) affect the well-being of some important species within criteria (a), (b), (c), or are critical to the structure and function of a valuable ecological system.

knowledge of important aquatic species and their migratory requirements is necessary so that a "zone of passage" may be established that will permit the free migration of these species.

Fish and other aquatic organisms can be killed or injured by impingement on condenser cooling water intake screens or by entrainment in discharge plumes. Furthermore, fish eggs, larvae, and small juveniles, which are not held back by the screens, can be killed or injured in their passage through the condenser. The reproductive capacity of important species populations may thus be impaired (Chapter 12).

To minimize the loss of eggs and young fish, the site characteristics should allow placement of intake structures where the abundance of important species is small. As a general rule, intake from deep waters, which are usually less productive, is preferred over that from shallow waters.

Fish are often attracted to condenser water discharge areas because of the warmer temperatures and the availability of food. Consequently, site characteristics should be considered relative to the possibilities of fish being confined in (or unable to escape readily from) a situation that will have adverse effects. For example, fish may remain in condenser (thermal) effluent mixing zones or discharge canals under summer-like temperatures; their migration, which would normally be triggered by a drop in the water temperature, would thus be inhibited. Competition of large numbers of fish for the available food supply could also have adverse effects. Another factor is that cessation of plant operation in winter could be lethal to fish because of an abrupt drop in temperature.

Water Supply and Quality

Water supplies are required for condenser cooling and for plant services. When a facility obtains service water from an aquifer, there must be assurance that the aquifer will not be destroyed or that too much water will not be drawn from an aquifer that is the primary or only source of water for a community. The amount and quality of the water available for cooling purposes and the quality standards for discharged water determine the type of cooling system suited to the proposed site.

As a general rule, the quality (e.g., temperature or chemical composition) of the available water supply is not a major consideration in assessing the suitability of a site. A satisfactory condenser-cooling water system can usually be designed for the existing condition. However, the quality of the available water supply can affect the cost of the power plant and of its operation.

The quality of the water discharged from steam-electric power plants, both conventional and nuclear, is governed by the Federal Water Pollution Control Act of 1972 (see Chapter 12). Certification from the state in which the proposed plant site is located that the discharge will comply with the applicable requirements of the Act (Section 401) is necessary before the NRC can issue a construction permit. However, a permit may be issued if the state waives the requirements or if it fails to act within a reasonable length of time.

For a site near a large river or lake, or near the ocean, an ample supply of cooling water will usually be available. If the quality standards for the discharge can be met with once-through cooling, then this can make the site attractive from the cost standpoint. If sufficient water is not available or if water quality standards cannot be met, then some form of closed-cycle cooling, with its associated higher costs, will have to be adopted. The choice among various types of condenser-cooling systems was discussed in Chapter 13, and some site-related environmental aspects will be considered below.

Even with so-called closed-cycle cooling, a certain amount of makeup water is required from a nearby river or lake to compensate for unavoidable losses. The consumptive use of water for this purpose may be restricted by local statutes, it may be inconsistent with water use planning, or it may lead to an unacceptable impact on the local water resources. The site and plant design should be such that the consumptive use of water will not impair the supply of other users especially at times of low flow of a small river. It may be necessary, for example, to include a water reservoir, at additional cost, to provide makeup water at such times.

In cold weather, fogging and icing can result from water vapor plumes from cooling towers, lakes, canals, or spray ponds (Chapter 13). To a certain extent these effects can be controlled by the proper choice of condenser heat dissipation system. In areas of unusually high humidity that are protected from large-scale air-flow patterns, there is an enhanced potential for fog and ice formation from low-level plumes. The impacts of these phenomena, especially on surface transportation corridors and electric transmission lines in the vicinity of the proposed site, must then be taken into consideration. Because the plume from a natural-draft wet cooling tower is formed at a considerable height, there is little danger of fogging or icing on the ground, but the high-level plume might be a hazard to aircraft.

Where cooling towers are used, the sensitivity of local crops or natural vegetation may require that there be little salt drift. The vulnerability of existing industries or other facilities in the area to corrosion from the tower drift should also be borne in mind. Although these factors are seldom critical in evaluating the suitability of a site, they could require special cooling system designs or a larger site to confine the salt drift within an acceptable limit. The effects of drift may demand particular attention when saline water or water with a high mineral content is used for cooling purposes.

Land Use and Esthetics

Certain land areas are not acceptable for nuclear (or other) power plant sites because of their current or planned use. For example, plant sites, transmission lines, or transportation corridors close to special areas administered by federal, state, or local agencies for scenic or recreational use may cause unacceptable impacts. Some areas of historical and archeological interest may also fall into this category. The acceptability of plant sites near special areas of public use should be determined by consulting the appropriate government agencies.[c]

Some hitherto undesignated areas may be unacceptable for power plant sites because of possible future dedication to scenic, recreational, or cultural use.

[c]Federal agencies include the National Park Service, the National Park Preservation Program, the Bureau of Sport Fisheries and Wildlife, and the Forest Service.

These areas may include rare land types, such as sand dunes, wetlands, or coastal cliffs. Official land use plans developed by governments at all levels and by regional agencies must be consulted for possible conflicts with power plant siting.

In certain cases the use of land for a plant site would preempt exceptionally important existing land use. An example would be land where specialty crops, such as cranberries or artichokes, are grown; such a land conversion may be considered to be unacceptable without good cause. If the site is proposed for a nuclear plant, a detailed evaluation will be required of the potential impact and justification for the use of the site.

The presence of power plant structures may introduce adverse visual impacts to residential, recreational, scenic, or cultural areas or other areas with significant dependence on desirable viewing characteristics. Such esthetic impacts can often be reduced by selecting sites where existing topography and forests can be used for screening plant buildings. Large structures, such as natural-draft cooling towers, can be located where they do not interfere with recognized views of scenic or historic interest.

Socioeconomic Factors

Nuclear power stations can require large land areas, especially when cooling ponds are planned. The land requirement issue is likely to be important in areas where the land is used for agriculture (or other purposes) important to the local economy. If a preliminary study of the local impact indicates a large dislocation, a detailed evaluation of the situation would be necessary. In some cases a partial compensation can be provided by the development of recreational facilities on part of the land preempted for the nuclear plant.

The construction of a nuclear facility in certain locations may have a serious impact on the socioeconomic structure of a community by placing severe stresses on the labor supply, housing, transportation, schools, and community services. Problems may arise concerning tax bases, cost of additional services, and compensation for persons who have to be relocated. When construction has been completed, however, the situation changes drastically, and this causes new problems. It is generally possible to resolve difficulties during phases of construction and subsequent operation by proper coordination with affected communities. Where local acceptance problems can be reasonably foreseen, evaluation of the suitability of a site should include consideration of the purpose and probable adequacy of plans intended to mitigate the socioeconomic impacts.

Certain communities in the neighborhood of a proposed site may have distinctive features that require special consideration. Among such communities are towns that, for historical or other reasons, have become tourist attractions. The acceptability of a site near a distinctive community requires demonstration that construction and operation of the plant, including transmission and transportation corridors, will not adversely affect the character of the community or cause disruption of the tourist trade.

Bibliography—Chapter 14

Burwell, C. C.; M. J. Ohanian; and A. M. Weinberg. "A Siting Policy for an Acceptable Nuclear Future." *Science* 204 (1979): 1043.

Crowley, J. H.; Doan, P. L.; and McCreath, D. R. "Underground Nuclear Plant Siting: A Technical and Safety Assessment." *Nuclear Safety* 15 (1974): 519.

International Atomic Energy Agency. *Proceedings of the Symposium on Containment and Siting of Nuclear Power Plants.* Vienna: International Atomic Energy Agency, 1967.

______. *Proceedings of the Symposium on Siting of Nuclear Facilities.* Vienna: International Atomic Energy Agency, 1974.

Klepper, O. H., and Anderson, T. D. "Siting Considerations for Future Offshore Nuclear Energy Stations." *Nuclear Technology* 22 (1974): 160.

National Academy of Engineering, Committee on Power Plant Siting. *Engineering for Resolution of the Energy-Environment Dilemma*, p. 267. Washington, D.C., 1972.

Piper, H. B., and Heddleson, F. A. "Siting Practice and Its Relation to Population." *Nuclear Safety* 14 (1973): 576.

U.S., *Code of Federal Regulations*, Title 10 (Energy), Washington, D.C.
Part 51. Licensing and Regulatory Policy for Environmental Protection;
Part 100. Reactor Site Criteria.

U.S., Energy Policy Staff, Office of Science and Technology, Executive Office of the President. *Considerations Affecting Steam Power Plant Site Selection.* Washington, D.C.: Government Printing Office, 1970.

U.S., Nuclear Regulatory Commission. "Design Basis Floods for Nuclear Power Plants" (Regulatory Guide 1.59). Washington, D.C.

______. "General Site Suitability for Nuclear Power Stations" (Regulatory Guide 4.7). Washington, D.C.

______. *Nuclear Energy Center Site Study—1975.* "Part I: Summary and Conclusions" (NUREG-001). Washington, D.C. See also *Nuclear Safety* 17 (1976): 411.

______. "Preparation of Environmental Reports for Nuclear Power Plants (Regulatory Guide 4.2). Washington, D.C.

______. *Report of the Siting Policy Task Force* (NUREG-0625). Washington, D.C., 1979.

______. "Seismic Design Classification" (Regulatory Guide 1.28). Washington, D.C.

Yadigaroglu, G., and Anderson, S. O. "Novel Siting for Nuclear Power Plants." *Nuclear Safety* 15 (1974): 651.

Index

About the Authors

Samuel Glasstone—PhD (1922), D Sc (1926), University of London, physical chemistry - is a well-known author of scientific materials. He has written more than 35 books, the best known of which is the *Sourcebook on Atomic Energy*, first published in 1950, revised in 1958 and 1967, and still a best seller. He has written about reactor theory, nuclear engineering, nuclear weapons, and controlled thermonuclear research, as well as various aspects of physical chemistry. Dr. Glasstone received the Worcester Reed Warner Medal from the American Society of Mechanical Engineers in 1959 and the Arthur Holly Compton Award from the American Nuclear Society in 1968.

Walter H. Jordan—PhD (1934), California Institute of Technology, physics —was formerly assistant director of Oak Ridge National Laboratory and a professor of nuclear engineering at the University of Tennessee. Currently he is a member of the Atomic Safety and Licensing Board Panel of the Nuclear Regulatory Commission and advisory editor to the journal *Nuclear Safety*. He is a fellow of the American Physical Society and a fellow emeritus of the American Nuclear Society.